Quantum Measurement and Chaos

NATO ASI Series

Advanced Science Institutes Series

A series presenting the results of activities sponsored by the NATO Science Committee, which aims at the dissemination of advanced scientific and technological knowledge, with a view to strengthening links between scientific communities.

The series is published by an international board of publishers in conjunction with the NATO Scientific Affairs Division

A	**Life Sciences**	Plenum Publishing Corporation
B	**Physics**	New York and London
C	**Mathematical and Physical Sciences**	D. Reidel Publishing Company Dordrecht, Boston, and Lancaster
D	**Behavioral and Social Sciences**	Martinus Nijhoff Publishers
E	**Engineering and Materials Sciences**	The Hague, Boston, Dordrecht, and Lancaster
F	**Computer and Systems Sciences**	Springer-Verlag
G	**Ecological Sciences**	Berlin, Heidelberg, New York, London,
H	**Cell Biology**	Paris, and Tokyo

Recent Volumes in this Series

Volume 156—Gravitation in Astrophysics: *Cargèse 1986*
edited by B. Carter and J. B. Hartle

Volume 157—The Physics of the Two-Dimensional Electron Gas
edited by J. T. Devreese and F. M. Peeters

Volume 158—Physics and Chemistry of Small Clusters
edited by P. Jena, B. K. Rao, and S. N. Khanna

Volume 159—Lattice Gauge Theory '86
edited by Helmut Satz, Isabel Harrity, and Jean Potvin

Volume 160—Super Field Theories
edited by H. C. Lee, V. Elias, G. Kunstatter, R. B. Mann, and K. S. Viswanathan

Volume 161—Quantum Measurement and Chaos
edited by E. R. Pike and Sarben Sarkar

Volume 162—Quantum Uncertainties: Recent and Future Experiments and Interpretations
edited by William M. Honig, David W. Kraft, and Emilio Panarella

Series B: Physics

Quantum Measurement and Chaos

Edited by

E. R. Pike

King's College
London, United Kingdom

and

Sarben Sarkar

Royal Signals and Radar Establishment
Malvern, United Kingdom

Springer Science+Business Media, LLC

Proceedings of a NATO Advanced Research Workshop on Quantum Chaos,
held September 22-26, 1986,
at the Alessandro Volta Center for Scientific Culture, Villa Olmo,
Como, Italy

Library of Congress Cataloging in Publication Data

NATO Advanced Research Workshop on Quantum Chaos (1986: Como, Italy)
Quantum measurement and chaos / edited by E. R. Pike and Sarben Sarkar.
p. cm.—(NATO ASI series. Series B, Physics; v. 161)
"Proceedings of a NATO Advanced Research Workshop on Quantum Chaos, held September 22-26, 1986, at the Alessandro Volta Center for Scientific Culture, Villa Olmo, Como, Italy"—T.p. verso.
"Published in cooperation with NATO Scientific Affairs Division."
Bibliography: p.
Includes index.
ISBN 978-1-4899-5384-1 ISBN 978-1-4899-5382-7 (eBook)
DOI 10.1007/978-1-4899-5382-7

1. Quantum theory—Congresses. 2. Chaotic behavior in systems—Congresses. I. Pike, Edward Roy, 1929- . II. Sarkar, Sarben. III. North Atlantic Treaty Organization. Scientific Affairs Division. IV. Title. V. Series.
QC173.96.N378 1986
530.1′33—dc19 87-18934
CIP

Originally published by Plenum Press, New York in 1987
Softcover reprint of the hardcover 1st edition 1987

PREFACE

We would like to express our deep appreciation of Dr Mario di Lullo, late of the NATO Scientific Affairs Division for every possible help in setting up our Advanced Research Workshop on Quantum Chaos held at Como, Italy in September 1986. The news of his untimely death came as a great shock. The annual programme with which he was so deeply concerned must be of enormous benefit to the NATO community and there is a substantial memorial in the innumerable personal scientific contacts as well as the published volumes which have resulted from these meetings.

It is perhaps now generally accepted that the subject of chaos has become a major new field of physical and mathematical interest. Many of us who attended the meeting and considered ourselves as reasonable experts, came away at the end of it realising that there were many ramifications that we only vaguely comprehended. Whether, as Joe Ford claims, the ongoing investigations of possible "classical" chaos in quantum mechanics is going to lead to a revolution in our scientific thinking, remains to be seen. However, the transition from quantum mechanics to classical mechanics is a subtle one and connections between the two promises to stimulate researchers for years to come.

A special feature of this meeting was the juxtaposition of the topics of quantum chaos and quantum measurement. Apart from the concern of both with basic quantum theory, the more serious intent was the cross-fertilisation of ideas on reversibility since both communities were interested in that from different viewpoints. The ARW was also preceded by a special seminar on "Chaos, Noise and Fractals", sponsored by the London Office of US Naval Research. The proceedings of this seminar are published separately by Adam Hilger, Bristol. Certainly the meetings were outstanding not only for the formal contributions, but for the lively discussion throughout. This book contains the papers of most of the talks given at the NATO Workshop together with the report of a working party on important future directions. Professor Ian Percival's contribution is not included as it is available substantially in another Plenum Press publication, "Order and Chaos in Nonlinear Physical Systems", eds Linqvist, March and Tosi (1987).

Although the participants deserve much credit for the atmosphere, it is hard to imagine a more congenial or appropriate place for such a meeting than Como and the Villa Olmo remembering the famous 1927 meeting there where Bohr announced his theory of complementarity and also the first meeting there on Quantum Chaos in 1983.

We thank Professor Guilio Casati and his staff at the Alessandro Volta Centre for Scientific Culture for all their generous support and assistance. We also would like to thank most warmly Gerda Wolzak and Jane Zeuli who made sure that both business and social affairs ran smoothly throughout. Their contribution to the success of the meeting was considerable. Jane Zeuli and Madelaine Carter of Plenum Press, have both worked efficiently in the production of these proceedings for which we are also most grateful.

ROY PIKE and SARBEN SARKAR

CONTENTS

STUDIES OF THE SINUSOIDALLY DRIVEN WEAKLY BOUND ATOMIC ELECTRON IN THE THRESHOLD REGION FOR CLASSICALLY STOCHASTIC BEHAVIOR

James E. Bayfield

Department of Physics and Astronomy
University of Pittsburgh
Pittsburgh, PA 15260 USA

INTRODUCTION

The investigation of classically stochastic quantum mechanics is beginning to include both numerical and experimental work on specific physical systems. Most of these are provided by atomic physics. These systems are the quantum analogues of members of a class of classical Hamiltonian nonlinear systems that are called near-integrable. Their classical counterparts exhibit intermingled small regions of regular (periodic and quasiperiodic) and irregular (stochastic or chaotic) motion in phase space. In addition, there are boundaries between large regions of phase space dominated by regular classical trajectories and other large regions dominated by stochastic ones. The boundaries move in phase space as the parameters of the system are changed. For a system with otherwise given initial conditions a transition from mostly regular behavior to mostly chaotic behavior can then occur.

One subset of the near-integrable atomic systems are the autonomous ones of two dimensions in configuration space, where the total energy is constant in time. Mathematically these are viewed as essentially two strongly coupled, sufficiently nonlinear one-dimensional oscillators in a state of high total internal energy. A rapid and strong competition of the submotions in the two different dimensions seems to characterize classically stochastic behavior in these systems. One example is the hydrogen atom in a strong static magnetic field (1-3), where the electron's motion along the field direction, dominated by the nuclear Coulomb interaction, competes with a motion in the perpendicular plane, dominated by the Zeeman interaction. Other examples of autonomous systems come from molecular physics (4,5), such as (a) the diatomic molecule simultaneously in highly excited vibrational and highly excited rotational states, and (b) a linear triatomic molecule with both bonds simultaneously highly excited. The work on all these systems has emphasized their energy eigenstates and related spectra. The statistical properties of spectral distributions and the sensitivity of spectra to small changes in system parameters are two areas of special interest.

A second class of simple atomic and molecular near-integrable systems includes the externally driven ones, where one of the oscillators of the autonomous system is replaced by a classical oscillator of fixed amplitude and frequency. Now we have just one sufficiently nonlinear oscillator, at high time-averaged internal energy and strongly coupled to the classical oscillator. The Hamiltonian for the driven nonlinear oscillator is explicitly time dependent. An extension of phase space to include time is believed to place the externally driven one-dimensional oscillator problems and the autonomous two-dimensional, coupled-oscillator problems in the same class of near-integrable systems (10). One example of an externally driven system is a diatomic molecule coupled to an intense external electromagnetic radiation field, the coupling being to either the vibrational (6) or the rotational (7) molecular degrees of freedom. A second example is the topic of this paper, the bound electron in a hydrogen atom again in an external radiation field. We shall see that an almost one-dimensional externally driven system can be achieved in the laboratory, when the atom is in a highly excited state and electrically polarized by a static electric field, and when the radiation field is linearly polarized and directed along the static field direction (8). This is a particularly simple manifestation of the more general externally driven bound electron problem, which we call the DBE problem.

Of all the quantum nonlinear systems mentioned above, only the DBE system has been experimentally investigated in any detail. The feasibility of studying the spectrum of this system has been established in experiments where the driving radiation field was a microwave field (9). Optical transitions in the atom were observed while it was in the driving field. The transitions were between a weakly excited quantum state essentially unperturbed by the microwaves and a highly excited state that was strongly perturbed. Some results are shown in Figure 1. Experiments at microwave field strengths producing classically stochastic behavior involve much more densely distributed spectral lines and need improvements in spectroscopic resolution. What the figure does emphasize is that at low driving fields the system is a weakly perturbed quantum system that generally cannot be described by classical particle mechanics. The wave effects are very important. Only at high driving fields does classical nonlinear dynamics have a possible role to play.

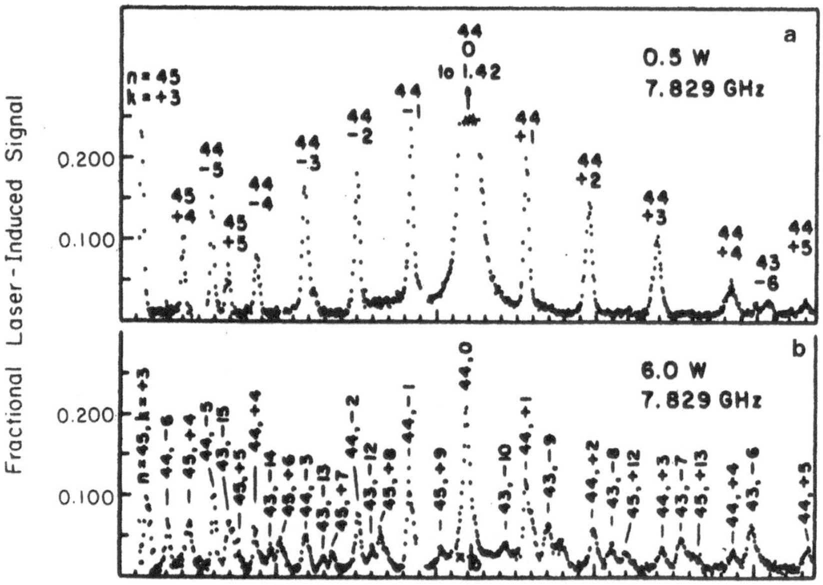

Fig. 1. Spectra of highly excited hydrogen in a microwave field F_{μ}, in the region near n=44. (a) $F_{\mu}=0.1F_c$; and (b) $F_{\mu}=0.4F_c$; where F_c is the stochastic threshold field near n=44. The energy spacing between adjacent n=44 sidebands in (a) is the microwave frequency.

The externally driven systems have an advantage over the autonomous ones in that one can focus on the time dependent changes in the observables of the nonlinear oscillator without being concerned with any changes in the driving oscillator. One can calculate probabilities for changes in these observables either classically or quantum mechanically. This includes transition probabilities for changes in the quantum numbers of an initially prepared quantum state. The time dependences of these probabilities have been of central interest. When the quantum numbers of the system are large and the number of photons absorbed from the radiation field is large, one might expect classical physics to be relevant. As we shall see, this does seem to be the case. It then becomes very interesting to see whether some of the phenomena of modern classical nonlinear dynamics (10), such as period-doubling bifurcations and transitions to stochastic behavior, are exhibited in some way in these quantum systems.

In this paper we shall discuss further only the highly excited atom in external electromagnetic fields. Experiments have already contributed signficantly to an understanding of this DBE system that exhibits the closest thing to "quantum chaos" yet uncovered. It exhibits many manifestations of classical chaos while not being known to have truly random underlying non-quantum statistical characteristics. We might call the system's behavior "quantum quasichaos," although, in the absence of a clear quantum picture of what is going on, we usually prefer the term "classically stochastic" behavior.

We shall review the experimental evidence for this quasichaos in the externally driven, weakly bound electron and discuss the connections between experiment, classical theory, and quantum theory.

THE DBE PROBLEM IN CLASSICAL NONLINEAR DYNAMICS

The classical electron orbits within an isolated hydrogen atom are ellipses with the atomic nucleus at one focus. The angular momentum is constant and is directed perpendicular to the plane of the ellipse. The Runge Lenz vector is directed from the nucleus to the aphelion of the orbit, lies in the orbit or xy plane, and is also constant. The motions in physical space and in phase space (x,y,v_x,v_y) are periodic with a frequency ω_n. Let us consider the changes in the orbit in physical space as the linearly polarized external time-oscillating electric field F_μ is increased from zero towards large values. Under certain conditions the period doubling sequence to chaos is expected to occur (10), as is qualitatively depicted in Figure 2. A part of a plane is used to sample the orbit, with the points of intersections being marked as the trajectory develops in time. At small values of F_μ, the orbit remains periodic with only a slight shift in frequency that grows with F_μ. The trajectory winding number W is defined as the ratio of the value T_n of the period of the electron orbit to the period τ of oscillation of the external field. It thus is a rational number equal to a frequency ratio, having the starting value unity in the figure. As F_μ is increased further, a point is reached where the periodic orbit cannot remain this simple. If a frequency shift is compensated for by adjusting the driving frequency $\omega = 2\pi/\tau$, then a period-doubled orbit is possible, with the trajectory coming back on itself after two ellipse-like periods. Then two stable intersection points are punctured in the sampling plane; the motion is "double valued" because of the stronger forcing of the external field along its direction, and the winding number becomes two. As F_μ is further increased, a second period-doubling occurs and the winding number becomes four. Further doublings occur for ever smaller increases in F_μ, leading to a chaotic limiting value of F_μ where the number of intersection points punctured in the sample plane becomes infinite and where the location of a next puncture point becomes unpredictable. Notice that even in this limit there is a remnant "ellipse-like" character to the trajectory; not everything

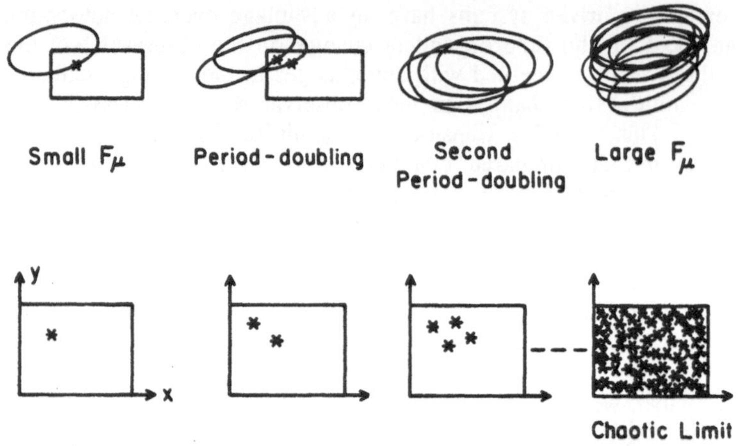

Fig. 2. Schematic of a hypothetical transition to classical chaos via the period-doubling mechanism. A plane in physical space is used to sample three-dimensional particle trajectories. Randomness is seen in the trajectory-plane intersection points when the motion is chaotic at large coupling parameter values.

about it is randomized. In particular, it is the partial motion along the external field direction that is most strongly randomized; it is therefore natural to focus upon this motion by considering one-dimensional models. Such models certainly pertain to the early-time motion of an electron initially started off in an orbit that is highly stretched along the direction of the microwave field.

Let us consider the surface-state electron or SSE model for our problem (13). It was originally developed for an electron bound outside a surface of liquid helium by its image charge. The motion of the electron remains outside the surface, or to one side of our atomic nucleus. This is accomplished by replacing these points of singularity by an infinite potential wall and by reversing the electron trajectory whenever it gets sufficiently close to the wall. Then our model Hamiltonian for the unperturbed "one-dimensional atom" becomes

$$H_0 = \frac{1}{2}p^2 + \begin{cases} -Z/x, & x > 0, \\ \infty, & x \leq 0, \end{cases} \qquad (1)$$

where here and below we use atomic units $m=e=\hbar=1$ (42). To take advantage of the fact that the energy E and action I of the electron are conserved, one transforms to action-angle variables (I,θ) according to

$$\theta = \begin{cases} 2[\sin^{-1}(\sqrt{x/a} - \sqrt{(x/a)(1-x/a)}], & p \geq 0 \\ 2\pi - 2[\sin^{-1}(\sqrt{x/a} - \sqrt{(x/a)(1-x/a)}], & p < 0 \end{cases} \qquad (2)$$

where $a = Z/E$.

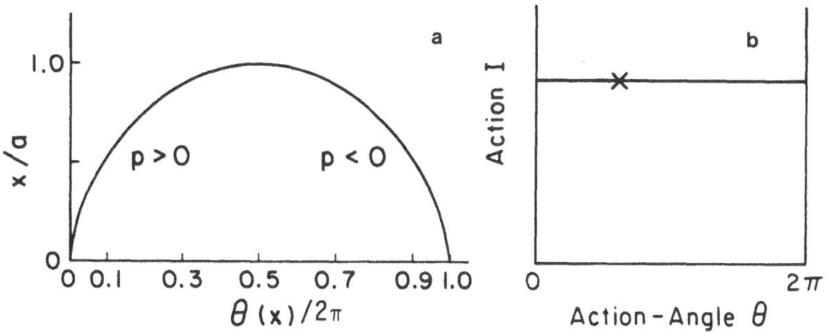

Fig. 3. (a) The SSE transformation function from electron-nucleus separation x to action-angle θ. For θ < 0.5, the electron is moving away from the nucleus. (b) A free-atom, periodic electron trajectory in action-angle phase space. Sampling at the natural frequency leaves a single point.

The Hamiltonian in terms of I is the constant $H_0 = -Z^2/(2I^2)$, the Bohr formula with the principal quantum number n being replaced by I. The transformation from electron position x to action angle θ is shown in Figure 3a. As the action angle increases linearly in time and by an amount 2π for each additional orbit period T_n, a sampling with this period of the trajectory in action-angle phase space leaves a single point determined by the relative phase of the electron orbit and the sampling sequence, see the phase space plot of Figure 3b. Now let us add to the Hamiltonian the interaction H' with the external oscillating field

$$H' = xF_{\mu} \sin(\omega t + \phi). \tag{3}$$

The problem becomes nonintegrable and phase space plots become very complicated. Numerical calculations of such a plot for parameter values relevant to the experiments give the results of Figure 4a (13,14). Classical scaling of parameters is used here, with initial action being normalized to one for the initial chosen value n_0 of quantum number; field strengths are multiplied by n_0^4 to scale them to the Coulomb field at the outside of the unperturbed Coulomb well, while frequencies are multiplied by n_0^3 to scale to the unperturbed electron orbit period. A static field is included in the calculations with a value typical of the experiments.

Three features of the phase space trajectory samplings of Figure 4a are notable. First of all, there are some trajectories that produce island structures such as that about the I=1.03, θ=0.2 point in the figure and that of the three-island chain near I=1.07; these features arise from resonances in this sinusoidally driven nonlinear oscillating system. There is a point within each island feature that lies within a periodic trajectory that has a rational winding number W. Several values of winding numbers are indicated in the figure. Trajectories started within an island generally are quasiperiodic, containing a number of incommensurate frequency components, with the nearby periodic trajectory orbit frequency being important. Island trajectories tend to keep the electron relatively localized in phase space. Secondly, at lower values of action there are in addition to the islands many other trajectories that span all values of θ; these are quasiperiodic trajectories containing a number of incommensurate

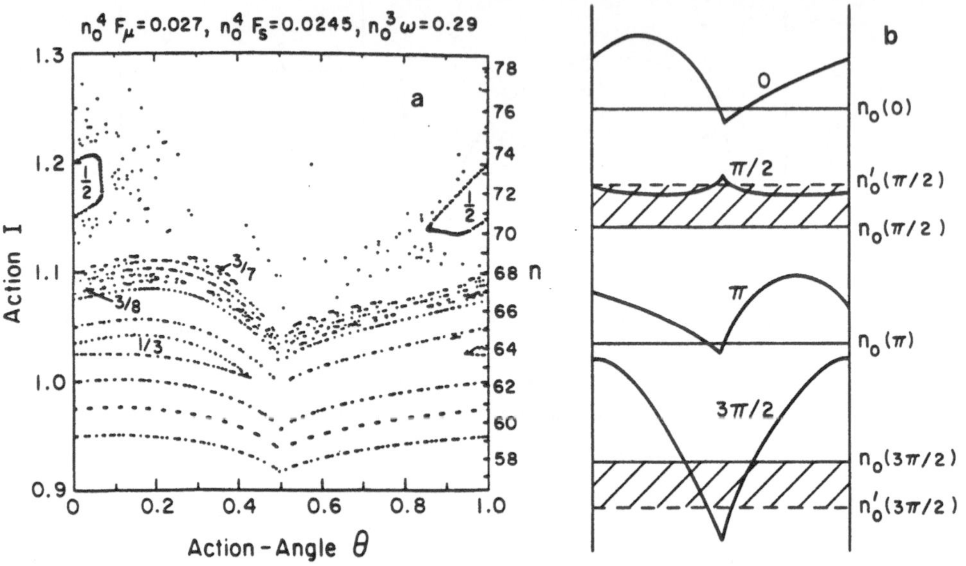

Fig. 4. (a) Some perturbed-atom, driven-electron trajectory samplings in phase space. (b) The boundary KAM surface at four different values of the instantaneous field, $F_\mu \sin(\omega t+\phi)$, having total phase values 0, $\pi/2$, π, and $3\pi/2$. Horizontal lines characteristic of typical initial action or quantum number values are also shown. n_0 values do not include the instantaneous energy shift of the external field interaction, while n_0' values do. Evidently, one is working close to the stochastic threshold.

frequencies and associated with an irrational winding number: Both kinds of curves in Figure 4a that correspond to quasiperiodic motion are called Kolmogorov-Arnold-Moser (KAM) surfaces; the KAM theorem (10) assures that the system when once on such a surface is quite unlikely to leave it. Yet it is the breakup of such surfaces that reflects true classical changes in the system that can correspond to transitions in the analogue quantum system. Finally, at higher values of action, most of the trajectories become unstable in the sense of not being quasiperiodic but rather stochastic; these produce randomly appearing points in the phase space sampling plots. There is a boundary region, here near $I=1.08$, where the instability becomes noticeable; this is the threshold region for stochastic behavior, where KAM surface breakup can be relatively rapid.

The location of the threshold region for KAM surface breakup can be roughly estimated using Chirikov's resonance island overlap criterion (10). After one obtains enough resonance island structure in a phase space sampling plot, then as one looks to higher regions of action, one sees that the islands have larger widths in action. This is a result of an increasing competitiveness of the electron's interaction with the external field compared to that with the Coulomb field. When the widths reach the stage where islands touch one another, all the horizontal-lying KAM surfaces are gone and non-island motion becomes stochastic rather than quasiperiodic. Approximate formulas for this have been derived (13) that give stochastic threshold field values to within a factor of two. We note in passing that in the related problem of a particle in a chain of sinusoidal potential wells, the classical resonance island overlap criterion has been shown to determine the threshold for field-induced barrier crossing and random walk motion along the chain (15).

It is important to realize that the detailed appearance of phase space sampling plots such as in Figure 4a depends upon the initial time chosen for the trajectory sampling; the KAM surfaces are "time dependent" as shown in Figure 4b (14).

In our system the period-doubled trajectories discussed qualitatively above are predicted to occur at high field strengths, where most of the nearby phase space is stochastic. The locations of some periodic trajectories in frequency-field strength space are shown in Figure 5 (16). The two plots refer to the different situations where (a) the electron is furthest from the nucleus when the external field is at a maximum and (b) instead when the field is near zero. The two situations are quite different. The experimental observability of these bifurcations as a splitting of resonances will depend upon the extent of any frequency locking of nearby orbits to periodic orbits to form

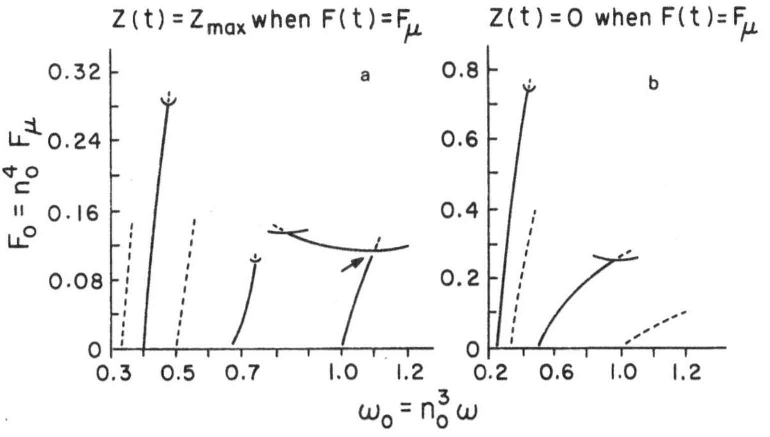

Fig. 5. The theoretical location of periodic trajectories in parameter space, the scaled field-strength/scaled frequency plane, for two different sets of initial conditions. Bifurcations are predicted for field strengths often above the average threshold field values for classically stochastic behavior, the averaging being over initial conditions, see Fig. 21.

large islands; the quantum mechanical enhancement of resonance effects also will affect their observability. At present, we have only classical theory for this effect. It suggests that the Feigenbaum period-doubling sequence is not the major pathway to classically stochastic behavior. Rather, it is the breakup of horizontal-lying KAM surfaces mentioned above that appears to be the major mechanism (14).

Associated with classically stochastic behavior is an exponential sensitivity of trajectories to initial conditions (10). If a system is started in turn on two stochastic trajectories infinitesimally close together in phase space, then at long times their locations in phase space are separated by a distance that on time-average grows exponentially with time. The exponent characteristic of this is called the (minimum positive) Lyapunov exponent. The observation of sensitivity to initial conditions and the determination of apparent Lyapunov exponent values are two important concerns of both quantum theory and experiment.

EXPERIMENTS ON MICROWAVE ABSORPTION BY HIGHLY EXCITED HYDROGEN ATOMS

The first observation of the microwave ionization of highly excited atoms occurred in 1974 (17,18). Figure 6 shows the early data for the ionization probability as a function of microwave field strength for three quite different microwave frequencies. The peak field required for ionization decreases with increasing frequency. If microwave ionization were only static-field ionization time-averaged over a field oscillation, then the peak field required would be higher than that of a static field and would be independent of frequency. A partial explanation of the ionization mechanism came a few years later with the observation of resonances in the ionization probability as a function of microwave frequency (19); this along with the observation of microwave absorption leaving the atom in higher excited bound states proved that the bound states of the atom were involved in a crucial way. The first theory for the ionization was a brute force three-dimensional classical calculation of the ionization probability, which successfully explained the first measurements (20). This led to considerable other classical theoretical work, including the establishment that the ionization classically was a stochastic process. The most recent of the sequence of experimental measurements using incompletely state-selected beams (21) has produced considerable ionization probability data that again is explained quite well by classical theory. This will be discussed further below.

The experiments have undergone a qualitative improvement with the introduction of electrically polarized atomic beam techniques (8). Now all the quantum numbers of the initially prepared atoms are well defined; in addition, one has the one-dimensionality already mentioned above. Let us see how such experiments are done.

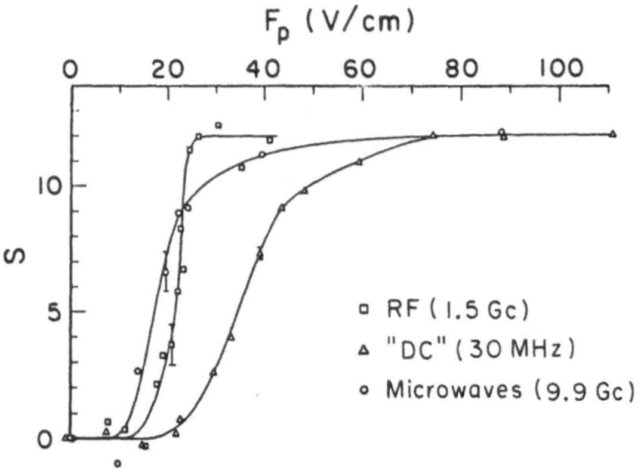

Fig. 6. First results for microwave ionization of highly excited hydrogen, at three quite different frequencies. Experimental ionizing values of field strength decrease with increasing frequency, not expected for time-average, static-field quantum tunneling.

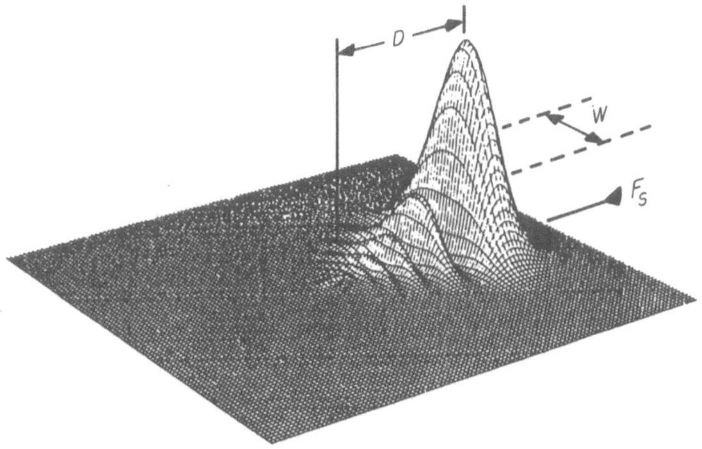

Fig. 7. The calculated electron spatial probability density for a hydrogen atom with parabolic quantum numbers (n, n_1, n_2, m) = (8, 0, 7, 0). A static field F_s determines the direction of the electron's displacement D from the nucleus. The electron wave packet is Stark-localized to a transverse width W.

Figure 7 shows an electron spatial probability density distribution for an electrically polarized hydrogen atom with quantum number n=8 (22). Such atoms are in a special one of the parabolic or Stark energy eigenstates, which are eigenfunctions of the components of the angular momentum and Runge-Lenz vectors taken along some direction in space, rather than of the magnitude and one component of the angular momentum (42). The special state shown has the parabolic quantum numbers, n, n_1, n_2, m = n, 0, n-1, 0. The electron distribution is seen to be shifted far to one side of the atomic nucleus, something that is maintained by application of a static electric field along the displacement direction. The electric dipole moment of such electrically polarized atoms is huge, 3/2 n(n-1). At large values of n, the half width of the Stark-localized electron in the direction perpendicular to the field direction is about a factor 5/n smaller than the mean distance between the electron and the nucleus; thus the electron is quite stretched out along a line when n is very large.

Electrically polarized hydrogen atoms with quantum numbers like n=60 can be produced using the optical double resonance technique (8,23) of Figure 8. Atoms collisionally produced in the n=7, n_1=m=0 state are passed in turn through two different static electric field regions. In the first, the field is very large, about 24 kV/cm. The second order Stark effect is then sufficient to separate energy levels of different quantum numbers, to within the sign of the azimuthal quantum number m. One can therefore select out of the atomic beam the electrically polarized n=7 atoms by inducing a resonant infrared laser transition to the n=10, n_1=m=0 state, while the atoms are in the static field. The atoms are stretched out a little bit in the process, the dipole moment increasing by a factor 100/49. The selected atoms then pass into the second static field region (field strength about 5 V/cm), where a second infrared laser transition resonantly induces the transition to the n=60, n_1=m=0 state. This stretches the atom out by a second factor of 36 to place the electron on time-average some 4000 Angstroms away from the nucleus.

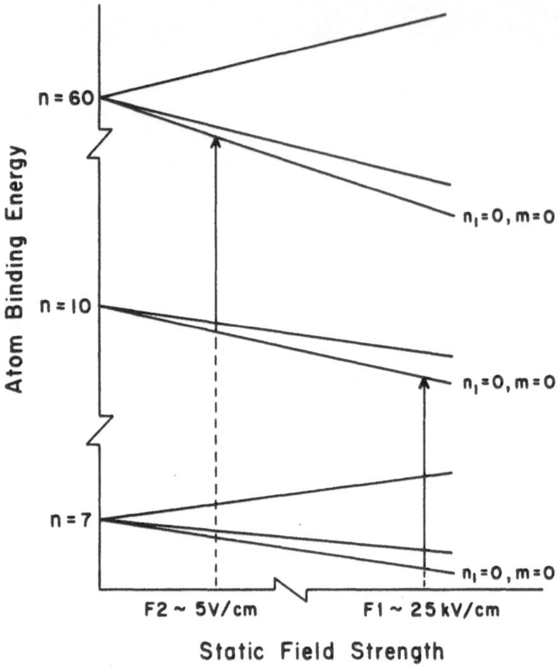

Fig. 8. The procedure for two-step resonant laser excitation of n=60 electrically polarized hydrogen atoms, using a pair of CO_2 laser transitions in appropriate static electric fields.

The n=60 atoms prepared by the laser excitation technique are passed through a microwave waveguide section, exposing them to a pulse of microwaves that is about 3000 microwave oscillations in duration, with 100 cycle pulse rise and fall times during atom passage through the fringe field regions of the waveguide section. The static electric field is maintained during this microwave exposure. The microwave electric field direction is the same as that of the static field, to within a few degrees or better. In this case, the matrix elements of the electric dipole interaction that enter are those of the displacement operator component z along the field direction. The off-diagonal or coupling elements are given by (24)

$$\langle n,0,0|z|n',\delta,0\rangle = \frac{(-1)^{n'+\delta+1}}{2^{\Delta n+2\delta+4}} \; \frac{\Delta n^{2\delta+\Delta n}}{(\Delta n+\delta+1)!} \; I \; \bar{n}^{2-\delta} \qquad (4)$$

where

$$I = 8\left\{\left[1 - \frac{(\Delta n)^2}{4(\Delta n+\delta+2)}\right] - 2\Delta n\left[1 - \frac{(\Delta n)^2}{4(\Delta n+\delta+1)}\right]\left[\frac{1}{\Delta n} + \frac{\delta+1}{(\Delta n)^2}\right]\right\} \qquad (5)$$

Here m=0, the change in m is zero, $\Delta n = n - n'$ is the difference in principal quantum number of the states being coupled, $\bar{n}$ is the average value of n, and δ is the change in parabolic quantum number n_1 from its initial value of zero. One sees from these equations that the coupling between states both having $n_1=0$ is a factor of about n larger than between states with $\delta=1$; coupling for $\delta=2$ is reduced a further factor of n. As microwave multiphoton transition rates are proportional to powers of the square of the coupling matrix element, we see that the relative probabilities for the system to remain on the ladder of $n_1=m=0$ quantum energy levels can be about 3000 times higher than for the system to go off this ladder. This propensity of the system to remain electrically polarized is the essence of the one-dimensionality of the system (25) that makes possible comparisons of experimental data with the results of both quantum and classical numerical calculations. The ladder of energy levels and the absorption of many microwave photons via their simultaneous coupling is schematically indicated in Figure 9.

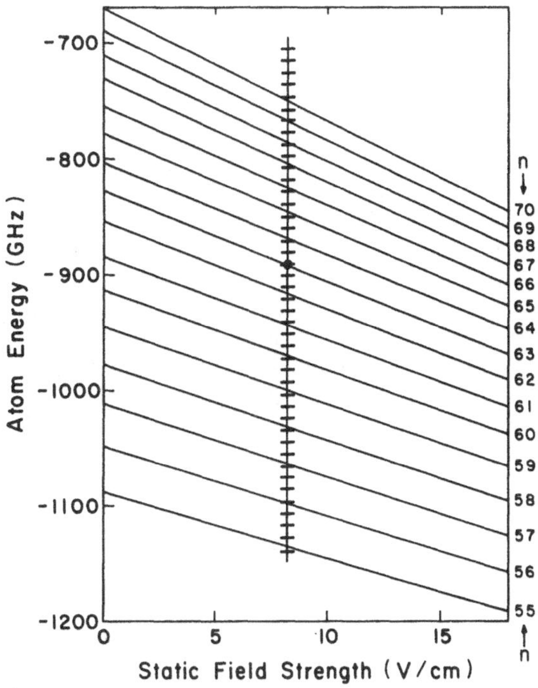

Fig. 9. A vertical chain of 7 GHz microwave photon energies superimposed upon a portion of the ladder of electrically polarized hydrogen atom energy levels. The ladder anharmonicity varies with the static electric field, F_s, here chosen to be 8 V/cm.

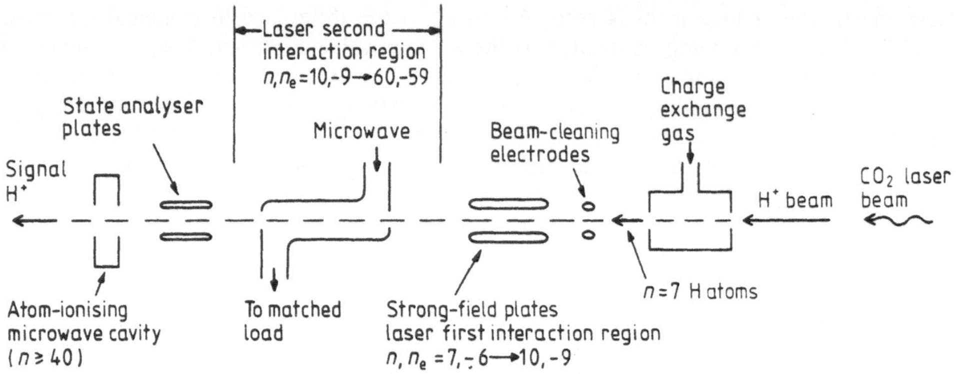

Fig. 10. Schematic of a fast atomic beam apparatus used to study microwave absorption by electrically polarized highly excited hydrogen atoms. See text and references 8, 26, and 28 for details.

Figure 10 gives the reader a feeling of how entire experiments have been carried out (26). A keV energy partially excited hydrogen atom beam is produced by charge exchange collisions during passage of a proton beam through a gas cell. The beam from a grating-tuned CO_2 laser is collinear with the atom beam. An intense static electric field between the beam-cleaning electrodes field-ionized n=10 and higher atoms, but not n=7 atoms. One laser photon was then absorbed by n=7 atoms in a transition within the strong field plates. Another photon was absorbed in the low field region after the strong field plates and before the microwave waveguide interaction region, producing the n=60 atoms. The waveguide was operated in the TE_{10} mode and had a length of about 70 cm. The static field was present within the waveguide, as it was produced by the motion of the atoms in a uniform magnetic field of a few Gauss generated by large coils outside the apparatus. The atom kinetic energy of motion, laser spectral line wavelength, and static field strengths were adjusted for simultaneous resonance between the desired Stark energy levels of Figure 8. The laser-excited atoms were detected by ionization during passage through a microwave cavity and subsequent measurement of the ion current. The state quantum-number analyzer shown in Figure 10 employed static field ionization to investigate the quantum numbers in the beam emerging from the waveguide interaction region. As the analyzer field strength was increased, the ion current dropped in a series of steps as atoms with each value of principal quantum number were removed in turn from the beam. An example of such steps is shown in Figure 11. The height of each step was a measure of the fraction of the initially laser-excited atom beam left in a quantum state with a particular quantum number after exposure to the waveguide microwaves. The locations of each step give the absolute value of the quantum number n when the other quantum numbers n_1 and m are zero. The quantum-number calibration of the analyzer was in excellent agreement with known static electric field ionization threshold values (27). Microwave ionization of the laser-excited atoms in the beam was observed as a drop in the ion current that was not associated with any step and occurred for all low values of the state analyzer field.

The experiments verified to a good approximation that the one-dimensional behavior of our atoms in our microwaves did occur. Figure 12 shows the sensitivity of the state-analyzer step location to the value of the parabolic quantum number n_1. The observed location of the steps generally ruled out values of n_1 larger than about two. Thus the microwave absorption left the direction of the atom's electric dipole moment fixed along the static field direction to within an angle of five degrees or less.

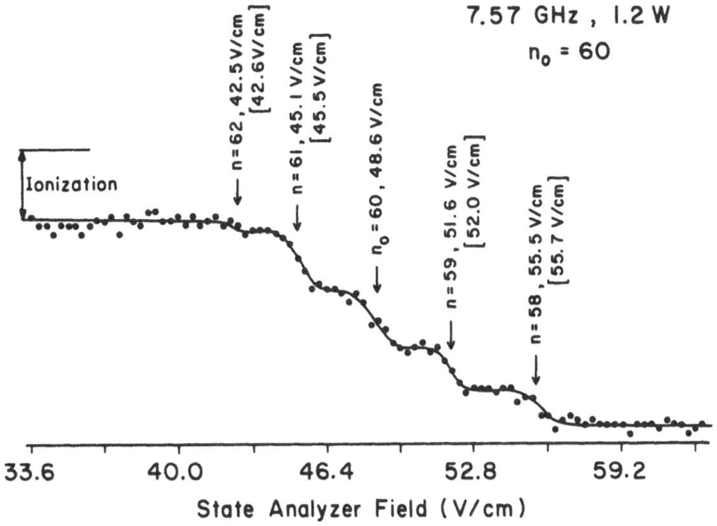

Fig. 11. A highly excited atom signal as a function of the field within the state-analyzer plates of Fig. 10. With the waveguide microwaves off, the signal is one large step starting from the $F_{\mu}=0$ level and occurring at the $n_0=60$ initial state field-ionization field value of 48.6 V/cm, see Fig. 12.

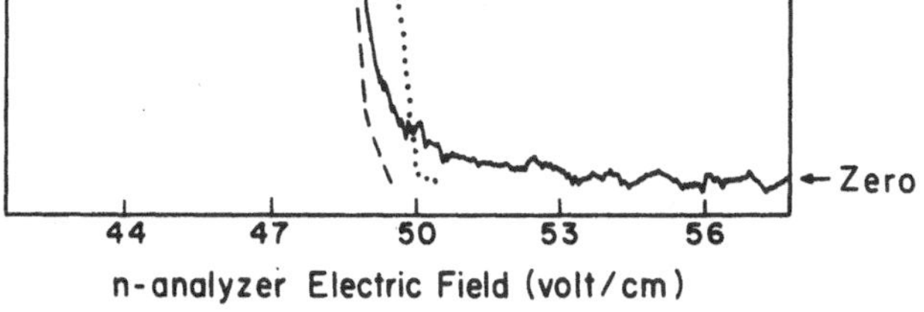

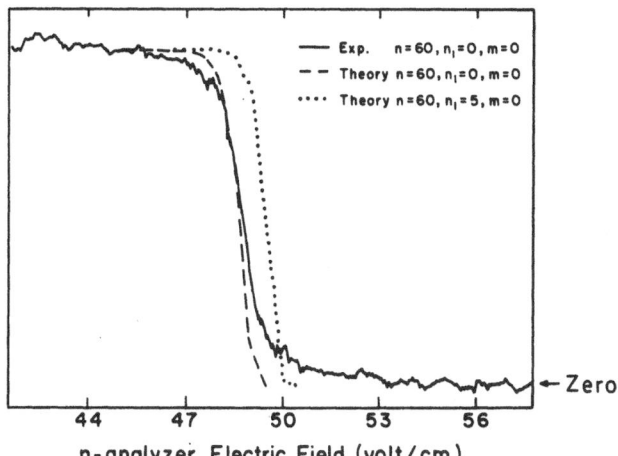

Fig. 12. An experimental $n_0=60$ state-analyzer field ionization curve, compared with theoretical curves for "one-dimensional" ($n_1=0$, $m=0$) and "somewhat two-dimensional" ($n_1=5$, $m=0$) atoms.

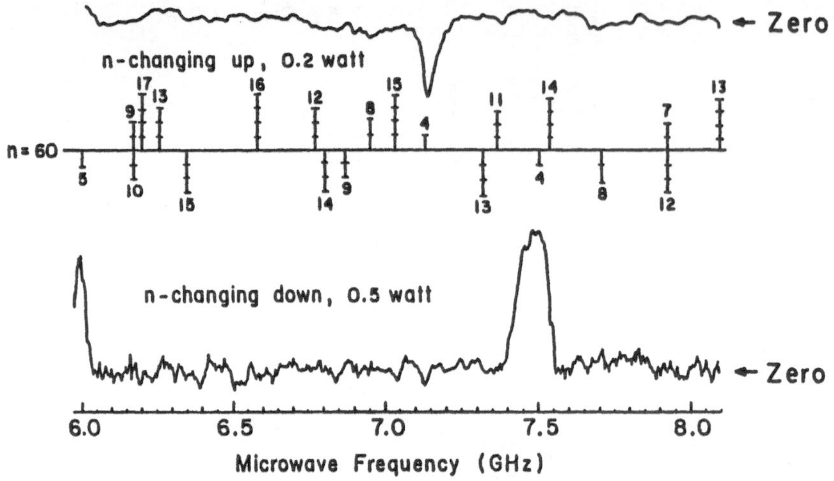

Fig. 13. Experimental $\Delta n = \pm 1$ n-changing signals as a function of microwave frequency, at relative low microwave fields. Along the center scale are indicated the expected frequencies for resonant multiphoton transitions, the number being the net number of photons absorbed, and the number of horizontal tics being the change Δn in n.

The experiments measured the fractional probabilities for microwave-induced n-changing (bound-bound transitions) and for microwave ionization, as a function of microwave frequency and microwave field strength. At low field strengths, only resonant microwave absorption producing changes in n of one unit were observed, see Figure 13. The expected resonant frequencies and numbers of microwave photons needed to be absorbed are indicated in the figure. These quantum multiphoton transitions were found to be in reasonable accord with two-state quantum calculations (24). The situation becomes more interesting when the microwave field strength is increased to higher values. Then numerous processes were found to simultaneously occur with observable probabilities (28,29). Probability variations with microwave field strength for two particular initial quantum number values and two particular microwave frequencies are shown in Figure 14. At these chosen parameter values, the microwave absorption goes into more and more states of higher quantum number, as the microwave field is increased. This occurs without much quantal state selectivity, the final state distributions at fixed value of field being smooth functions of n. As the mean value of the final state quantum number was increased beyond a certain point, ionization of the atom began to occur with observable probabilities. In these experiments the ionization involved both the static and microwave fields.

A great deal of information about the microwave photon absorption processes occurring in the experiments has been obtained. We discuss in the next section those features that bear closely upon the connections with classical nonlinear dynamics.

EXPERIMENTAL EVIDENCE FOR PARTIAL CLASSICALLY STOCHASTIC BEHAVIOR IN THE DRIVEN BOUND ELECTRON PROBLEM

Let a free electrically polarized atom be in a quantum energy eigenstate with quantum number n_0 at time $t=0$. We especially consider subsequent microwave exposures at frequencies away from the prinicpal resonances, which are those connecting states with n very near n_0. There are three observed effects that seem to relate to classically stochastic behavior. We refer to Fig. 4 for a theoretical indication of this behavior.

The first effect is the frequency-independent n-changing process (30). This has observable probability at fairly low microwave field strengths and follows resonant n-changing as the second type of transition observed as the field strength is increased. The superposition of these two processes is shown in the experimental frequency scan of n-changing down by one unit presented in Figure 15. Frequency independent n-changing has also been observed for n-changing up by one and by two units, the latter

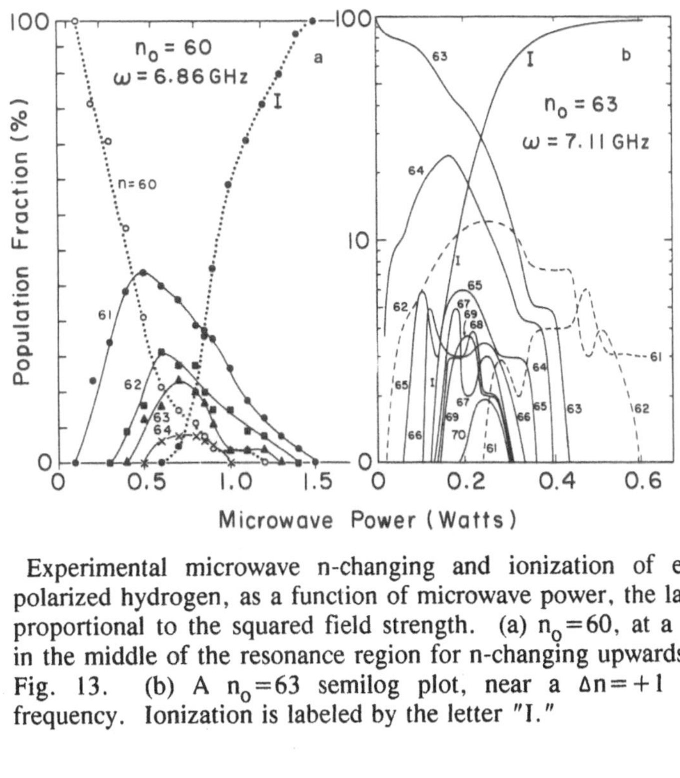

Fig. 14. Experimental microwave n-changing and ionization of electrically polarized hydrogen, as a function of microwave power, the latter being proportional to the squared field strength. (a) $n_0=60$, at a frequency in the middle of the resonance region for n-changing upwards in n, see Fig. 13. (b) A $n_0=63$ semilog plot, near a $\Delta n=+1$ resonance frequency. Ionization is labeled by the letter "I."

being seen at somewhat higher field strengths. This absorption of microwave energy that disregards the value of frequency implies the effective development of a quasicontinuum. We show below that the quasicontinuum arises from the presence of one-photon resonances that are microwave field-broadened to the point of almost total overlap with one another. One is always on-resonance for one-photon absorption with a change in n of one unit and at the higher fields also for two units. Embedded in this quasicontinuum are the multiphoton resonances, which still have fairly small widths, see Figure 15. Photon absorption via the quasicontinuum most closely corresponds to classically stochastic behavior and absorption via the multiphoton quantum resonances to motion on classical resonant islands in action-angle space.

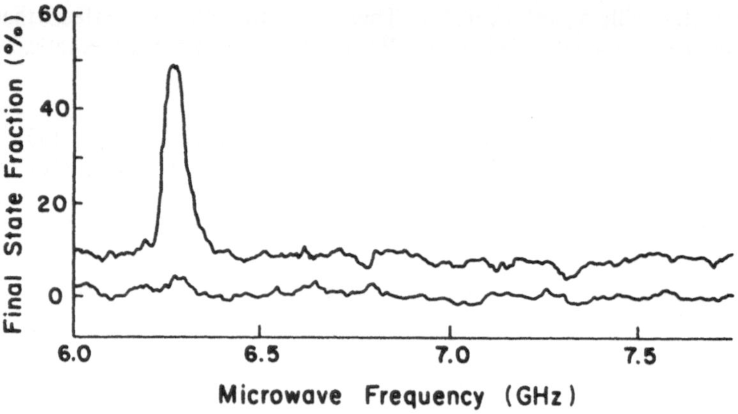

Fig. 15. The frequency dependence of $n_0=63 \rightarrow n=62$ n-changing, at a field strength near the threshold for classically stochastic ionization. Note the frequency-independent signal, the difference between the two curves away from the resonance.

Support for the above interpretation of the frequency-independent n-changing lies in its dependence upon microwave field strength. Figure 16 shows that this is quadratic, that is, the dependence on microwave power is linear (28). This is the signature of a one-photon process.

We have not thought of a viable alternative to the above quasicontinuum explanation for the frequency-independent n-changing. One might initially suppose that the effect might be due to laser excitation of atoms while they are in the microwave field, see Fig. 10; the atom energies might then be modulated in some continuous way, making photon absorption resonant at some point in the microwave field oscillation, no matter what the frequency. This mechanism is ruled out, however, by several facts. Primary among these is the power dependence of Figure 16; if the continuous modulation mechanism were in effect, then the n-changing signal would drop with increasing power, as the modulation amplitude increases and the atom's time spent in resonance within the bandwidth of the laser transition decreases, opposite to what is observed. It is also clear from Figure 1 that the modulation of the atom is not slow enough for the atom to be in resonance for a specific change in n for essentially all frequencies; said another way, the microwave frequency of about 7 GHz is too high for the sideband spacing of Figure 1a to be smaller than the laser resonance bandwidth of 0.2 to 0.3 GHz achieved in the measurements of Figures 15 and 16.

Further support for our interpretation of the frequency-independent n-changing comes from the numerical quantum calculations (30). Although we discuss these later, we show in Figure 17 some results for n-changing down by one unit, for frequencies near a four-photon resonance. The presence of frequency-independent n-changing is evident.

The superposition of resonance effects upon the absorption in the quasicontinuum as seen in Figure 17 has also been found in numerical quantum calculations of final state quantum number distributions (31). Resonant population of final states tends to favor those separated from the initial state by energies equal to integral numbers of microwave photon energy. The situation is very similar to that of the above threshold ionization (ATI) phenomenon that occurs within the true continuum (32).

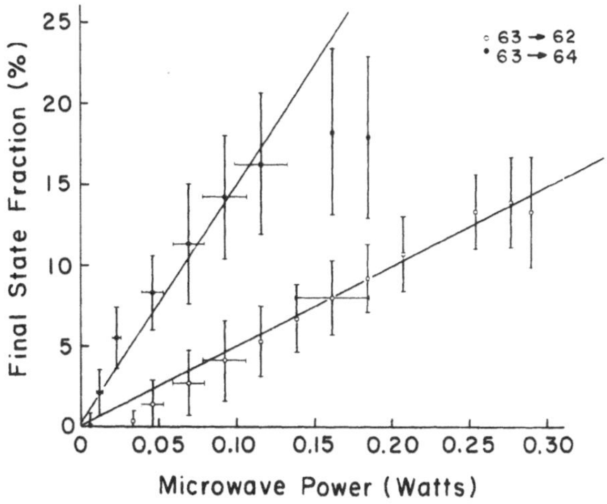

Fig. 16. Dependence upon microwave power of the frequency-independent signals for Δn=-1 (see Fig. 15) and Δn=+1 n-changing. At higher powers, the signals would saturate and then drop, as processes involving larger changes in n become important, see Figs. 14 and 18.

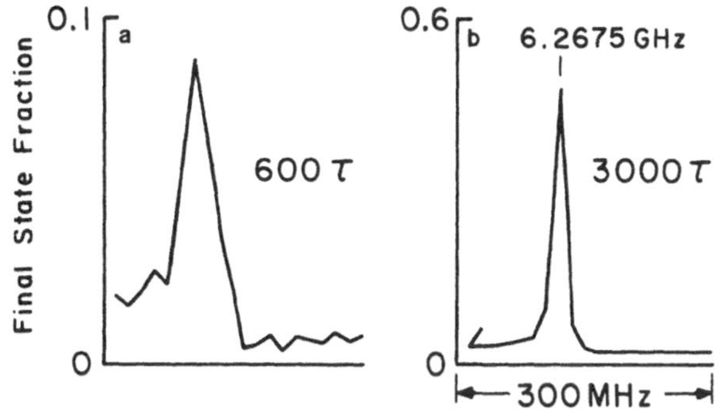

Fig. 17. Many-state quantum numerical results for the $n_0=63 \to n=62$ n-changing of Fig. 15, (a) after 600 microwave oscillations and (b) after the full 3000 oscillations. $F_\mu=7.2$ V/cm and $F_s=8.0$ V/cm. Only a 300 MHz range of frequencies near the resonance was considered, and only the case $\phi=0$, see equation 3 and Fig. 4b, is shown. Including other, more important values of ϕ increases the final resonance width, quasienergy resonance shift, and level of frequency-independent n-changing. The accord with experiment is then reasonable, see reference 30.

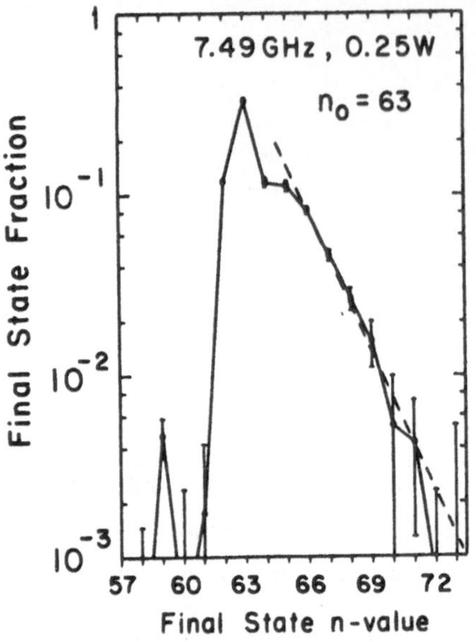

Fig. 18. A measured n-changing final state distribution exhibiting an exponential dependence at high n.

One reason for the observation of only partial classically stochastic behavior at the partially ionized fields used here is that it occurs only for some atoms, those with correct timing of their introduction into the microwaves, see Figure 4b. Even then, it occurs mostly for a part of the initial state wavefunction that corresponds to the electron being far from the nucleus.

The second sign of classically stochastic behavior is the smooth exponential final state distributions observed for some values of n_0 and for some values of frequency (30). Such distributions are unlikely to be the result of processes that depend upon the unbroadened resonance structure of the system. The smooth exponential distributions arise as shoulders on the final state distributions occurring on the high-n side, see Figure 18. The measured exponent of about 0.5 per n-value implies a partial delocalization (10,33,34) of the electron via microwave absorption within the quasicontinuum. The delocalization is in both action space (quantum number space) and in physical space, since the mean value of the electron-nucleus separation and the quantum number are monotonically related. The numerical quantum calculations of (10,35). This is convincingly shown in the results of Figure 19 (35). Once the experimental conditions of field switching on-and-off and random timing of the atom entrance into the microwaves are accounted for, the quantum calculations of final state distributions are in satisfactory agreement with the experimental data, see Figure 20. Final state distributions also find such exponential distributions, and furthermore show that they arise when the atom is initially within a stochastic region of phase space.

The third sign of classically stochastic behavior is the agreement of experimental data with classical calculations for ionization probabilities at high microwave field strengths and at short microwave exposure times, a few hundred cycles. Again, this agreement is for frequencies away from resonances, particularly the quantum resonances for n-changing down. Figure 21 shows this ionization data (21), compared

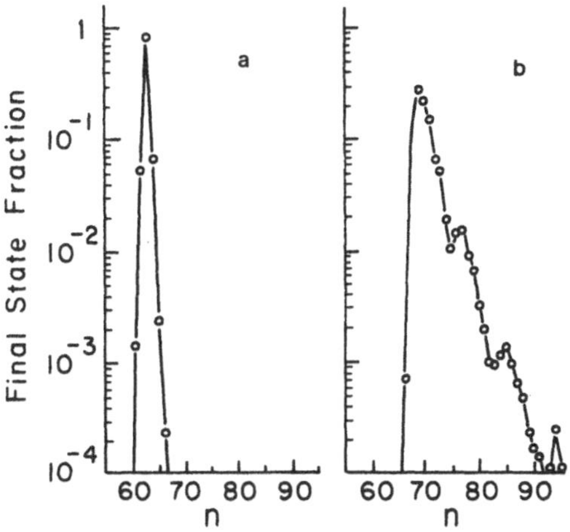

Fig. 19. Numerical quantum calculations of n-changing final state distributions starting at (a) $n_0=63$, below and (b) $n_0=70$, above the quantum number band corresponding to the classically stochastic threshold.

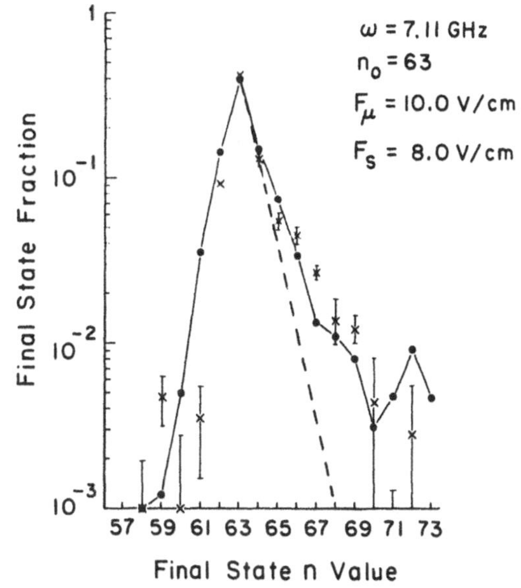

Fig. 20. A comparison of experimental n-changing data, ⌶ with quantum results that average over ϕ and include experimental field switching.

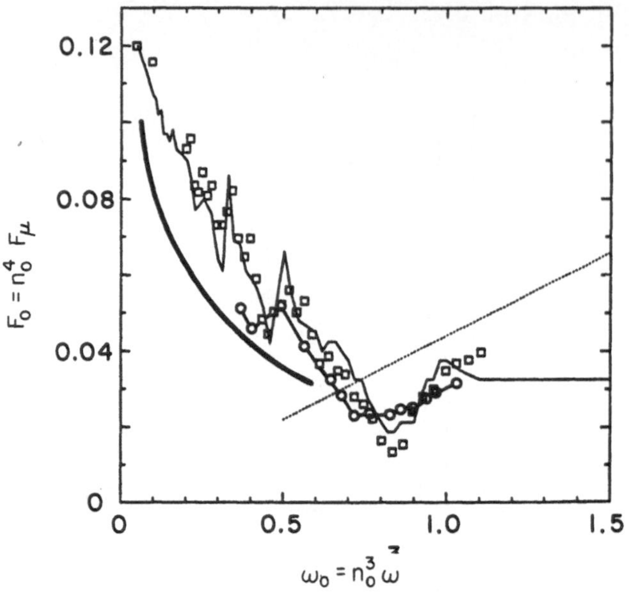

Fig. 21. The curve for 10% ionization probability, plotted in the parameter space of the scaled microwave field strength/scaled microwave frequency plane. Squares are experimental data for a mixture of atoms of varying n_1, n_2, and m, reference 21; the solid line is a one-dimensional classical calculation, reference 36; the open dots are a one-dimensional quantum calculation, reference 37; the heavy solid line is the prediction of the quantum Chirikov resonance overlap criterion, equation 7. The dotted line is an analytic estimate of the threshold field for quantum delocalization, reference 31.

with the classical calculations (16,21,36) and also with recent numerical quantum calculations (37). There is complete agreement about the general feature that indeed the microwave field strength needed for a given ionization probability does very significantly decrease from the static field value as the (scaled) microwave frequency is increased, nicely confirming the early results of Figure 6. The results of Figure 21 also exhibit the expected resonance effects near rational values of the winding number that correspond to large resonance islands in the phase space plots like in Figure 4a; these along with the other curves shown in Figure 21 will be discussed below.

We should point out that the microwave field strengths shown here for observable ionization are much larger than those required for microwave ionization of non-hydrogenic Rydberg atoms with low values of angular momentum (39-41). There is a different state-coupling mechanism in such non-hydrogenic systems that arises from the additional Coulomb interactions of the Rydberg electron with the other electrons in the atom.

The experimental data we have shown have uncertainties in reproducibility indicated either by error bars or by the jitter in the curves. An interesting question is concerned with how reproducible one should expect data to be when it is expected to be perhaps exponentially sensitive to initial conditions and to parameter values! Partly, we expect some reproducibility because quantum calculations exhibit more stability than classical ones (11). Still, ever since the first ionization experiments in 1974, we have consistently seen an unusual amount of sensitivity of experimental signals to parameter values such as microwave frequency and field strength. Efforts are under way to try and quantify this.

THE DBE PROBLEM IN QUANTUM MECHANICS

Let us now look at the quantum picture of our hydrogen-in-microwaves system. Consider the response of the atom to the microwaves as the field strength is increased. We recall that the parabolic atomic energy eigenstates have mixed parity. As a consequence, there are nonzero diagonal matrix elements of the electric dipole operator (42) that imply permanent dipole moments and a first order interaction with the microwave field. This is responsible for the low-field modulation of the wavefunction that produces the sideband spectrum of Figure 1a; this is an effect associated with a single parabolic state that begins to be observabe at microwave fields typically 20 times lower than that for observable ionization. As the field strength is increased a factor of 5 or 10, adjacent modulated states begin to couple via the off-diagonal matrix elements of the dipole interaction, and resonant multiphoton transitions of order 2 to 5 between states with adjacent values of n are observable (8,19). This requires the larger fields because the order of the multiphoton process is one unit larger, because the off-diagonal matrix element is 5 times smaller than the diagonal one with the same n, and because there is destructive interference between a large number of the terms in the formula for the transition probability (24). Increasing the field strength another factor of 2 or so makes the frequency-independent n-changing transitions of Figure 15 observable. Finally, another relatively small field increase brings in the higher order resonant and frequency-independent n-changing transitions accompanied by observable ionization probabilities, see Figure 14.

It is clear that the phenomena occurring at the higher field strengths involve strong coupling of many free-atom quantum states, a situation amenable only to numerical calculations as is appropriate for a nonintegrable system. Even the $\Delta n = \pm 1$ microwave absorption via the quasicontinuum discussed above is a multistate phenomenon, since at least four free-atom states must be included in numerical calculations in order to roughly obtain the frequency-independent n-changing probabilities. Nevertheless, one may try to understand the many-state problem by focusing on two-state interactions. One way to do this for the quasicontinuum absorption is to consider the two-state, one-photon. Rabi-flopping probability given by

$$P(t) = \left[\frac{(z_{12}F_\mu)^2}{(\omega-\omega_{12})^2+(z_{12}F_\mu)^2}\right] \sin^2\{\frac{1}{2}\,[(\omega-\omega_{12})^2 + (z_{12}F_\mu)^2]^{1/2}t\} \qquad (6)$$

Here z_{12} is the off-diagonal coupling matrix element, F_μ the microwave field strength, ω the microwave frequency, ω_{12} the resonant frequency for a transition from state 1 to state 2, and t is the time. The transition probability is seen to oscillate in time at an off-resonance Rabi frequency given by the square root in Equation 6. The amplitude P_0 for the probability oscillation is given by the coefficient in front of the time-oscillating function. Let us now calculate this amplitude for typical values of the quantities in the equation, fixing them except for the quantum numbers of the initial and final states. For a given initial state, we select out the final state with largest value of P_0 and plot its value as a function of initital state quantum number n. Some Rabi-flopping population-changing probability amplitude curves obtained this way are shown in Figure 22. We find empirically that when the value of P_0 is 0.5 or larger for the initial

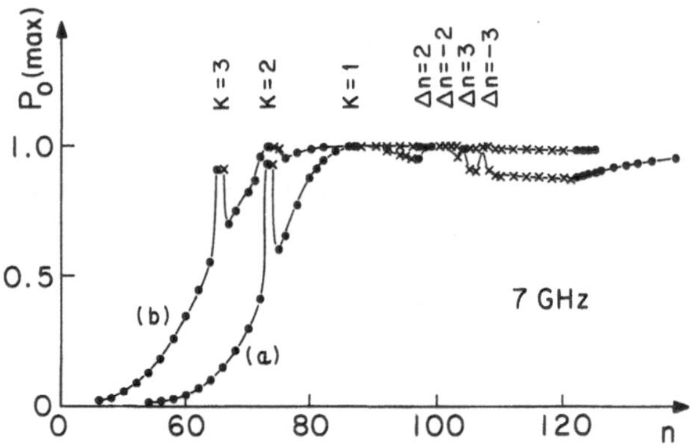

Fig. 22. Maximum two-state Rabi-flopping population changing probability amplitudes as a function of the quantum number of the chosen state, for (a) a field strength F_μ=3.1 V/cm, weakly ionizing at n_0=63, and (b) a field strength F_μ=10.3 V/cm, strongly ionizing at n_0=63. Case (b) is characteristic of the effective formation of an energy quasicontinuum for quantum numbers down to n_0=63.

state and all higher states, one has observable probabilities for ionization and for the classically stochastic, many-photon absorption that leads to ionization. Thus, we have the simple rough estimate for the threshold for classically stochastic behavior given by P_0=0.5, implying a threshold microwave field strength of

$$F_\mu(\text{threshold}) = \frac{3}{n^5}\ \text{a.u.}, \qquad \omega < \omega_{12} \tag{7}$$

This quantum condition produces the $F(\omega_0) \sim \omega_0^{-1/3}$ curve shown in Figure 21; the agreement with the ionization threshold curves are satisfactory, given that the latter is for a few hundred field oscillations while the quantum condition is based upon n-changing data for 3000 oscillations that is additionally enhanced by the presence of a static electric field. This agreement further supports the idea that even quantum mechanically the ionization is the result of many-photon absorption within a quasicontinuum, i.e. it is classically stochastic.

When the microwave frequency is increased to the point where one-photon transitions jump over one or more energy levels in the ladder of Figure 9, the maximal

Rabi-flopping probability amplitudes begin to drop from what they otherwise might be. How large the drop is depends upon the field strength. It can be as large as the ratio of squares of the matrix elements of Equation 4 for the two values of Δn, the actual value and the value unity. This effect tends to produce an upper-n cutoff to the classically stochastic region, but the effect is weakened by one-photon ionization linking the microwave absorption within the bound states to the continuum. If an additional static electric field is present, static field ionization would also do this. If the upper-n bottleneck is effectively absent, then the quasicontinuum and true continuum can be expected to act as one continuum, with the quasicontinuum region becoming part of a truly classically stochastic region.

What we have meant in this paper as quasichaos or classically stochastic behavior requires quantum many-photon absorption via the one-photon quasicontinuum. True chaos presumably would at least require absorption via k-photon quasicontinua for all k.

We point out that while a net number of order 100 microwave photons are required for microwave ionization of $n=60$ hydrogen atoms, still the Rabi-flopping frequencies or rates of Equation 6 can reach as high as one to five photon absorptions or emissions per microwave field oscillation or per electron orbit period. (Classically the picture of this situation is that the field-free character of the electron orbit changes one to five times during a field-free electron orbit time, a rather chaotic situation when the orbit change can occur in many competing different ways.) If ionization occurs after 3000 cycles, the accumulated number of photon absorption and emissions can be 10,000, 100 times the net number of photon absorptions. Thus the total number of photons involved is very large indeed, as needed for quantum mechanics to give classical results.

Let us now look at the equations of one of the simpler many-state quantum models used in the numerical calculations for the DBE system. In the first-order stretched-atom model, any static field is accounted for using field-free parabolic states and atom energy levels shifted by the first order static field Stark effect. Exact electric dipole matrix elements for real parabolic states of hydrogen atoms are used, which are essentially the same as those matrix elements for the truly one-dimensional quantum SSE model when n is 30 or larger (30). The stretched atom model can include the static field to higher order or exactly. It has the advantage over truly one-dimensional quantum models in that one can include off-ladder transitions in a systematic way, accounting for the small deviations from one-dimensionality that are expected. In the first-order version of the model, Schroedinger's equation reduces to a set of coupled differential equations for the time-dependent amplitudes of the wavefunction expanded in the parabolic basis:

$$i\,C_n(t) = E_n C_n + F_\mu \sin(\omega t+\phi) \sum_{n'} z_{nn'} C_{n'}(t) \quad (8)$$

The matrix elements and energies are given by

$$Z_{nn} = 1.5\,n(1-n) \quad (9)$$

$$Z_{nn'} = 0.32\,n^2, \qquad n' = n\pm 1$$

$$Z_{nn'} = 0.11\,n^2, \qquad n' = n\pm 2, \text{ etc;}$$

$$E_n = -\frac{1}{2n^2} + \frac{3}{2} n(1-n)F_s$$

One is interested in the time development of the vector C(t) of amplitudes. If the microwave field strength is constant, then one can utilize Floquet theory (30,52,53) and propagate the system one microwave period at a time by multiple matrix multiplication of C(t) by the one-period, time-evolution operator T:

$$C(\tau) = T(n_i, F_\mu, \omega, \phi)C(0) \tag{10}$$

once the operator is obtained numerically by integration of equation (8) over a first microwave oscillation. Some of the results obtained this way already have been shown in Figures 17, 19, 20, and 21. We later turn to further evidence for quantum quasichaos that is revealed by such numerical quantum calculations.

QUALITATIVE RELATIONSHIPS BETWEEN THE QUANTUM AND CLASSICAL PICTURES OF THE DBE PROBLEM: SOME EVIDENCE FOR A QUANTUM NONLINEAR DYNAMICS

We expect correspondences between the quantum and classical pictures discussed above, since the (coarse grained) Wigner transform of wavefunctions produces a correspondence between classical and quantum densities of states in phase space (43-45). Thus we seek quantum analogues of the principal features of the phase space structure, see Figure 4. These are the resonance islands, the horizontal KAM surfaces, and the stochastic region.

We can make a simple connection between the classical and quantum resonances that is accurate in the limit of large n and when the quantum structure of the atom in the field is not too different from that of the free atom. We recall that the winding number of the periodic trajectory at the center of a classical resonance island is the frequency ratio

$$W = \omega/\omega_n \tag{11}$$

At large n, the free-atom correspondence between quantum energy level separation ΔE_n and classical orbit frequency ω_n is

$$\Delta E_n = \frac{\Delta n}{n^3}, \tag{12}$$

The condition for a quantum resonance is that an integral number k of microwave photons must equal the energy level separation:

$$k\,\omega = \Delta E_n. \tag{13}$$

Combining these equations, we find that the winding number characterizing the classical resonance is the ratio of the two integers characterizing the quantum resonance:

$$W = \Delta n/k. \tag{14}$$

Thus, insofar as the quantum resonances for n-changing up and down can be considered as one resonance, the classical and quantum resonances are in one-to-one correspondence.

Classical period doubling splits one island in phase space into two. Thus, in quantum nonlinear dynamics this should be a kind of splitting of a resonance state into two.

We have earlier mentioned the classical resonance overlap criterion for the threshold for classically stochastic behavior. But the quantum estimate for this threshold that was given in the previous section is just a rough quantum resonance overlap criterion, since the on-resonance Rabi-flopping probability is just the resonance width. An improved quantum resonance overlap criterion would include the widths of two adjacent quantum resonances rather than that of just one. The idea of a quantum resonance overlap criterion has also been introduced in theoretical work on the related problem of an externally driven particle in an infinite square well (47).

From the above, we expect a correspondence in the threshold value of classical action in phase space needed for stochastic classical ionization with the threshold value in quantum number space for quantum quasichaotic ionization via quasicontinuum photon absorption. This correspondence exists and will be discussed in the next section.

N-changing absorption via the quasicontinuum (see Figure 18) and short-time ionization (see Figure 21) both are processes that can be inhibited by the interference of the resonances. Classically, this is because some trajectories at a given level of initial action are of the resonance island type, which are stabilized. Quantum mechanically, it is the coherent interference of the down n-changing transitions (26) that is at the root of this stabilization.

All this leads us to construct a semi-empirical proposed list of classical-quantum correspondences (48,49) that appears to apply for strongly-perturbed highly excited hydrogen atoms when the external driving frequency is less than the fundamental resonance orbit frequency of the atom's initial state. These are enumerated in Table I.

There is not yet any semiclassical theory to put these correspondences on a rigorous footing, but progress in this area appears promising (50,51,54).

Table I. Proposed correspondences between features of the classical and quantum nonlinear dynamics of the DBE system.

CLASSICAL	QUANTUM
Periodic motion and quasiperiodic motion on island KAM surfaces	Resonant n-changing with interfering restricted quasicontinuum n-changing
KAM surface breakup ("chaos")	Quasicontinuum n-changing to high n, and ionization
Quasiperiodic motion on non-island KAM surfaces	Restricted quasicontinuum n-changing

NUMERICAL EVIDENCE FOR CONNECTIONS BETWEEN THE QUANTUM AND CLASSICAL PICTURES: FURTHER SUPPORT FOR A QUANTUM NONLINEAR DYNAMICS

In this section we consider comparisons between quantum calculations and classical calculations that are not directly comparable with existing experimental data. Comparisons between the results of the two theories can be made without averaging over certain parameter values or initial conditions. In addition, convenient switching on and off of the microwave field can be selected; this is usually taken to be sudden in the sense that at t=0 the system is started off in the field, and at the final time t, it is still in the field.

We need to introduce the idea of the quasienergy states of our system, which for such time-dependent problems are the closest things to energy eigenstates. They are time independent in a way reminiscent of Fourier amplitudes. The quantum dynamics of a system driven with a periodic force of period τ can be discussed in terms of the time-independent quasienergy operator G defined by (52,53)

$$T(t,t_0) = P(t-t_0)\exp[-(i/\hbar)G(t-t_0)]. \tag{15}$$

$T(t,t_0)$ is the system's time propagator from time t_0 to time t. Here P(t) is some time dependent unitary operator periodic with period τ and having the value unity at t=0. Sine P(t) is also unity at times $t_0 + N\tau$ with N integral, the one-period time evolution operator $T(t_0 + \tau, t_0)$ that propagates the system over one oscillation of the field is diagonal in the quasienergy representation defined by the basis set $|\phi_\alpha>$ that satisfies

$$G\,|\phi_\alpha> = \omega_\alpha\,|\phi_\alpha>. \tag{16}$$

A wavefunction initially given by ψ_0 at $t=t_0$ will be mapped at time $t=t_0 + N\tau$ into

$$|\psi_N> = \sum_\alpha \exp[-(i/\hbar)\omega_\alpha N\tau] < \phi_\alpha\,|\psi_0>\,|\phi_\alpha>. \tag{17}$$

Thus, the time evolution of the wavefunction is entirely determined by the quasienergy mixing amplitudes $<\phi_\alpha|\psi_0>$ and the quasienergies ω_α. The spectral properties of G can be found by diagonalizing the one-period time evolution operator directly

$$T(\tau)|\phi_\alpha> \equiv T(t_0 + \tau, t_0)\,|\phi_\alpha> = \exp[-(i/\hbar)\omega_\alpha\tau]\,|\phi_\alpha> \tag{18}$$

to obtain the quasienergy states and quasienergies. The quasienergy states are a form of dressed atom states in that they characterize the system of the atom plus external field. In systems of low symmetry such as ours, the quasienergy variations with a parameter such as microwave field strength usually exhibit numerous pseudocrossings, a situation linked by some with the idea of quantum chaos (54).

The quasienergy states for parameter values of the data shown in Figure 20 have been obtained and their projections onto the physical parabolic states calculated (30). In Figure 23 we show these projections for the case of 21 states in the parabolic basis set, the value of n ranging from 55 to 75 with $n_0=63$. Ten eigenvectors are almost completely dominated by a single parabolic state, accounting for at least 98% of the state population. Five eigenvectors have strong mixing between two or three neighboring levels. The remaining six provide strong coupling among a group of states that extends up to the highest n value included in the

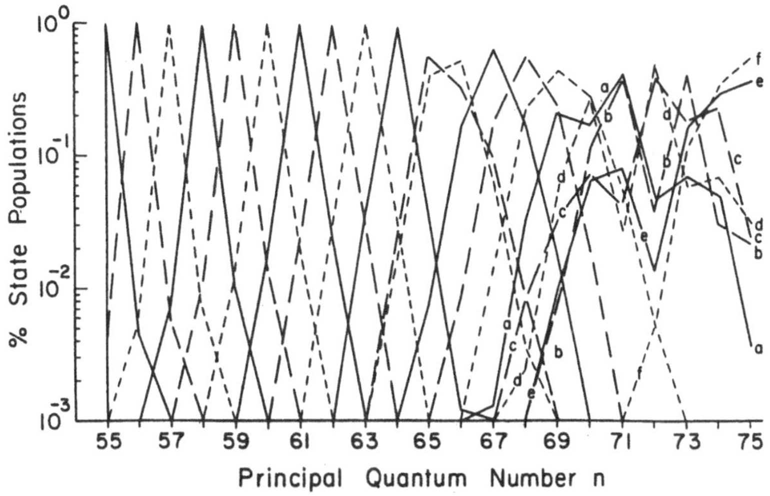

Fig. 23. An example of the unperturbed-state composition of 21 time-evolution or quasienergy eigenstates. To aide the eye, straight-line segments of different types connect the values defined at the discrete quantum numbers. A curve for one eigenstate has a fixed segment type and ends with a point of population 10^{-3}, larger than the calculated value. Three different zones of behavior can be seen, the regular zone 1 on the left, the transition zone 2, and the irregular or "chaotic" zone 3 on the right. Letters are used to indicate the six vectors of type 3; the remaining fifteen eigenvectors are localized with substantial coefficients from only a few free-atom states.

basis set. When the calculation is repeated with a higher upper limit to the basis set, the eigenvectors in the first two groups remain essentially unchanged, whereas those in the third group are very sensitive to such changes. A calculation with 41 basis states showed that there is strong mixing between the level $n=69$ and all higher levels in the basis set, up to $n=95$. Thus above a quantum threshold there is an explosion of physical state content of the quasienergy states; they become more or less delocalized in quantum number (action) space. Below the threshold, the quasienergy states are perturbatively localized, i.e. essentially localized to one physical state. This fact was first sensed as a large increase in basis set instability of those quantum-calculated final state probabilities having quantum number values above the classical stochastic threshold (29).

The variations in the above (partial) quantum delocalization thresholds with microwave frequency, microwave field strength, static field strength, and value of n_0 have all been explored (35). Similarly, the variations in the location of the top horizontal KAM surface in classical phase space have been studied for the same set of parameter values (14). Several different definitions of comparable thresholds for the quantum and classical phenomena have been introduced, with the conclusions being the same: the quantum delocalization thresholds and the classical stochastic behavior thresholds are the same to within one unit of quantum number. Table II shows such a comparison for a few sets of parameter values. The classical definition of the threshold

TABLE II. Comparison of (a) classical action thresholds for onset of stochastic motion and (b) quantal quantum number thresholds for basis-unstable delocalized behavior.

F_μ(V/cm)	F_s(V/cm)	ω(GHz)	n_0	(a)	(b)
9.1	8.2	7.1047	63		
		$\phi=0$:		68	69
		$\phi=\pi/2$:		62	62
9.1	0.0	7.1047	63	70	70
17.3	0.0	7.1047	63	67	66
16.0	5.45	7.0061	60	64	65
11.0	0.0	57.0	66	58	58
9.8	0.0	31.58	63	55	54

was chosen to be the lowest point of the top KAM surface, while the quantum definition was the uppermost parabolic state which had 1% or more quasienergy state content of the third or high-n type. It is this correspondence in quantum and classical thresholds that is responsible for the quantum final-state behavior shown in Figure 19.

Other quantum calculations have shown that final state distributions are the same as those obtained classically when the quasienergy states are completely delocalized (11). This occurs at very high microwave field strengths where the resonance islands have essentially disappeared, and the classical motion is completely stochastic. Further evidence has recently been obtained showing that the range of delocalized quantum quasienergy state n-delocalization corresponds more generally to the classical range of action of the motion (36).

Lastly, we point to the good agreement of recent microwave ionization quantum calculations with classical theory and with experiment, as is shown in Figure 21. Although many details remain to be worked out, it seems clear that quantum mechanics contains much of the classical dynamics in our system, at least for high microwave fields and short microwave exposure times. It seems reasonable to believe that quantum nonlinear dynamics that is parallel to classical nonlinear dynamics does exist, at least transiently.

DIFFERENCES IN THE PREDICTIONS OF QUANTUM AND CLASSICAL THEORY

The predictions of quantum mechanics differ from those of classical particle dynamics because of wave interference effects, both constructive and destructive, and because of classically unallowed tunneling within regions of phase space. The observable results of these wave effects will depend upon the coherence of and correlations in the amplitudes and phases of the component waves used to build up the wavefunction. A truly chaotic wavefunction would be expected to have random values for at least one of these.

The wave effects seem to particularly alter the classical role of resonances. Although classical physics exhibits the resonances, it doesn't treat them properly because of interference. In the quantum system there are two resonances for each classical one, corresponding to n changing up or down by equal amounts; these have somewhat different frequencies and yet can coherently interfere because of their finite widths. Furthermore, the resonant multiphoton absorption may coherently interfere with the quasicontinuum absorption.

Additional deviations from classical physics arise because the absorption of energy by the atom from the microwave field is quantized in its amount; the field is a quantized field and satisfies photon statistics.

The Heisenberg uncertainty principle requires that the system be largely distributed over a non-zero volume of phase space at any one time; thus quantum mechanics averages over collections of classical trajectories in phase space, with the results of the averaging not always easily predictable.

There is considerable experimental evidence for quantum modifications of classical microwave ionization probabilities. The classical results work best at the higher microwave field strengths; these fields correlate with shorter microwave exposure times because of the constraint that the ionization signal be observably large. At longer exposure times, ionization occurs at unexpectedly low fields (14). Even at the shorter times, such low field ionization occurs for some microwave frequencies (21, 55) with ionization dependences on field strength showing quantum step structure and sometimes not even being monotonic (8,28). Recent data for $n_0=63$ electrically polarized atoms in combined static and microwave fields is shown in Figure 24 (28). A

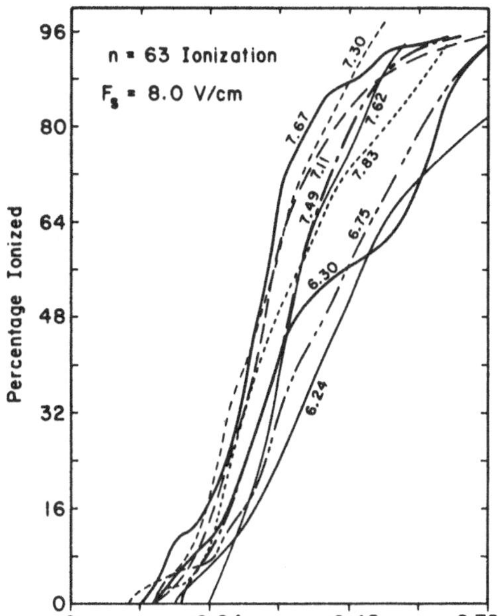

Fig. 24. Ionization probability curves for $n_0=63$ electrically polarized hydrogen atoms, for the indicated microwave frequencies in GHz. Many of the curves exhibit steps believed due to microwave power-shifted atomic resonances being swept past the applied microwave frequency.

physical explanation for the steps is the microwave field-induced shifting of resonances through the frequency of the applied field. Since quantum mechanics is needed to get the resonance structure effects correctly, the steps can be considered a quantal effect, although field-shifted resonances do also occur classically, see Figure 5.

Some numerical quantum calculations at high microwave frequencies and short exposure times have shown a quantum suppression of the classical time development of the microwave absorption, resulting in greatly reduced final state populations at large n (31). An example of this is shown in Figure 25. An approximate formula for the upper limit on the field strength range for the suppression to occur has been obtained; the result is the dotted line in Figure 21. One sees that the suppression should occur when the microwave frequency is larger than the electron orbit frequency, at the edge of the region where n-changing by more than one unit plays a larger role.

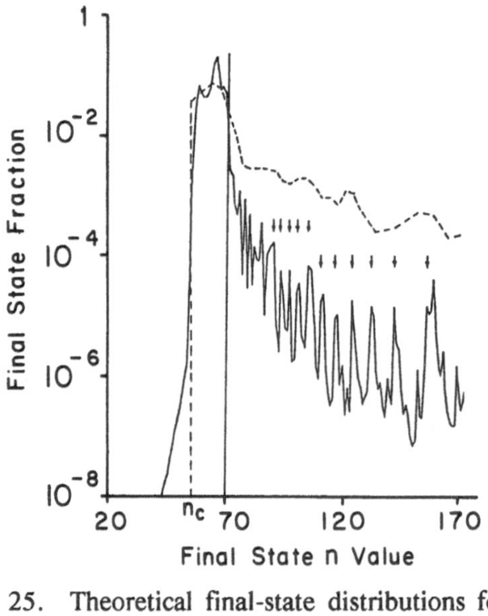

Fig. 25. Theoretical final-state distributions for $n_0=66$, a microwave frequency of 28 GHz and a microwave field strength of 8 V/cm. The atom is started out at n_0 well above the stochastic threshold value n_c near 58. The dashed line is a one-dimensional classical result, while the solid line is a one-dimensional quantum result displaying the quantum inhibition of the classical diffusive microwave energy absorption. Peaks separated by the microwave photon energy are seen for the quantum curve.

One mechanism for the quantum suppression is concerned with the effects of cantori, remnants of KAM surface structure in otherwise stochastic regions of phase space. Unless the KAM surface is totally destroyed over lengths of at least h, the motion of quantum waves past the remnants can be hindered via wave interference effects. This has been demonstrated in numerical studies of wavepacket motion in such regions (56) and is known to occur for the quantum kicked rotor (57).

A second possible quantum supression mechanism can occur in stochastic regions of phase space. It is an analog of an effect in condensed matter physics call Anderson localization (58), an effect determining the electrical conductivity of disordered solids and expected to produce metal-insulator phase transitions. In a perfect crystal at zero temperature, the electronic wave functions are the periodic Bloch states, and the electron density is delocalized throughout the crystal. If, however, the lattice site potential energies are random variables distributed according to a uniform distribution with a width W, and if W is much larger than the nearest-neightbor intersite interaction, then a spatial destructive quantum interference effect created by the strong disordering produces strongly localized wavefunctions about random centers in the lattice. The probability densities decay exponentially with the distance from these centers, the exponent being called the delocalization length. Now it has been established for a few model quantum driven oscillator problems that a mapping onto the Anderson localization, solid state problem exists (33,59). The action in the driven oscillator problems corresponds to the lattice sites in the solid state problem. Thus, a localization of quasienergy state wavefunctions in action space (quantum number space) is predicted, which should be reflected in a localization in final state quantum number probability distributions when primarily one quasienergy state is initially populated. What otherwise would be a wide final state probability distribution over a strong coupling region, characteristic of the region of classically stochastic behavior being populated during the exposure time, is greatly reduced in width by destructive interference, as is seen in Figure 25 (31). The delocalization length in quantum number space may be related to the Lyapunov exponent for the associated classical motion (34). The centers of localization depend upon the quasienergy, with states belonging to nearby quasienergies being centered in general around quantum number values that are far apart and states centered at nearby quantum numbers having, in general, quite different quasienergies. The amplitudes of the quasienergy wavefunctions are sensitive functions of the initial conditions. All this is a possible explanation of the data of Figure 18, as the possibility of initial preparation of a number of perturbatively localized QES along with a partially delocalized one exists, see Figure 23 (30). A detailed analysis remains to be carried out.

Ongoing experiments in our laboratory are searching for the Anderson localization at higher frequencies and are studying the effects on final state quantum number distributions of any further destruction of phase coherence produced by added broadband microwave noise (60).

REFERENCES

1. A. Hareda and H. Hasegawa, J. Phys. B. 16:L259 (1983).
2. D. Delande, A. Steitz, and J. C. Gay, in "Highly Excited States of Atoms and Molecules," S. S. Kano and M. Matasuzawa, ed., Fuji-Yoshida, Japan (1986); D. Delande and J. C. Gay, J. Phys. B. 19:L173 (1986); Phys. Rev. Lett. 57:2010 (1986).

3. G. Wunner, A. Steitz, U. Woelk, W. Schweizer, H. Herold, and H. Ruder, "Abstracts Tenth Int. Conf. Atomic Physics," H. Narum and I. Shimamura, ed., Tokyo, Japan, paper I-94 (1986); G. Wunner, U. Woelk, I. Zech, G. Zeller, T. Ertl, F. Geyer, W. Schweizer, and H. Ruder, preprint, (1986).
4. R. E. Wyatt, G. Hose, and H. S. Taylor, Phys. Rev. A 28:815 (1983).
5. D. Carter and P. Brumer, J. Chem. Phys. 77:4208 (1982).
6. J. R. Ackerhalt, H. W. Galbraith, and P. W. Milonni, Phys. Rev. Lett. 51:1259 (1983).
7. A. D. Bandrauk and L. Claveau, Proc. SPIE 664:paper 31 (1986).
8. J. E. Bayfield and L. A. Pinnaduwage, Phys. Rev. Lett 54:313 (1986).
9. J. E. Bayfield, L. D. Gardner, Y. Z. Gulkok, and S. D. Sharma, Phys. Rev. A 24:138 (1981).
10. A. J. Lichtenberg and M. A. Liberman, "Regular and Stochastic Motion," Springer-Verlag, New York (1983).
11. G. Casati, B. V. Chirikov, I. Guarneri, and D. L. Shepelyansky, Phys. Rev. Lett. 56:2437 (1986).
12. G. Casati, B. V. Chirikov, D. L. Shepelyansky, and I. Guarneri, Phys. Rev. Lett. 57:823 (1986).
13. R. V. Jensen, Phys. Rev. Lett. 54:2057 (1985); Phys. Rev. A 30:386 (1984).
14. R. V. Jensen, private communication.
15. W. A. Lin and L. E. Reichl, Phys. Rev. A. 31:1136 (1985).
16. J. G. Leopold and D. Richards, J. Phys. B 18:3369 (1985).
17. J. E. Bayfield and P. M. Koch, Phys. Rev. Lett. 33:258 (1974).
18. P. M. Koch and J. E. Bayfield, "Abstracts Fourth Int. Conf. Atomic Physics," J. Kowalski and H. G. Weber, ed., Heidelberg, Germany (1974).
19. J. E. Bayfield, L. D. Gardner, and P. M. Koch, Phys. Rev. Lett. 39:76 (1977).
20. J. G. Leopold and I. C. Percival, J. Phys. B 12:709 (1979).
21. K. A. H. van Leeuven, G. V. Oppen, S. S. Resnick, J. B. Bowlin, P. M. Koch, R. V. Jensen, O. Rath, D. Richards, and J. G. Leopold, Phys. Rev. Lett. 55:2231 (1985).
22. D. Kleppner, in "Radio Recombination Lines," P. Shaver, ed., (1980).
23. P. M. Koch and D. R. Mariani, J. Phys. B 13:L645 (1980); Phys. Rev. Lett. 46:1275 (1981).
24. J. N. Bardsley and B. Sundaram, Phys. Rev. A 32:689 (1985).
25. D. L. Shepelyansky, in "Chaotic Behavior in Quantum Systems," G. Casati, ed., Plenum Press, New York (1985).
26. J. Bayfield and L. A. Pinnaduwage, J. Phys. B 18:L49 (1985).
27. R. J. Damburg and V. V. Kolosov, in "Rydberg States of Atoms and Molecules," R. F. Stebbings and F. B. Dunning, ed., Cambridge University Press, Cambridge, UK (1983).
28. L. A. Pinnaduwage, Ph.D. Dissertation, University of Pittsburgh (1986).
29. J. E. Bayfield, in "Fundamental Aspects of Quantum Theory," V. Gorini and A. Frigerio, ed., Plenum Press, New York (1986).
30. J. N. Bardsley, B. Sundaram, L. A. Pinnaduwage, and J. E. Bayfield, Phys. Rev. Lett. 56:1007 (1986).
31. G. Casati, B. V. Chirikov, and D. L. Shepelyansky, Phys. Rev. Lett. 53:2525 (1984).
32. H. G. Muller, A. Tip, and M. J. Van der Wiel, J. Phys. B 16:L679 (1983).
33. D. R. Grempel, R. E. Prange, and S. Fishman, Phys. Rev. A 29:1639 (1984).

34. D. L. Shepelyansky, Phys. Rev. Lett 56:677 (1986).
35. B. Sundaram, Ph.D. dissertation, University of Pittsburgh (1986).
36. R. V. Jensen, in "Highly Excited States of Atoms and Molecules," S. S. Kano and M. Matsuzawa, ed., Fuji-Yoshida, Japan (1986).
37. J. N. Bardsley and M. J. Comella, J. Phys. B 19:L565 (1986).
38. J. E. Bayfield, Proc. SPIE 664:paper 18 (1986).
39. P. Pillet, H. B. Van Linden van den Heuvell, W. W. Smith, R. Kachru, N. H. Tran, and T. F. Gallagher, Phys. Rev. A 30:280 (1984).
40. H. B. van Linden van den Heuvell and T. F. Gallagher, Phys. Rev. A 32:1495 (1985).
41. D. R. Mariani, W. van de Water, P. M. Koch, and T. Bergeman, Phys. Rev. A 50:1261 (1983).
42. H. A. Bethe and E. E. Salpeter, "Quantum. Mechanics of One- and Two-Electron Atoms," Springer-Verlag, Berlin, (1957).
43. E. J. Heller and M. J. Davis, J. Phys. Chem. 86:2118 (1982).
44. S.-J. Chang and K.-J. Shi, Phys. Rev. Lett. 55:269 (1985).
45. K. Nakamura, Y. Okazaki, and A. R. Bishop, Phys. Rev. Lett. 57:5 (1986).
46. B. V. Chirikov, Phys. Rpt. 52:263 (1979).
47. L. E. Reichl and W. A. Lin, Phys. Rev. A 33:3598 (1986).
48. J. E. Bayfield, in "Highly Excited States of Atoms and Molecules," S. S. Kano and M. Matasuzawa, ed., Fuji-Yoshida, Japan (1986).
49. J. E. Bayfield and L. A. Pinnaduwage, in "The Physics of Phase Space," Lecture Notes in Physics xxx, Springer-Verlag (1986).
50. J. R. Klauder, Phys. Rev. Lett. 56:897 (1986); "Global, Uniform Semiclassical Approximations for Quantum Systems on the Half Line," preprint (1986).
51. I. B. Bernstein, Phys. Rev. A 32:1 (1985).
52. A. G. Fainshtein, N. L. Manakov, and L. P. Rappoport, J. Phys. B 11:2561 (1978).
53. K. Yajima, J. Math. Soc. Japan 29:729 (1977).
54. M. V. Berry, in "Chaotic Behavior of Quantum Systems," G. Iuoss, R. H. G. Helleman, and R. Stora, ed., North-Holland, Amsterdam (1983).
55. P. M. Koch, J. Physiq. Colloq. 43:C2-187 (1982).
56. R. C. Brown and R. E. Wyatt, Phys. Rev. Lett. 57:1 (1986).
57. T. Geisel, G. Radons, and J. Rubner, "KAM-barriers in the Quantum Mechanics of Chaotic Systems," submitted for publication.
58. R. Abou-Chacra, P. W. Anderson, and D. J. Thouless, J. Phys. C 6:1734 (1973).
59. R. Blumel, S. Fishman, and U. Smilansky, J. Chem. Phys. 84:2604 (1986).
60. E. Ott, T. M. Antonsen, Jr., and J. D. Hanson, Phys. Rev. Lett. 53:2187 (1984).

RELEVANCE OF CHAOS IN QUANTUM MECHANICS:
THE HYDROGEN ATOM IN THE MICROWAVE FIELD

Giulio Casati

Dipartimento di Fisica dell'Università, Via Celoria 16
20133 Milano, Italy

Abstract

We discuss some recent analytical and numerical results which show that, for highly excited hydrogen atoms irradiated by microwaves, a large ionization peak occurs at frequencies much below those required for the conventional one-photon photoelectric effect. A striking property of this ionization peak, which can be much higher than the usual photoelectric effect, is that its frequency width is jointly determined by two independent effects: the classical chaotic threshold and the quantum delocalization border. Moreover a simple reversibility test shows that the quantum motion proves to be perfectly stable in sharp contrast to the high instability and unpredictability of classical chaotic motion.

Introduction

More than three years ago the first international conference on Quantum Chaos was held here in Villa Olmo. Later, this subject was discussed in several national and international meetings, a fact which indicates the rapid growth of interest in the field. A similar rapid development took place in the study of classical dynamical systems after the first meeting held in Como in 1977.

On the other hand it is quite natural to expect that the manifestations of chaos at the microscopic level, in the domain of quantum mechanics, are of great relevance not only for atomic, molecular and nuclear physics, but also for some fundamental problems in quantum mechanics. In this last respect a main contribution to our present understanding of quantum mechanics has been given here several years ago in 1927. After two years of silence, Niels Bohr announced here the principle of complementarity and presented his interpretation of quantum mechanics which became known as the "Copenhagen interpretation"(1). The audience included, Born, Bose, Bragg, de Broglie, Brillouin, A.H. Compton, Debye, Duane, Fermi, Frank, Frenkel, Gerlach, Hall, Heisenberg, Kramers, Lande, Langmuir, Levi-Civita, Lorentz, Majorana, Marconi, Millikan, von Neumann, Paschen, Pauli, Planck, Saha, Smekal, Sommerfeld, Stern, Tolman, Volterra, Wentzel, Wigner, Zeeman.

Turning now to specifics, we will consider in the following the problem of excitation and ionization of an hydrogen atom under an external, linearly polarized, monochromatic electric field. The motivation of the choice of this model is related to the results obtained from the analysis of a simple quantum system, the so-called δ-kicked rotator[2] . This analysis shows that quantum mechanics, typically, appear to suppress the classical chaotic motion. More precisely, the classical δ-kicked rotator exhibits a "stochastic transition" for a certain value of the perturbation strength[3]; above this value, the system behaves in a quite disordered way, its motion being almost the same as if the perturbation were a random and not a deterministic one. In particular, its energy grows indefinitely, according to a diffusive law. On the contrary, in the corresponding quantum system this chaotic diffusion is suppressed[2,4]: apart for an exceptional set of values of the kicks period[5], the quantum interference effects lead to a complete arrest of the diffusive growth of energy after a finite time. As a result, only a finite number of unperturbed levels are significantly excited during the whole course of quantum evolution.

The interesting question arises whether such quantum suppression is a property only of the particular model studied or, instead, is a general occurence in quantum mechanics. Also one may inquire whether such suppression of classical chaos can be observed in a laboratory experiment.

In order to answer the above questions we consider the ionization mechanisms for the hydrogen atom when a linearly polarized monochromatic electric field induces transitions from initial states having principal quantum number $n_0 \gg 1$. For simplicity, we restrict ourselves to the study of very elongated quantum states having parabolic quantum numbers $n_1 = n_0 - 1$, $n_2 = 0$, and magnetic quantum numbers $m = 0$. Since to a good approximation these wave functions have nonzero values only along the direction of the applied field, we are at liberty here to treat the hydrogen atom as if it were one dimensional, having the Hamiltonian:

$$H(I, \phi, t) = -1/2I^2 + \epsilon\, x(I, \phi) \cos \omega t; \quad x > O \qquad \omega = 2\pi/T \tag{1}$$

Where ϵ and ω are the microwave electric field strength and frequency, respectively, in atomic units. Here, $x(I, \phi)$ is the x coordinate of the electron, expressed as a function of action angle variables of the unperturbed atom. The validity of this one-dimensional approximation is due to the small value of matrix elements for transition having $\Delta n_2 \neq 0$. As a consequence, the atom remains one dimensional during the relevant interaction times[6,7]. This important fact has also been verified in laboratory experiments which produce such states and excite them by microwave fields[8].

The Quantum Delocalization Border

The classical system (1) exhibit a transition to chaos when the coupling parameter ϵ exceeds some "stochasticity threshold", that can be estimated by means of Chirikov's resonance overlapping criterion[9,10]. This transition occurs for

$$\epsilon_0 > \epsilon_s = 1/(50\omega_0^{1/3}); \quad \epsilon_0 = \epsilon n_0^4, \quad \omega_0 = \omega n_0^3 \tag{2}$$

for $\omega_0 > 1$. (When $\omega_0 < 1$, the analysis of the chaotic transition is more involved). Above this threshold, the motion is more conveniently described in statistical terms. Specifically, considering an ensemble of trajectories leaving with a fixed value I_0 of the action and with randomly distributed phases ϕ, one finds that the distribution function f(I, t) in action space is well approximated by the solution of the Fokker-Planck equation:

$$\partial f/\partial t = 1/2\, \partial/\partial I\, (D(I)\, \partial f/\partial I). \tag{3}$$

The diffusion coefficient D(I) depends on action I : $D(I) \approx \epsilon^2 I^3/(\pi\omega^{4/3})$. A quite remarkable feature of the diffusion ruled by eq. (3) is that the moment $<I> = \int |f(I,q\ t)|\, dI$ grows to infinity-i.e., the atom ionizes - in a finite time $t_I \approx 2\omega^{4/3}/(\epsilon^2 n_0)$.

The quantized version of model (1) involves the solution of the Schroedinger equation:

$$i\, \partial\psi/\partial t = H_0\psi + \epsilon\, V(t)\psi \tag{4}$$

with V(t) an operator depending periodically on time.

A particularly convenient way to study equations of this type is introducing the Floquet operator S = U(0, T) where U(s, s+t) is the unitary operator which gives the evolution of states ψ over the time t according to $\psi(t+s) = U(s, s+t)\psi(s)$. Indeed, in order to analyze the long-time behaviour of the solutions of eq.(4), it is sufficient to study the iterates S of this Floquet operator [11].

A first qualitative classification of various types of behaviour that solutions of (4) can show is provided by the nature of the quasy-energy spectrum, which is by definition the spectrum of the self-adjoint operator G such that S = exp(iG). As a matter of fact the hydrogen atom unperturbed Hamiltonian H_0 possesses a discrete spectral component, and one is interested in the time evolution of states initially coinciding with some unperturbed eigenstate. Then, it can be shown [11] that a continuous quasi-energy spectrum would enforce an indefinite spreading of such wave packets over the unperturbed spectrum. Instead, a pure point quasi-energy spectrum would be associated with a recurrent behaviour of the wave packets.

In the case of the hydrogen atom model (1) the question about the nature of the quasi-energy spectrum is clear. Indeed, it is known that the q.e. spectrum is absolutely continuous [12]. This means that, unlike the classical atom, the quantum atom will eventually ionize, no matter how small ϵ. The corresponding ionization mechanism, however, at least for small ϵ has nothing to do with the classical chaotic phenomenon, and involves very long time scales. From the mathematical point of view, this entails that identifying the nature of the q.e. spectrum is not enough for answering the question of the existence of "quantum chaos". Indeed, this question calls for the analysis of time evolution of wave packets over the time scale in which quantum and classical evolution may be expected to agree to some extent.

In the present case, the short time behaviour of the atom is dominated by resonances (in the sense of scattering theory). In other words, the q.e. spectrum is absolutely continous, but there are poles of the resolvent operator lying very close to the real axis [12]. For practical purposes, it can still be assumed that the q.e. spectrum has a pure point component, in which each level has a small width. Therefore, we can expect that the same mechanism which leads to quantum localization of the classical dynamical chaos in the δ-kicked rotator case will be working here, too. There is an essential difference, however; indeed, if it is possible for the hydrogen atom case to choose such parameter values that the spread $\Delta n(t)$ of the wave packet over the unperturbed levels follows the classical diffusion for a time larger than the classical chaotic ionization time, then we may expect "diffusive" ionization also in the quantum case. Conditions for this <u>delocalization</u> phenomenon have been derived and it was found that, for $\omega_0 > 1$ delocalization will occur for [7,13]

$$\epsilon_0 > \epsilon_s \quad , \quad \epsilon_0 > \epsilon_q = \omega_0^{7/6}/\sqrt{(6n_0)} . \tag{5}$$

The model (1) was the object of extensive numerical investigations, which fully confirm the main results of the above discussion and in particular the estimate (5) of the <u>quantum delocalization border</u> ϵ_q. Namely, three distinct quantum regimes can be observed according to the particular choice of ϵ_0, ω_0 and of the initial state n_0. If ϵ_0 lies below both the classical stochasticity border ϵ_s and the delocalization border ϵ_q, then both the classical and the quantum model will exhibit localization in action space; this is the case of Fig. 1 where $\omega_0 = 1.5$, $n_0 = 100$, $\epsilon_0 = 0.1$ and where the quantum and classical distributions over unperturbed actions are shown after 120 periods of the external field.

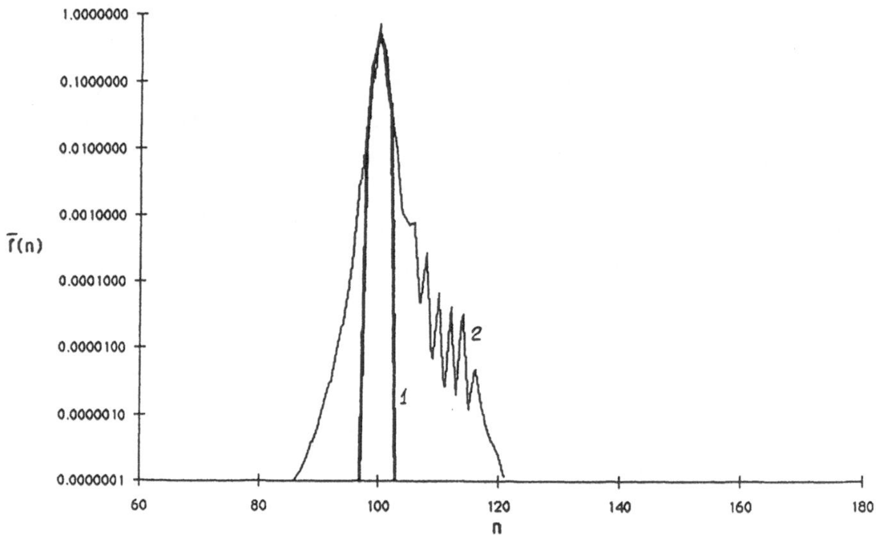

Fig. 1 Classical (1) and quantum (2) distribution function $\bar{f}_n$ averaged over 40 values of $\tau = \omega t/(2\pi)$ within the interval $80 < \tau < 120$. Here $\omega_0 = 1.5$, $\epsilon_0 = 0.1$, $n_0 = 100$. For these parameter values, both classical and quantum packets are localized. Notice the small quantum tunneling through the classical Kolmogorov invariant curves.

If ϵ_0 lies in between the threshold ϵ_s and ϵ_q, we observe classical chaotic motion and quantum localization; here the classical atom would ionize diffusively, but the quantum one does not (Fig. 2 : $n_0 = 100$, $\omega_0 = 1.5$, $\epsilon_0 = 0.03$; distributions after 480 periods). In Fig 3 we show the comparison of classical and quantum probability of excitation above $n = 1.5\, n_0$, as functions of time. Quantum localization of classical chaos is here quite evident.

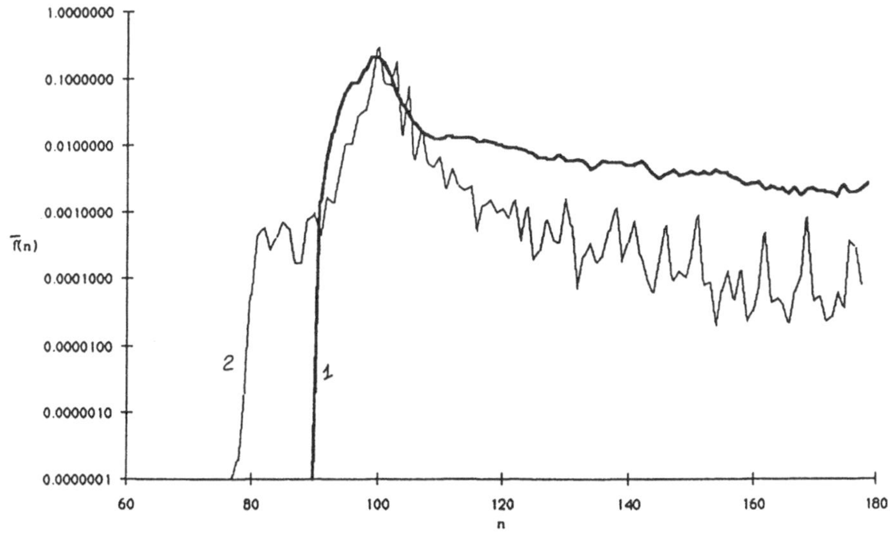

Fig. 2 Same as Fig. 1 but with $\epsilon_0 = 0.03$, the classical (1) and quantum (2) distribution functions are averaged over 40 values of τ within the interval $440 < \tau < 480$. In this case $\epsilon_s < \epsilon_0 < \epsilon_q$, and the quantum packet is localized as expected (besides a small resonant plateau). On the contrary, the classical packet is strongly diffusing as it is also shown in Fig. 3.

Finally if ϵ_0 exceeds both threshold, we observe a qualitative agreement between classical and quantum results (Fig. 4, $n_0 = 100$, $\omega_0 = 1.5$, $\epsilon_0 = 0.15$). Here a quantum ionization mechanism is at work, that cannot be understood in terms of standard perturbation theory. Indeed, this mechanism is effective at much lower frequencies than those ($\omega_0 > n_0/2$) predicted by the perturbative theory of the photoelectric effect.

Underthreshold ionization

Due to the delocalization phenomenon, the dependence of the ionization probability W_I on frequency at fixed field intensity displays a number of unexpected features. In Fig. 5, we plot W_I against ω_0 for $n_0 = 66$, $\epsilon_0 = 0.05$. First of all, since the classical diffusion is due to resonances between the external field and the harmonics of the electron motion, it takes place only for frequencies ω_0 above some critical value ω_c of the order of 1, when $\epsilon_0 > 1/50$ (for $\epsilon_0 < 1/50$, the value ω_c has to be determined from eq. (5)). Notice that, since the diffusion coefficient D that determines the rate of diffusive excitation is decreasing with ω_0 at fixed ϵ, the critical value ω_c for frequency has classically a marked threshold character for diffusive excitation.

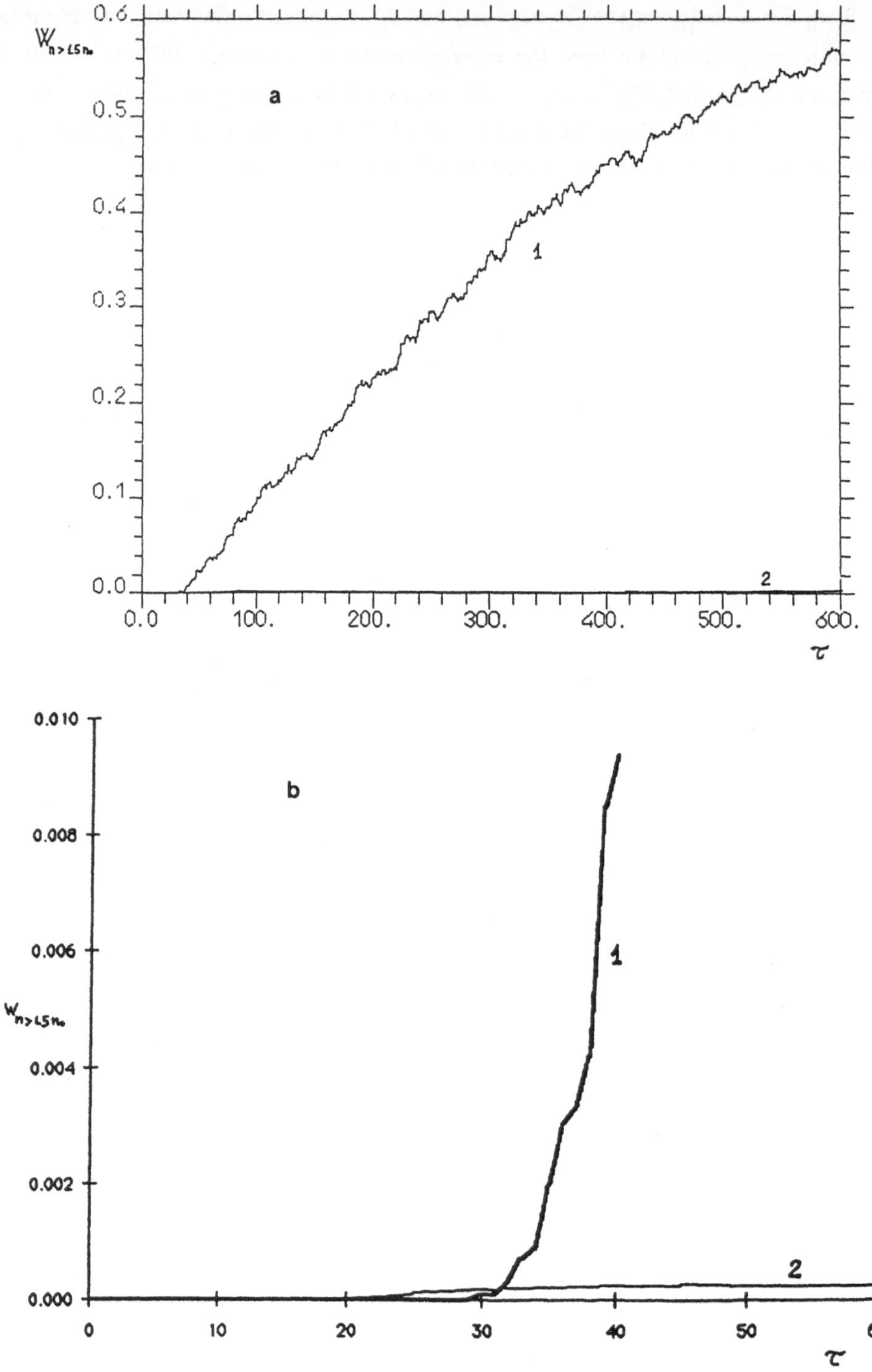

Fig. 3 Classical (1) and quantum (2) total probability $W_{n>1.5n_0}$ above level $n = 1.5n_0$ or the case of Fig. 2. The classical ionization probability is order of magnitudes higher than the quantum one (Fig. 3a). The comparison of the two ionization probabilities for short times is shown in Fig. 3b.

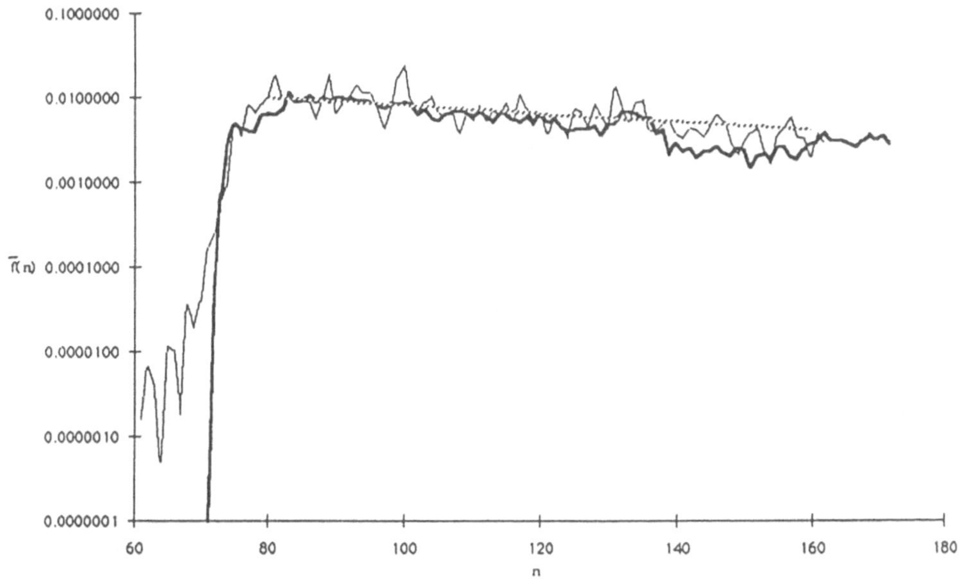

Fig. 4 Same as Fig. 1 but with $\epsilon_0 = 0.15$. Here, the classical (▬) and quantum (_) distribution functions are averaged over 40 values of τ within the interval $0 < \tau 40$. In this case $e_0 > \epsilon_q$ and both classical and quantum motion obey a diffusion law given by Fokker–Planck equation. The dotted line gives the solution of the Fokker–Planck equation.

In Fig. 5 we see that for the given field the quantum atom is delocalized at $\omega > \omega_c$, and quantum and classical excitation go in approximately the same way. However, since $\epsilon_q > \epsilon_s$, at $\omega_d \approx (6n_0\epsilon_0^2)^{3/7}$ one falls below the delocalization border; then localization occurs, and the ionization probability drops as compared to the classical value. Upon further increasing ω_0, the one photon threshold is reached; beyond that, the numerically computed ionization probability is in perfect agreement with standard theoretical predictions.

As is apparent from Fig. 5, diffusive ionization occurs at very low frequencies as compared to the one photon threshold. Even more strikingly, its rate is much higher.

Dynamical Stability of Quantum Motion

In the previous section we have seen that under certain conditions (field intensity $\epsilon_0 > \epsilon_q$, $\omega_0 > 1$) the quantum hydrogen atom mimics the classical diffusive behaviour which is ruled by the Fokker-Planck equation. The question has been frequently asked whether or not, in such situations, the quantum motion is as truly chaotic as the classical one. In this respect an impressive demonstration of the practical unpredictability of classical motion is the fact that Liouville equation, though exactly reversible in principle, cannot be time-reversed in practice on computer experiments. Indeed due to inevitable numerical errors, the memory of the initial distribution is completely lost after a while.

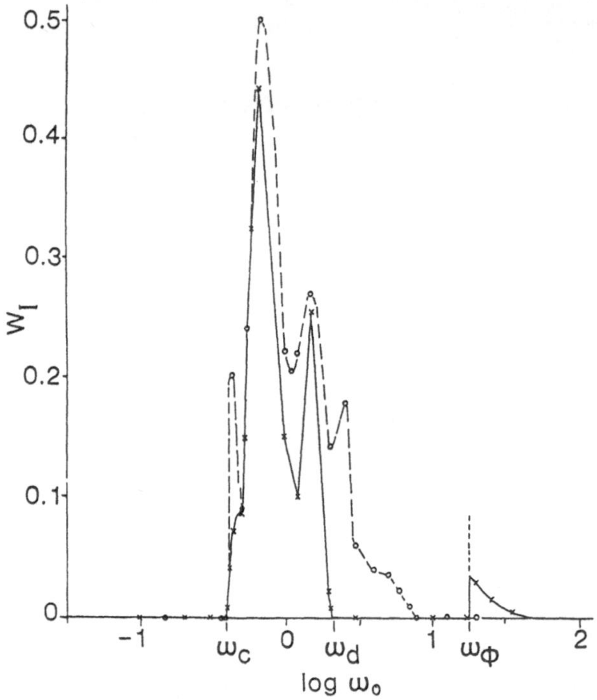

Fig. 5 Ionization probability $W_I = \Sigma_{n>\bar{n}} |c_n|^2$ versus field frequency ω_0 after a number of periods $\tau = 40\omega_0$ which corresponds to the same real physical time t for all frequencies. Here $n_0 = 66$, $\epsilon_0 = 0.05$, $\bar{n} = 99$. (x) Quantum case, (o) Classical case. The position of the threshold ω_0 is due to the definition of W_I adopted here.

Thus the numerically computed time evolution will not reproduce backwards the history of the system, except for a short time; afterwards, approach to equilibrium will again show up, and the initial distribution will be lost forever. Needless to say, the exactly computed evolution would in any case find its way back to the original state. We stress again however that the lack of "practical" reversibility in computer experiments unambiguously hints at a quite complex and sensitive nature of orbits and is a distinctive mark of true dynamical chaos. We may therefore inquire whether or not such instability character is a feature of the quantum motion also, at least in the delocalization regime. The fact that the quantum evolution is linear and unitary is of no relevance for the above question since the same property is shared by the classical evolution governed by Liouville equation.

In order to answer the above question [14] we performed the reversibility test with parameters and intial conditions above the delocalization border: $n_0 = 100$, $\epsilon_0 = 0.08$, $\omega_0 = 1.5$. In this way we provide the maximal chaos possible in a quantum system. Similarly to the quantum case where the single unperturbed state $n_0 = 100$ was excited, in the classical computations we chose the same parameters and analogous initial conditions. Namely, we computed 1000 trajectories with the same initial action $n_0 = 100$ and phases uniformly distributed within the interval $[0,2\pi]$.

In Fig. 6 the excitation probability is shown as a function of time τ. At time $\tau = 60$ we reversed the velocities in both the classical and quantum systems and followed the evolution for another sixty field periods.

The specular symmetry of the quantum curve about the time of reversal $\tau = 60$ demonstrates the reversibility and stability of the diffusive quantum motion. In contrast, the strong instability of the classical motion leads to a continuation of the diffusion process after velocity reversal (obviously except for a short time interval). The reversibility of the quantum motion is even more spectacular if we notice that part of the recovered initial state comes back from the continuum (see Fig. 6). The latter process is a peculiar kind of a coherent recombination.

We emphasize that the reversibility phenomenon does not depend on the particular initial condition which we used in this paper simply as an example.

In conclusion due to the stable character of quantum "chaotic" motion ***quantum dynamics has a much more deterministic character than classical dynamics*** and, unlike classical chaotic motion, the past history of a quantum system can always be recovered from its present state.

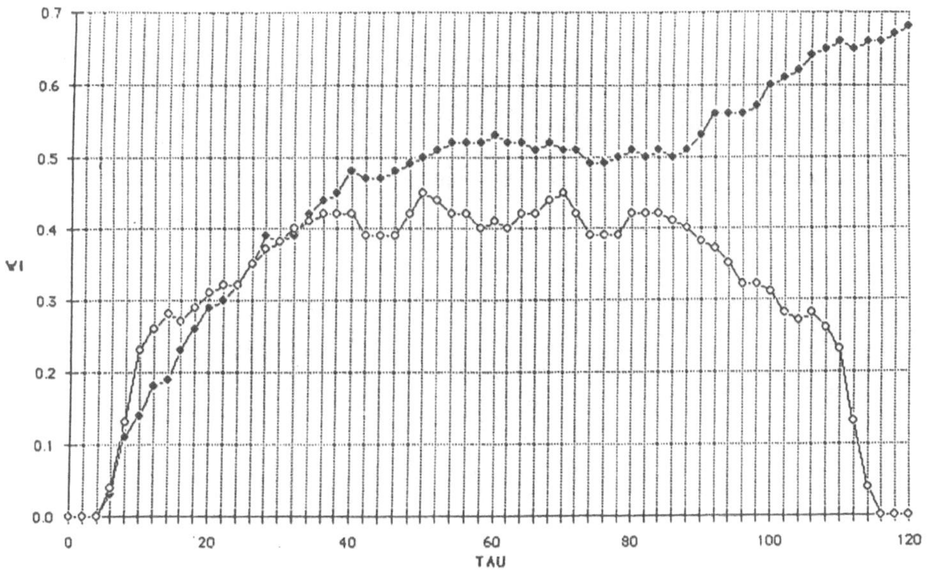

Fig. 6 Classical (solid lozenges) and quantum (open lozenges) ionization probability (excitation above the unperturbed level $n = 150$) as a function of time τ. Here $n_0 = 100$, $\epsilon_0 = 0.08$, $\omega_0 = 1.5$ and at $\tau = 60$ the velocities are reversed in both the classical and quantum system. Notice the perfect specular symmetry of the quantum curve about time of reversal $\tau = 60$.

References

1. Atti del Congresso internazionale dei Fisici (Zanichelli, Bologna, 1928) Vol. 2, pp. 565–588.
2. G. Casati, B.V. Chirikov, J. Ford, F.M. Izrailev, Lectures Notes in Physics Springer 93 (1979) 334.
3. B.V. Chirikov, Phys. Rep. 52 (1979), 263.
4. S. Fishman, D.R. Grempel, R.E. Prange, Phys. Rev. Lett. 49 (1982), 509; Phys. Rev. A29 (1984), 1639.
5. G. Casati I. Guarneri, Comm. Math. Phys. 95 (1984) 121.
6. D.L. Shepelyansky, Inst. of Nuclear Physics, Novosibirsk, Report No. 83–61 1983 (unpublished), and in "Chaotic Behaviour in Quantum Systems" edited by Giulio Casati (Plenum, New York, 1985), p. 187.
7. G. Casati, B.V. Chirikov, and D.L. Shepelyansky, Phys. Rev. Lett. 53, (1984) 2525.
8. J.E. Bayfield and L.A. Pinnaduwage, Phys. Rev. Lett. 54, (1985) 313, and J.Phys. B 118, (1985) 449.
9. N.B. Delone V.P. Krainov, and D.L. Shepelyansky, Usp. Fiz. Nauk. 140, (1983) 355 [Sov. Phys. Usp. 26, (1983) 551].
10. R.V. Jensen, Phys. Rev. A 130, (1984) 386.
11. J. Bellissard, in 'Trends and developments in the Eighties', Bielefeld Symposium 1983, Eds. S. Albeverio and Ph. Blanchard. World Publishing Company Singapore (1985).
12. S. Graffi, V. Grecchi, H.J. Silverstone, Ann. Inst. Poincare 42 (1985) 215.
13. G. Casati, B.V. Chirikov, I. Guarneri, D.L. Shepelyansky, "Relevance of Classical Chaos in Quantum Mechanics: the Hydrogen Atom in a Monochromatic Field". To appear in Physics Reports.
14. G. Casati, B.V. Chirikov, I. Guarneri, D.L. Shepelyansky, Phys. Rev. Lett. 56 (1986) 2437.

THE MEASUREMENT OF QUANTUM CHAOS

Sarben Sarkar and J. S. Satchell*

Centre for Theoretical Studies
Royal Signals and Radar Establishment
Malvern, UK

INTRODUCTION

The topics of quantum Hamiltonian, dissipative systems and measurements[1] are usually treated separately with little interplay between the fields. It is sometimes argued that it is possible to discuss Hamiltonian systems in ideal experimental situations where dissipation can be neglected. However, it is perhaps less widely appreciated that the very act of measurement on a Hamiltonian system introduces dissipation. Although the effect of measurements on quantum systems has been studied over the years, these studies have been almost exclusively for simple systems with linear Heisenberg equations of motion. Potentially chaotic systems are usually nonlinear. Although we do not have a definition of 'quantum chaos', in this paper it will denote, somewhat loosely, the quantum mechanical behaviour of systems which, at least, in the classical limit show complicated irregular time evolution.

In order to illustrate the effect of measurements on quantum chaotic systems we will consider two examples which between them contain the aspects that we wish to emphasise. One is the much studied kicked rotator[2] and the other is a two state system driven by a classical field with incommensurate frequencies[3]. Before we go into details we will remind ourselves of some elementary facts concerning classical and quantum mechanics which highlight the differences between the two dynamics. Let us first consider the problem of measuring (albeit in a very ideal and impractical way) the 3-momentum $\underline{p}_1$ of a classical particle of mass M moving on a frictionless table containing three identical balls of mass m (which constitute the measuring apparatus). The first initially stationary measuring ball is placed (somehow) in the path of the mass M. There is an elastic collision and the resultant measuring ball's momentum $\underline{q}_1$ is noted to arbitrary accuracy. As a result of the collision the mass M acquires a momentum $\underline{p}_2$. Since the collision is assumed to be elastic we have conservation of momentum and energy. This gives

$$\begin{aligned} \underline{p}_1 &= \underline{p}_2 + \underline{q}_1 \\ \frac{\underline{p}_1^2}{2M} &= \frac{\underline{p}_2^2}{2M} + \frac{\underline{q}_1^2}{2m} \end{aligned} \tag{1}$$

*Also at: Clarendon Laboratory, Oxford University, Oxford OX1 3PU, UK

Since $\underline{P}_1$ and $\underline{P}_2$ are unknown and we have only 4 equations determining them, there is not enough information to find $\underline{P}_1$. Fortunately we have two other particles of mass m which can collide in a similar fashion with the mass M. As a result we have for the second collision

$$\underline{P}_2 = \underline{P}_3 + \underline{q}_2$$

$$\frac{\underline{P}_2^2}{2M} = \frac{\underline{P}_3^2}{2M} + \frac{\underline{q}_2^2}{2m} \tag{2}$$

and for the third collision

$$\underline{P}_3 = \underline{P}_4 + \underline{q}_3$$

$$\frac{\underline{P}_3^2}{2M} = \frac{\underline{P}_4^2}{2M} + \frac{\underline{q}_3^2}{2m} \tag{3}$$

(with the obvious notation). Now there are twelve equations for twelve unknowns, ie $\underline{P}_1$, $\underline{P}_2$, $\underline{P}_3$ and $\underline{P}_4$. In order to find $\underline{P}_1$ we also have to calculate $\underline{P}_2$, $\underline{P}_3$ and $\underline{P}_4$. We would like to be able to have this measurement in such a way that $\underline{P}_1$, $\underline{P}_2$, $\underline{P}_3$ and $\underline{P}_4$ are not very different from each other, ie the system is only slightly disturbed by the measurement. From Eqn. (1) we deduce that

$$(\underline{P}_2 - \underline{P}_1)^2 = \underline{q}_1^2 = \frac{m}{M}(\underline{P}_1^2 - \underline{P}_2^2) < \frac{m}{M}\underline{P}_1^2 \tag{4}$$

and by making m/M arbitrarily small we can have $\underline{P}_2$ and $\underline{P}_1$ equal each other as closely as we wish. This is just an example of our ability in classical mechanics, in principle, to measure quantities accurately and with negligible disturbance.

We will now examine the effect of coupling together two quantum mechanical systems. For simplicity we consider the independent but identical harmonic oscillators of frequency ω in the absence of any friction (ie damping). The Hamiltonian H for the system is given by

$$H = \hbar\omega\, a^+a + \hbar\omega\, b^+b + g(a^+b + ab^+) \tag{5}$$

where (a^+,a) and (b^+,b) are the standard harmonic oscillator[4] creation and annihilation operator pairs for the two oscillators. For definiteness the 'a' harmonic oscillator will be regarded as the system being measured and the 'b' harmonic oscilator the apparatus (or meter) which is required to do this. The coupling of the oscillators is represented by the term involving g. If the 'a' oscillator happens initially to be in a number state with n quanta (a fact that we would like to discover) and we couple to it the 'b' oscillator in a state containing no quanta then the initial state of the system is

$$|n\rangle_a\ |o\rangle_b$$

where the subscripts 'a' and 'b' identify the oscillators to which the kets refer. Working in the Heisenberg representation it is easy to show that

$$a(t) = \tfrac{1}{2}(a(o)+b(o))\left(\exp\left(-\frac{i2gt}{\hbar}\right) - 1\right)\exp(-i(\omega - g/\hbar)t)$$
$$+\ a(o)\exp(-i(\omega - g/\hbar)t) \tag{6}$$

$$b(t) = (a(o) + b(o)) \exp(-i(\omega + g/\hbar)t)$$
$$- \tfrac{1}{2}(a(o) + b(o)) \left(\exp\left(-\frac{i2gt}{\hbar}\right) - 1\right) \exp(-i(\omega - g/\hbar)t)$$
$$- a(o) \exp(-i(\omega - g/\hbar)t) \qquad (7)$$

Hence we can deduce that

$${}_b\langle o|\, {}_a\langle n|\; a^+(t)\; a(t)\; |n\rangle_a\; |o\rangle_b \;=\; n \cos^2 \frac{gt}{\hbar} \qquad (8)$$

and

$${}_b\langle o|\, {}_a\langle n|\; b^+(t)\; b(t)\; |n\rangle_a\; |o\rangle_b \;=\; n \sin^2 \frac{gt}{\hbar} \qquad (9)$$

These quantities are respectively the expectation values of the number of quanta in oscillators 'a' and 'b'. As expected at $t = 0$ the mean number of quanta in oscillator 'a' is n whereas it is zero in oscillator 'b'. In time t of order $\pi\hbar/2g$ the situation is just the opposite. If we do not want to disturb oscillator 'a' in a short time, g has to be small, which is not surprising. However that which is surprising (from classical reasoning) is that

$${}_b\langle o|\, {}_a\langle n|\; (b^+(t)b(t))^2\; |n\rangle_a\; |o\rangle_b - ({}_b\langle o|{}_a\langle n|b^+(t)b(t)|n\rangle_a|o\rangle_b)^2$$

$$= n \sin^2 \frac{gt}{\hbar} \cos^2 \frac{gt}{\hbar} \qquad (10)$$

and so the standard deviation of the number of quanta in oscillator 'b' is $n^{\frac{1}{2}} \sin gt/\hbar \cos gt/\hbar$. This is large compared to the mean $n \sin^2 gt/\hbar$ (for n fixed). In fact the ratio of the two goes to infinity as g tends to 0. This suggests that if we are to measure the state of oscillator 'a' without appreciably disturbing if then the meter 'b' will give readings which are unreliable. As usual a better and better estimate of n can be obtained by considering a larger and larger assembly of systems together with their meters. The coupled oscillator system is not a measurement scheme since there is no mechanism given for extracting information out of oscillator 'b'. However, it, nonetheless, highlights the need for ensembles in order to perform good (ie reliable) measurements. We will now turn to the systems which have irregular behaviour classically.

The kicked rotator, in suitable units, has the Hamiltonian H_1

$$H_1 = \tfrac{1}{2} P_\theta^2 - k \cos\theta \sum_n \delta(t-n) \qquad (11)$$

where n is an integer, θ is the angle made by the rotator with the horizontal, and P_θ is the momentum conjugate to θ. This system has been influential in the study of quantum chaos since the observation that diffusive behaviour (present in the classical system for $k > 1$) is suppressed in the quantum mechanics. An interesting connection[4] has been found with the Anderson model for localisation in pseudo-random media; the intricate quantum mechanical interference responsible for localisation is now believed to be the reason for the lack of diffusive behaviour of the energy in the kicked rotator. Recently it has been noticed[5] that the addition of noise to the kicked rotator destroys this interference and the long time behaviour is again diffusive. It is thus interesting to examine the effect of measurement on this diffusive behaviour.

The second system that we will study does not have a classical analogue and is not chaotic in the sense of showing mixing. However it illustrates in an interesting way how the various parts of a complicated system can provide both a subsystem whose dynamics is of interest as well as the meter. We will introduce first the subsystem with Hamiltonian H_2

$$H_2 = \tfrac{1}{2}\hbar\omega\sigma_3 + i\hbar(B(t)\sigma^- - B^*(t)\sigma^+) \tag{12}$$

where B(t) is a classical field

$$B(t) = B' e^{i\omega' t} + B'' e^{i\omega'' t} + c.c.$$

and (ω, ω', and ω'') is an incommensurate set of frequencies. The σ_3 and $\sigma^\pm$ are 2 x 2 matrices

$$\sigma^+ = \begin{pmatrix} 0 & 1 \\ 0 & 0 \end{pmatrix}, \quad \sigma^- = \begin{pmatrix} 0 & 0 \\ 1 & 0 \end{pmatrix}, \quad \sigma_3 = \begin{pmatrix} 1 & 0 \\ 0 & -1 \end{pmatrix}.$$

The Hamiltonian H_2 is physically realised by a classically driven two state atom (with the energy difference between the two states $\hbar\omega$). The states of the two state atom are represented by

$$\begin{pmatrix} 1 \\ 0 \end{pmatrix} \quad \text{and} \quad \begin{pmatrix} 0 \\ 1 \end{pmatrix}.$$

It has been observed that the population difference between the two states has an irregular aperiodic time evolution.

MEASUREMENTS

The most straightforward way of examining the effect of measurements on a system is to explicitly incorporate a coupling to a meter[6], including any other systems that the meter may be coupled to. We shall adopt this procedure since we expect our qualitative conclusions to be valid for any bona fide measuring scheme. The relevant Hamiltonians with the coupling to the meters will be denoted by H_i^M (i = 1,2). We will take for H_1^M

$$H_1^M = H_1 + \tfrac{1}{2}\hbar\omega\sigma_3 + i\hbar\,(\varepsilon^* e^{i\omega t}\,\sigma^- - \varepsilon\, e^{-i\omega t}\,\sigma^+)\, p^2\, U(t) \tag{13}$$

where

$$U(t) = \sum_n (\theta(t_1+n-t) - \theta(t_o+n-t))$$

The quantities t_o and t_1 satisfy

$$0 \leqslant t_o \leqslant t_1 \leqslant 1$$

U(t) determines the time of coupling of the rotator to the meter in between kicks. The meter is taken to be a two state system, ie the 'pointer' can be either up or down. Here the meter is a microscopic object (eg a two state atom) and if we wish we can couple it to an amplifier. An amplifier however cannot improve the signal to noise ratio of the meter and so we still need to consider an ensemble of systems in order to improve the estimation of the signal. Given that we can construct a meter with the properties embodied in H_1^M, we still have to determine the quantity being measured. The Heisenberg equations of motion are

$$\frac{d}{dt}\sigma_3 = -2(\varepsilon^* e^{i\omega t}\sigma^- + \varepsilon e^{-i\omega t}\sigma^+) P_\theta^2 U(t) \quad (14)$$

$$\frac{d}{dt}\sigma^- = \varepsilon e^{-i\omega t} U(t) \sigma_3 P_\theta^2 \quad (15)$$

$$\frac{d}{dt}P_\theta^2 = 0 \quad (16)$$

It can be shown for $\omega^2 >> \varepsilon^2 P^4$ (an inequality holding strictly for matrix elements of the operators) that

$$\sigma_3(t) \sim -\cos(2\varepsilon p_\theta^2(t-t_o-n)) \text{ for } t \in (t_o+n, t_1+n) \quad (17)$$

On making an expansion based on the smallness of ε we have

$$\begin{aligned} \tfrac{1}{2}(1 + \langle\sigma_3(t_1+n)\rangle) &\simeq \varepsilon^2(t_1-t_o)^2 \langle p_\theta^4\rangle \\ &= \langle p_\theta^4\rangle/2\beta \end{aligned} \quad (18)$$

The quantity on the left hand side is just the probability that in an ensemble the meter will be showing an 'up' reading. In the analysis it has been assumed that the atom (or meter) before a particular period of coupling to the rotator is in its 'down' state. This is natural once we have explained how we perform a measurement since so far only a coupling of the meter to the system has been described. The first measurement procedure is similar to the field ionisation technique used in Rydberg atom studies. By some mechanism, after the period of interaction between kicks, those members of the ensemble with the atomic meters in the excited state, have the atoms ionised. The electrons from these 'destroyed' meters are collected together to give a macroscopic current. The strength of this current is a direct measure of the number of members of the ensemble with meters in the upper state. If we start off with a very large ensemble then this procedure will allow us to follow the behaviour of $\langle P_\theta^4\rangle$ for some time even allowing for the depletion of the ensemble at each measurement. A less wasteful form of measurement can be devised. It is unrealistic to state that an atom is decoupled from its environment, eg the electromagnetic radiation field which is responsible for the lifetime of the levels of an atom. The Hamiltonian H_1^M has to have added to it a term

$$i\int d^3k\, g(\underline{k})\, (a^+(\underline{k})\sigma^- - a(\underline{k})\sigma^+)$$

where $a(\underline{k})$ and $a^+(\underline{k})$ are annihilation and creation operators associated with photons of momentum $\underline{k}$ (and for simplicity we have suppressed helicity labels). $g(\underline{k})$ gives the appropriate coupling, and the integral accounts for the continuum of modes of the electromagnetic field in the space. Any of the wavefunction for the system of rotator, meter and radiation field modes could be expanded in terms of states

$$|n\rangle\ |\alpha\rangle \prod_{\underline{k}} |m_{\underline{k}}\rangle$$

where $|n\rangle$ is an eigenstate of P_θ, $|\alpha\rangle$ in an eigenstate of σ_3 and $|m_{\underline{k}}\rangle$ is an eigenstate of $a^+(\underline{k})\, a(k)$. A Schrödinger equation can in principle be written down for the combined system, but for this to be useful it is necessary to know an initial state. We have incomplete knowledge of the state of the radiation field and so a density matrix description of the system is forced on us. If we are interested in the expectation value of operators for the rotator and atom alone then we can trace over the radiation degrees of freedom to give

$$\frac{\partial\rho}{\partial t} = \frac{1}{i\hbar}[H_1^M, \rho] + \Lambda\rho \tag{19}$$

where

$$\Lambda\rho = \frac{\gamma}{2}([\sigma^-\rho, \sigma^+] + [\sigma^-, \rho\sigma^+])$$

Here γ is a decay rate of the upper state of the atom. The rotator is not directly coupled to the environment, but indirectly through the meter. By measuring the light emitted by the meter it can be shown that we measure the probability that a member of the ensemble has a meter showing a reading in the 'up' position. From Eqn. (19) we immediately see the effect of the measurement on the rotator and meter. Since there is no destruction of a part of the ensemble at each measurement we can proceed with the measurements as long as required. The effect of the measurement on the rotator can be found only after we solve Eqn. (19) and evaluate $\langle P_\theta^4\rangle$. (By choosing a somewhat different interaction between rotator and meter, we could have designed a meter to measure other quantities such as $\langle P_\theta^2\rangle$.) Before further analysis we will consider measurements related to H_2^M.

In a more realistic description of an atom often there are more than two states which are important, whereas in our discussion of the kicked rotator we added a meter to the rotator it would be more convenient if a system allowed an apparently redundant part of its structure to act as a meter. If an atom has two states $^2S_{\frac{1}{2}}$ and $^2P_{\frac{1}{2}}$ and each level consists of two states then the $^2S_{\frac{1}{2}}$ doublet can be excited by an external magnetic field B(t). The $^2P_{\frac{1}{2}}$ doublet will provide part of the meter. We make the meter couple to the ground state (or system) by a resonant right circularly polarised laser field $\underline{E}(t)$, which can be treated classically to a good approximation. If necessary a static magnetic field can be introduced to break any degeneracy of the levels. $\underline{E}$ excites population from the lower level in the ground state to the upper level in the excited state. By this procedure there is a net transfer of population from the ground to the excited level in the ground state and this probes the population difference. The electric field is described by

$$\underline{E}(t) = E_o(\hat{\underline{e}}\exp(-i\omega t) + \hat{\underline{e}}^*\exp(i\omega t))$$

with

$$\hat{\underline{e}} = -\frac{1}{\sqrt{2}}(\hat{\underline{x}} + i\hat{\underline{y}})$$

where

$$\hat{\underline{x}}^2 = \hat{\underline{y}}^2 = 1$$

and $\hat{\underline{x}}\cdot\hat{\underline{y}} = 0$

It is convenient to expand the density matrix ρ in terms of the irreducible tensor basis[7], ie

$$\rho = \sum_{\substack{\alpha,\beta=u,\ell \\ J,m}} \rho_m^J(\alpha\beta)\, T_m^{(J)}(\alpha\beta) \tag{20}$$

where u and ℓ refer to the excited and ground states respectively and

$$T_m^{(J)}(\alpha\beta) = \sum_{\substack{m',m'' \\ J',J''}} (-1)^{G-N} \langle J',J'',m',-m''|J,m\rangle \; |\alpha J'm'\rangle\langle\beta J''m''| \tag{21}$$

and $\langle J',J'',m',-m''|J,m\rangle$ is a Clebsch-Gordan coefficient. It is known[9] that the density matrix elements satisfy

$$\dot{\rho}_o^o(\ell\ell) = \gamma\, \rho_o^o(uu) - iv_1\, \rho_1^1(\ell u) + iv_1^*\, \rho_1^1(\ell u)^* \tag{22a}$$

$$\dot{\rho}_o^o(uu) = -\rho_o^o(\ell\ell) \tag{22b}$$

$$\dot{\rho}_o^1(\ell\ell) = -\frac{1}{3}\gamma\,\rho_o^1(uu) - \frac{i\omega_L(t)}{\sqrt{2}}\,(\rho_1^1(\ell\ell) + \rho_{-1}^1(\ell\ell)) - iv_1\,\rho_1^1(\ell u) + iv_1^*\,\rho_1^{1*}(\ell u) \tag{22c}$$

$$\dot{\rho}_1^1(\ell\ell) = -\frac{1}{3}\gamma\,\rho_1^1(uu) - \frac{i\omega_L(t)}{\sqrt{2}}\,\rho_o^1(\ell\ell) - i\omega_\parallel\,\rho_1^1(\ell\ell) + iv_1^*\,(\rho_o^o(\ell u)^* - \rho_o^1(\ell u)^*) \tag{22d}$$

$$\dot{\rho}_{-1}^1(\ell\ell) = -\frac{1}{3}\gamma\,\rho_{-1}^1(uu) - \frac{i\omega_L(t)}{\sqrt{2}}\,\rho_o^1(\ell\ell) + i\omega_\parallel\,\rho_{-1}^1(\ell\ell) + iv_1(\rho_o^o(\ell u) - \rho_o^1(\ell u)) \tag{22e}$$

$$\dot{\rho}_o^1(uu) = -\gamma\,\rho_o^1(uu) - iv_1\,\rho_1^1(\ell u) + iv_1^*\,\rho_1^{1*}(\ell u) \tag{22f}$$

$$\dot{\rho}_1^1(uu) = -\gamma\,\rho_1^1(uu) - iv_1^*(\rho_o^{o*}(\ell u) + \rho_o^{1*}(\ell u)) \tag{22g}$$

$$\dot{\rho}_{-1}^1(uu) = -\gamma\,\rho_{-1}^1(uu) - iv_1(\rho_o^o(\ell u) + \rho_o^1(\ell u)) \tag{22h}$$

$$\dot{\rho}_o^o(\ell u) = -\gamma_\perp\,\rho_o^o(\ell u) - iv_1^*(\rho_{-1}^1(uu) - \rho_{-1}^1(\ell\ell)) \tag{22i}$$

$$\dot{\rho}_o^1(\ell u) = -\gamma_\perp\,\rho_o^1(\ell u) - iv_1^*(\rho_{-1}^1(uu) + \rho_{-1}^1(\ell\ell)) \tag{22j}$$

$$\dot{\rho}_1^1(\ell u) = -\gamma_\perp\,\rho_1^1(\ell u) + iv_1^*(\rho_o^o(uu) - \rho_o^o(\ell\ell) - \rho_o^1(uu) - \rho_o^1(\ell\ell))$$

$$\dot{\rho}_{-1}^1(\ell u) = -\gamma_\perp\,\rho_{-1}^1(\ell u)$$

v_1 is proportional to the Rabi frequency associated with E_o. ω_L and $\omega_\parallel$ are the Larmor frequencies corresponding to B(t) and the static longitudinal field required to get splittings of the doublets. If we take $\gamma_\perp$, $\gamma >> \omega_L$ then adiabatic elimination is possible and we obtain[10]

$$\dot{\rho}_o^1(\ell\ell) = -\frac{4}{3}\frac{|v_1|^2}{\gamma_\perp}\,\frac{1+\rho_o^1(\ell\ell)}{1+\dfrac{6|v_1|^2}{\gamma\gamma_\perp}} - \frac{i\omega_L(t)}{\sqrt{2}}\,(\rho_1^1(\ell\ell) + \rho_{-1}^1(\ell\ell)) \tag{23}$$

$$\dot{\rho}_1^1(\ell\ell) = \frac{2|v_1|^2}{\gamma_\perp}\,\rho_{-1}^1(\ell\ell)^* - \frac{i\omega_L(t)}{\sqrt{2}}\,\rho_o^1(\ell\ell) - i\omega_\parallel\,\rho_1^1(\ell\ell) \tag{24}$$

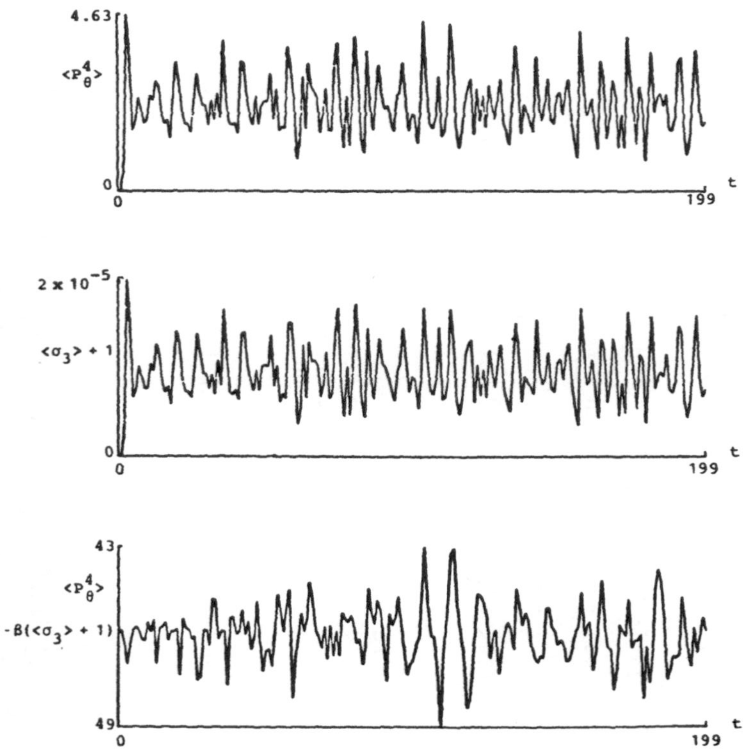

FIGURE 1
Destructive measurement of the kicked rotator with $\varepsilon = 3 \times 10^{-3}$ in a non-resonant case

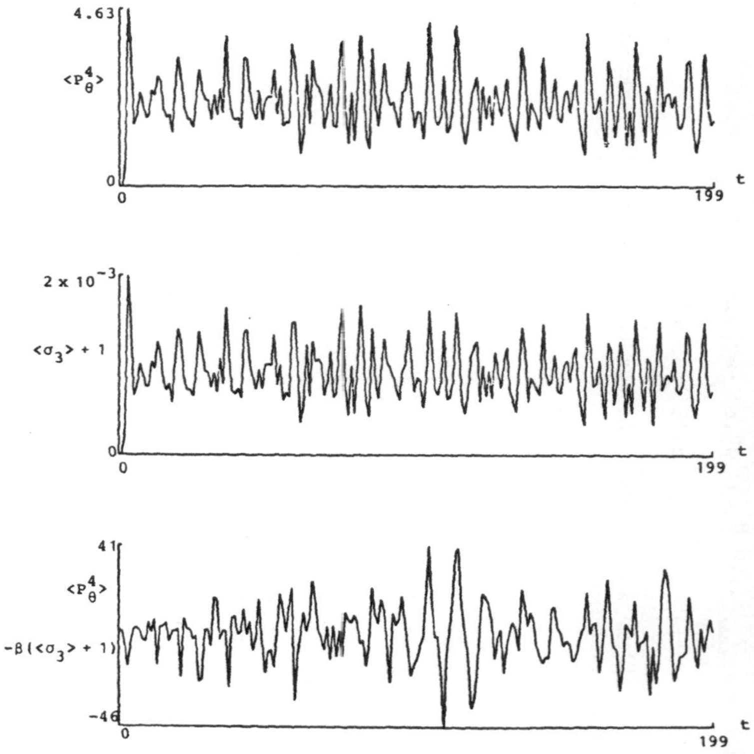

FIGURE 2
Destructive measurement of the kicked rotator with $\varepsilon = 3 \times 10^{-2}$ in a non-resonant case

$$\dot{\rho}^{1}_{-1}(\ell\ell) = -\frac{2|v_1|^2}{\gamma_\perp}\rho^{1}_{-1}(\ell\ell) - \frac{i\omega_L(t)}{\sqrt{2}}\rho^{1}_{o}(\ell\ell) + i\omega_\parallel \rho^{1}_{-1}(\ell\ell) \qquad (25)$$

Here $\rho^1_o(\ell\ell)$ is the population difference in the ground state doublet and $\rho^1_1(\ell\ell)$ are the Zeeman coherences. In the absence of measurement the equations for the density matrix are of course obtained by putting $v_1 = 0$. As far as $\rho^1_o(\ell\ell)$ is concerned from Eqn. (23) we can see that measurement brings in a certain amount of dissipation. The actual measurement consists of monitoring the absorption of the light field. This absorption is proportional to $\rho^1_1(\ell u)$ and

$$\rho^1_1(\ell u) = \frac{iv_1^*}{\gamma_\perp}(1+\rho^1_o(\ell\ell))\left(-1+\frac{6|v_1|^2}{\gamma\gamma_\perp\left(1+\frac{6|v_1|^2}{\gamma\gamma_\perp}\right)}\right) \qquad (26)$$

Hence the absorption directly measures $\rho^1_o(\ell\ell)$. In principle a similar scheme could go through for more complicated systems like a driven hydrogen atom.

We are now in a position to analyse the effects of measurement quantitatively on quantum chaotic systems. In Figure 1 we consider the destructive measurement of a kicked rotator with $\hbar = 0.1$, $k = 1.1$, $\varepsilon = 3 \times 10^{-3}$, $\omega = 31$, $t_o = 0$, and $t_1 = 0.5$. (We recall that the units are rescaled and so $\hbar$ is not the usual Planck's constant.) In the absence of measurement $\langle P^4_\theta \rangle$ saturates quickly and has a quasi-periodic behaviour around a steady value. Since the coupling to the meter is very small it would be hoped that $\langle P^4_\theta \rangle$ is essentially unchanged from the result in the absence of measurement. This is indeed the case for the 200 time units of evolution. It is necessary now to check that the meter is estimating $\langle P^4_\theta \rangle$ well. In the second graph of Figure 1 we see that $(\langle \sigma_3 \rangle + 1)$ is showing behaviour similar to that of $\langle P^4_\theta \rangle$. A quantitative comparison is made in the third graph and shows that the meter is estimating $\langle P^4_\theta \rangle$ at worst to about 10%. This is in quite good agreement with Eqn. (18); by making ω larger we would get better agreement. Figure 2 shows the effect of increasing the coupling by an order of magnitude. The coupling is sufficiently weak that in 200 time units there is negligible change from the first figure. We increase the coupling by another order of magnitude (see Figure 3). The change is dramatic. $\langle P^4_\theta \rangle$ and $\langle S_z \rangle + 1$ both show an initial rise but then a pronounced decay. The measurement has severely changed the behaviour of the system being measured. It is thus no surprise that the third graph of Figure 3 shows that the quality of the measurement in terms of estimating $\langle P^4_\theta \rangle$ is extremely poor. We will now turn down the coupling ε to 10^{-4} which is smaller even than the value used in Figure 1. Instead of having $\hbar = 0.1$ we will consider $\hbar = \pi/2$ (see Figure 4). When h is a rational multiple of π a resonance condition is satisfied. In the absence of measurement, $\langle P^4_\theta \rangle$ for large time increases as t^4. However even a weak coupling with the meter alters the situation significantly. $\langle P^4_\theta \rangle$ increases quite fast but then decreases to a smooth limit which is at most slowly varying. Although $(\langle \sigma_3 \rangle + 1)$ shows a similar behaviour, the quality of the estimate for $\langle P^4_\theta \rangle$ is wrong by 100%. This should not be surprising since Eqn. (18) was derived on the basis of the inequality $\omega^2 \gg \varepsilon^2 P^4_\theta$. In resonant cases this cannot hold except for short times. The results that we have presented are not qualitatively dependent on the nature of the destructive measurement. The form of measurement embodied in Eqn. (19) shows similar behaviour. It is difficult to do a satisfactory calculation of the resonant case though for the non-destructive measurement, owing to the size of the basis involved.

We will now present the analysis of Eqns. (23)-(25) for $\omega = 5.0$, $\omega' = \sqrt{2} + \sqrt{3}$ and $\omega'' = \sqrt{3} - \sqrt{2}$. Figure 5 shows the irregular

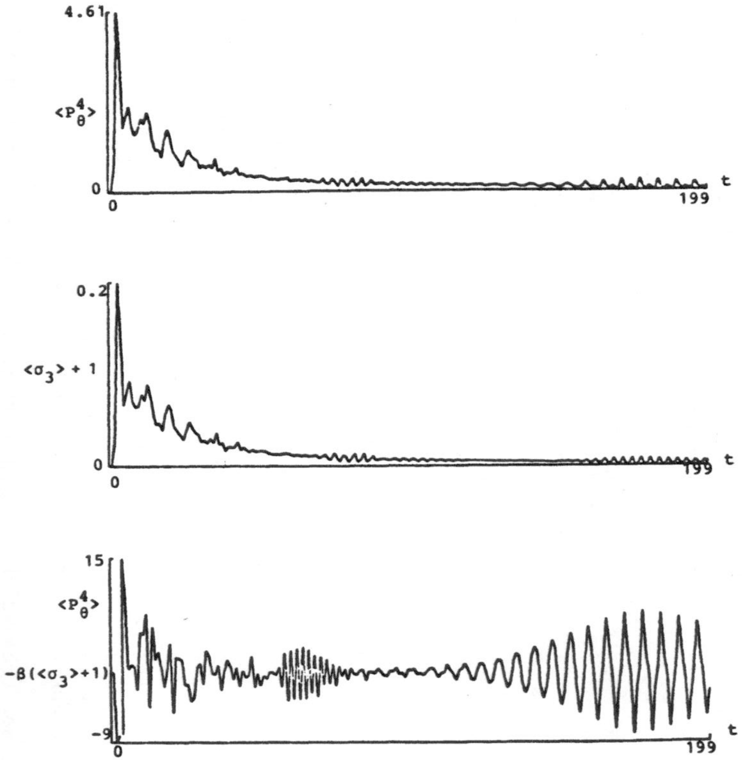

FIGURE 3
Destructive measurement of the kicked rotator with $\varepsilon = 0.3$ in a non-resonant case

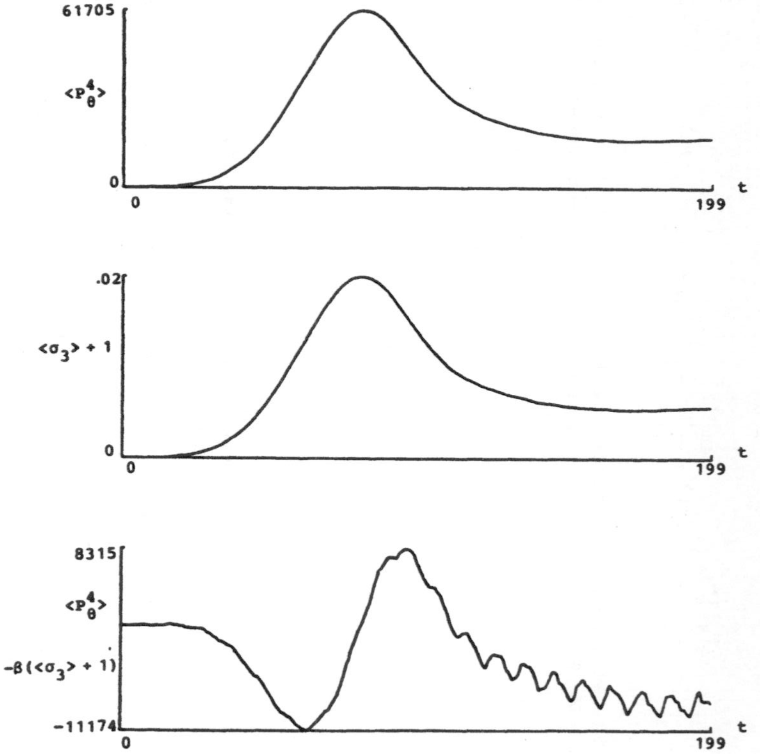

FIGURE 4
Destructive measurement of the kicked rotator with $\varepsilon = 10^{-4}$ in a resonant case

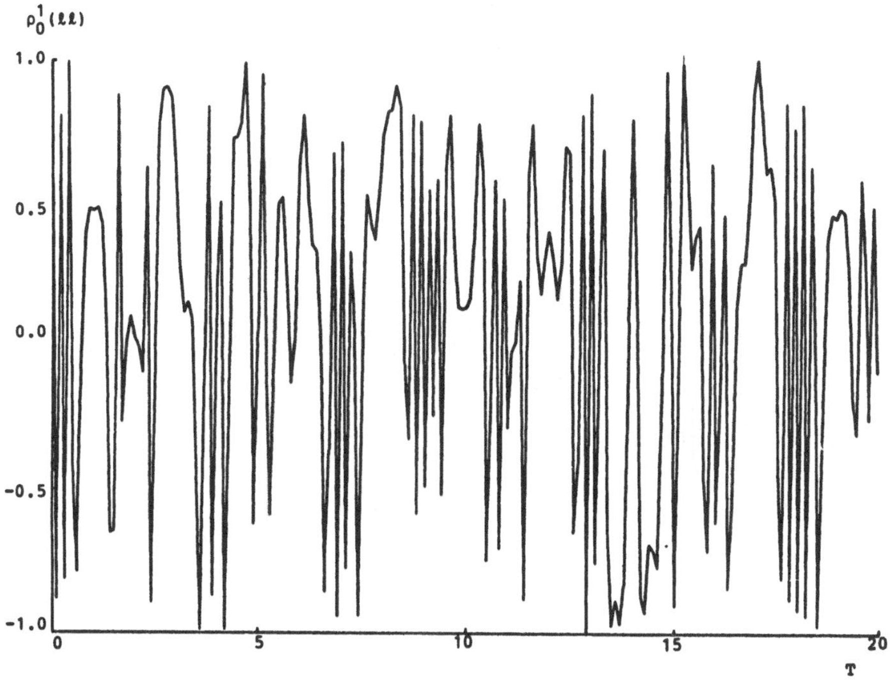

FIGURE 5
Time evolution of the population difference of the driven two state atom for S = 0

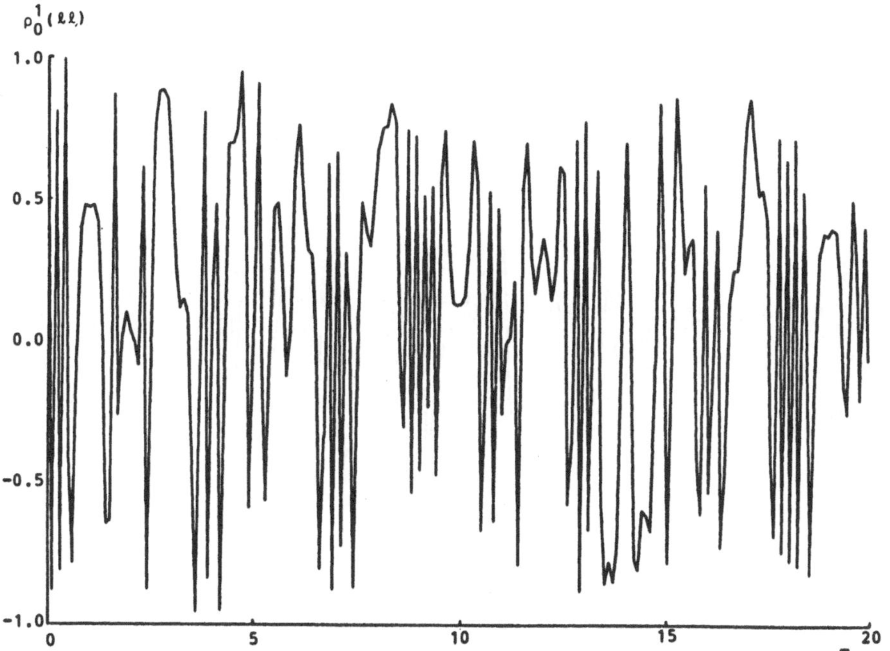

FIGURE 6
Time evolution of the population difference of the driven two state atom for S = 0.01

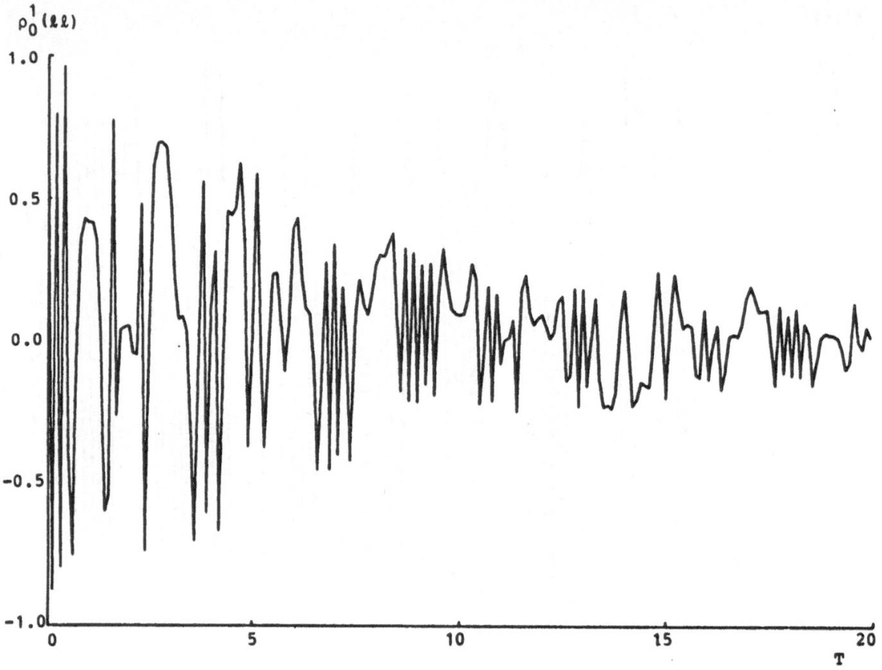

FIGURE 7
Time evolution of the population difference of the driven two state atom for S = 0.1

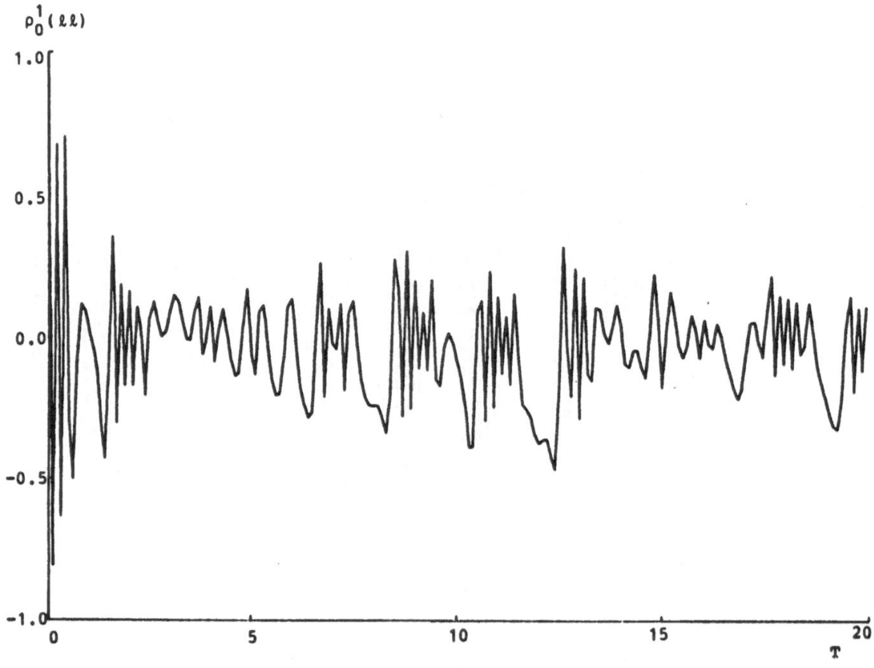

FIGURE 8
Time evolution of the population difference of the driven two state atom for S = 1.0

(multi-periodic) motion of $\rho_0^1(\ell\ell)$ in the absence of measurement. It is convenient to introduce the analogue of ε in the previous example to be $s = 2|v_1|^2/\gamma_\perp$. In Figure 6, s = 0.01 and there is a slight damping of $\rho_0^1(\ell\ell)$ which is perceptible. When s is increased by an order of magnitude then the damping of $\rho_0^1(\ell\ell)$ is substantial (see Figure 7). We might expect that by increasing s by a further order of magnitude to 1 we would obtain a further damping of the motion. This is however not the case. Figure 8 shows that in this case the disturbance of the measurement on the system is so great that the nature of the motion has changed from seemingly irregular to quasi-periodic.

We have shown that the principal effect of measurement on quantum chaotic systems is to introduce dissipation. Too strong a coupling to the meter gives not only a poor measurement but drastically alters the behaviour being studied. Our considerations show that conceptually the introduction of measurement for quantum Hamiltonian systems brings in dissipative dynamics.

REFERENCES

1. See for example:
'Chaotic Behaviour in Quantum Systems' ed. G. Casati (Plenum, N.Y., 1983).
'Frontiers in Quantum Optics' ed. E. R. Pike and S. Sarkar (Adam Hilger, Bristol, 1986).

2. G. Casati, B. V. Chirikov, F. M. Izraelev and J. Ford, in 'Stochastic Behaviour in Classical and Quantum Hamiltonian Systems', eds. G. Casati and J. Ford (Springer-Verlag, Berlin, 1979).

3. Y. Pomeau, B. Dorizzi and B. Grammaticos, Phys. Rev. Lett. 56 681 (1986).

4. D. R. Grempel, S. Fishman and R. E. Prange, Phys. Rev. Lett. 49 833 (1982).

5. E. Ott, T. M. Antonsen, Jr. and J. D. Hanson, Phys. Rev. Lett. 53 2187 (1984).

6. C. M. Caves, K. S. Thorne, R. W. P. Drever, V. D. Sandberg and M. Zimmermann, Rev. Mod. Phys. 52 34 (1980).

7. S. Sarkar and J. S. Satchell, The effect of measurement on the quantum kicked rotator, RSRE Malvern preprint, August 1986.

8. W. Happer, Rev. Mod. Phys. 44, 169 (1972).

9. W. J. Sandle and M. W. Hamilton, in 'Laser Physics', page 54, eds. J. D. Harvey and D. F. Walls (Springer-Verlag, Berlin, 1983).

10. J. S. Satchell and S. Sarkar, Quantum measurement as dissipation in chaotic atomic systems, RSRE Malvern preprint, August 1986.

QUANTUM CHAOS AND THE MEASUREMENT PROBLEM

Asher Peres

Department of Physics
Technion—Israel Institute of Technology
32 000 Haifa, Israel

1. INTRODUCTION AND SUMMARY

Ever since its inception, quantum theory has been haunted by the measurement problem.[1] While there is no disagreement on how to use the quantum formalism in practical calculations (such as predicting energy levels, cross sections, etc.) the *ontological* meaning of this formalism is the subject of a perennial and lively debate. In recent years, an additional, more technical source of controversy is quantum chaos.[2] There is no agreement yet on how to *define* this term, and the very existence of quantum chaos is subject to numerous claims and counterclaims.

The purpose of this paper is to review and clarify both problems, which are intimately related. In the next section, the "measurement problem" is posed in unambiguous terms. This problem originates in the ambivalent nature of the "observer:" Although the observer is *not* described by quantum theory, it should nevertheless be possible to "quantize" him and include him in the wavefunction, if quantum theory is universally valid.[3] *The problem is to prove that no contradiction may arise in these two conflicting descriptions.* The proof involves the notion of quantum chaos, which is introduced in Section 3. It is shown that, in quantum physics as well as in classical physics, there are questions having a unique solution (from the point of view of pure mathematics) but this solution requires an inordinate amount of computing work and is effectively unattainable. It is then argued in Section 4 that *informational complexity* is a real physical limitation, which must be considered as seriously as the second law of thermodynamics (to which it is not unrelated, as a matter of fact). Finally, Section 5 integrates all these notions into a coherent solution of the measurement problem.

Most of this paper is frankly pedagogical and claims no originality. Clarity of exposition has been my main concern. On the other hand, I have not hesitated to follow my personal bias and to take sides on controversial issues. I have also taken the liberty to advance some unproven conjectures (which are clearly labelled so) with the expectation that they will eventually turn out to be correct.

2. THE MEASUREMENT PROBLEM (AND RELATED PSEUDOPROBLEMS)

The numerous paradoxes[4] associated with quantum theory can conveniently be plotted on a complex plane (Fig. 1). Some of these paradoxes are merely due to the naive belief that the time evolution of the wavefunction ψ represents what is actually happening in the real world. The classic example is Schrödinger's cat,[5] doomed to be killed by an automatic device triggered by the decay of a radioactive atom. Initially, the combined state of the cat and the atom is $|L, u\rangle$, namely a living cat with an undecayed atom. After one half-life of the atom, their state becomes

$$\psi = 2^{-1/2}(|L, u\rangle + |D, d\rangle), \tag{1}$$

with obvious notations. Nobody has ever seen a cat in such a strange situation.

At this point, the most reasonable attitude is to be pragmatic and to renounce the illusion that every physical system *has* at every instant a well defined state ψ (or, possibly, a density matrix ρ) and that the evolution of $\psi(t)$, or of $\rho(t)$, represents what actually happens in the real world.[6] Quantum theory can make no such claim. It is only a mathematical formalism, allowing us to compute *probabilities* for the occurence of events of a specified kind, following a specified preparation.[6] The wavefunction ψ (or the density matrix ρ) represents statistical *information.*[7–11] For example, Eq. (1) tells us that if Schrödinger's gedankenexperiment is repeated many times, roughly one half of the cats will be found alive and one half will not. The information contained in $\psi(t)$, or $\rho(t)$, can *evolve* because of known dynamical interactions (such as those causing the decay of the radioactive atom); it can also get *degraded*, and ultimately *obsolete*, with the passage of time.[11] This loss of information is related to *quantum chaos*, as will be shown below.

There are some physicists (perhaps the majority of them) who have been bred in realistic doctrines and reject the above pragmatic attitude. They attempt to save the objectivity of the wavefunction by arguments such as the following one: Nobody

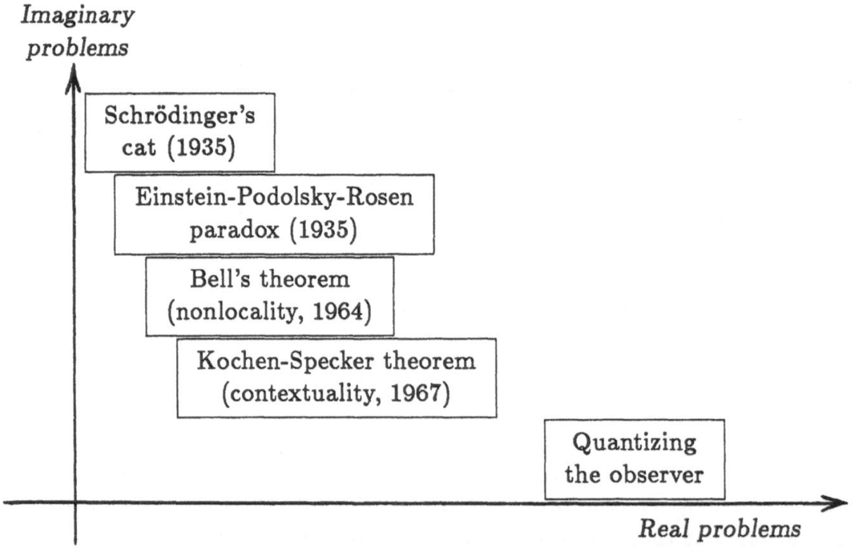

Figure 1: Real and imaginary problems in the interpretation of quantum theory.

indeed has ever *seen* a cat in state (1), but this is because the mere *observation* of the cat causes the wavefunction to jump[12] into either a $|L,u\rangle$ or a $|D,d\rangle$ state (this is called the "collapse"[13] of the wavefunction). There is nothing inconsistent in this story, but it is nevertheless incredible, as it implies a powerful and mysterious interaction between the brain of the observer and the entire body of the cat. A measurement, after all, is not a supernatural event. It is a physical process, involving ordinary matter, and whatever happens ought to be explained by the ordinary physical laws.[3]

A more radical attempt to salvage realism without invoking a "collapse" of the wavefunction is Everett's interpretation of quantum theory,[14] in which Eq. (1) *does* represent the real situation. More precisely, Eq. (1) represents the situation if the cat is considered as the observer of the atom. Alternatively, if we are reluctant to use a dead observer, we may include in the wavefunction an additional observer who does not die with the cat. Now, Everett boldly assumes that the two "branches" of the wavefunction (1) correspond to different worlds, between which there can be no communication. One may distinguish several variants of this approach,[15] which have been called the "relative state interpretation" and the "many worlds interpretation." None is satisfactory because they merely replace the arbitrariness of the "collapse" postulate by that of the "no-communication" dictum.

It is nevertheless possible to reinterprete Everett's work in a way quite compatible with the pragmatic approach that I advocated earlier, according to which the wavefunction only represents statistical information: In that approach, effectively, each observer imagines that he is dealing with a *Gibbs ensemble*[11] of quantum objects and apparatuses, and the wavefunction represents all the information compatible with that ensemble. The inclusion of the observer in the wavefunction, as proposed by Everett, simply means that now one has to consider a *Gibbs ensemble of observers,*[11] each one of them experimenting with a single quantum object and a single apparatus. This twist of Everett's interpretation may not be what Everett or his disciples had in mind, but it can be used to solve the "measurement problem," as shown in the final section of this paper.

Both the Everett interpretation, in its original form, and the naive "realistic" attitude (namely, each system has a well defined state, which may or may not be known to us) run into additional difficulties when there are *several* independent observers, especially if the latter are in relative motion.[10] Consider the situation illustrated by Fig. 2, which involves three events. The first one (E) is the emission of a pair of photons in opposite directions and with *opposite polarizations,* due to the the decay of a spinless system (for example, an atomic SPS cascade). The two other events are measurements of the linear polarizations of these photons by observers O′ and O″, respectively. These observers recede with uniform velocities from the emission point (and thus from each other) along the z axis, which is the line of flight of the photons. Therefore, by virtue of the Lorentz transformation, each observer considers his experiment as occurring *earlier than that of his colleague.* Although there is no disagreement about what really happens (what is actually observed) this set of events *cannot* be described by a wavefunction with a relativistic transformation law. If one tries to do that, the wavefunction becomes ambiguous and noncovariant[16] and no relativistically satisfactory version of the collapse postulate can be found.[17,18]

The events described in Fig. 2 (but with the observers possibly at rest) can also be used to illustrate Bell's theorem.[19] The latter is based on the Einstein-Podolsky-Rosen hypothesis that, if two systems are well separated and do not interact, "no real change can take place in the second system in consequence of anything that may be done to the first system."[20] (Here, obviously, Einstein, Podolsky and Rosen had in mind that each physical system had a well defined state which could, or could not, be affected by the actions of various observers. The purpose of the EPR paper[20] was to investigate whether that state was adequately described by quantum theory.) Bell showed that the EPR hypothesis implies a set of inequalities[19,21] for correlations between the outcomes of various experiments, such as those described in Fig. 2. These inequalities are grossly violated by quantum statistics. Experimental tests[22] clearly favor quantum theory and violate Bell's inequalities. It can in fact be shown that in at least 41% of cases, the result obtained by each one of the observers cannot be assumed independent of the *choice* of the experiment performed by the other observer.[23]

An even stronger result was proved by Kochen and Specker[24] and, less formally, in an earlier paper by Bell[25] (ref. 25 was obviously written before ref. 19): Consider three operators, A, B and C, such that $[A,B] = [A,C] = 0$, but $[B,C] \neq 0$. The measurements of B and C are therefore incompatible, but none of them, by itself, disturbs that of A, so are we told by textbooks on quantum theory.[12,13] We can therefore measure A alone, or A and B, or A and C, as we please. Now, assume tentatively that the result of a measurement of A is independent of whether we measure A alone, or A and B, or A and C. It can then be shown[24–26] that this assumption is, with a suitable choice of operators, incompatible with the fact that the outcomes of quantum measurements are *discrete* and, in some cases, *correlated*. This result is *stronger* than the experimental violation of Bell's inequalities, because

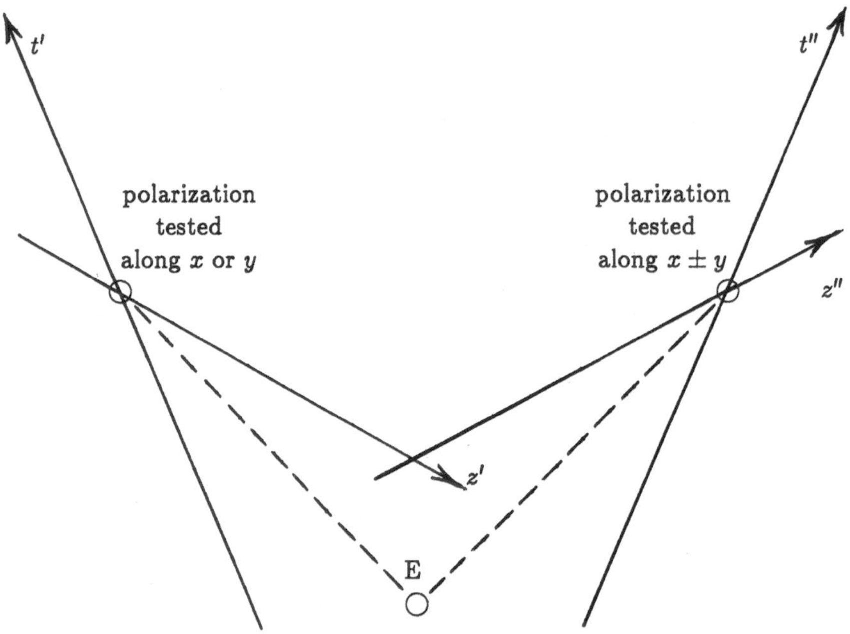

Figure 2: The t' and t'' axes are the world lines of two observers who are receding from each other, while they test different polarization components of a photon pair. Along their z' and z'' axes, these observers have t'=const and t''=const , respectively.

it does not involve quantum *statistics*: Even the results of *individual* measurements are "contextual." They cannot be assumed independent of which other measurements are involved in the *complete* experimental setup.

This apparent nonlocality is considered by some physicists as a serious intellectual challenge[27] while others take a more relaxed attitude[28] and say that the nonlocality of a photon pair is not essentially different from that of a single photon, which can pass through two widely separated slits and interfere with itself. The photon *pair* then is a single, indivisible, nonlocal object. If anything here is curious and ought to be explained, it is that macroscopic phenomena do appear to be localized, pronouns such as "I" and "you" appear to be meaningful[29] and, in particular, the *instantaneous transfer of information is impossible.* I have discussed these questions in another publication.[29] Their solution is closely related to that of the measurement problem, which is the subject of the present paper.

What is then the "measurement problem?" As already stated, it is to show that *no* inconsistency can arise if the observer is included in the quantum dynamics, and then observed by someone else, a "superobserver" who in turn has *no* dynamical description. For example, it should not matter whether Schrödinger's cat is considered as the observer, or is part of the dynamical system and is observed by someone else. The reason that one could fear an inconsistency is the following:

If the observer were not a cat or some other, possibly inanimate but utterly complicated measuring apparatus, the evolution leading to Eq. (1) would be reversible. Simple, highly idealized models of measuring apparatuses can easily be constructed, which have that property.[30] After the act of measurement is consummated and its outcome is definite, it is still possible to undo the whole process. Can this apply to *real* apparatuses? In that case, a superobserver, capable of fully analyzing the dynamical behavior of the measuring apparatus (e. g., capable of writing explicitly the exact Hamiltonian of Schrödinger's cat) could cause the original observer (or apparatus—this makes no difference) to "unlearn" the result of his measurement[31] and return to his original state of ignorance (for example, the cat would be resurrected). And then, the superobserver, by measuring the *same* system, could obtain a *different* result for *his* measurement.

If such a scenario were indeed possible, the notion of measurement would become meaningless, as no measurement could be conclusive (note that a supersuperobserver could likewise override the measurements of the superobserver, an so on). Consistency thus requires the measurement process to be *irreversible.*[32–34] There are no superobservers in our physical world.

Our task thus is to explain how irreversibility occurs in quantum systems, when the basic dynamical law—Schrödinger's equation—is reversible. This is the problem of quantum chaos, which is the subject of the next section.

3. CHAOS IN CLASSICAL AND QUANTUM PHYSICS

The intuitive meaning of "irreversibility" is that there are some dynamical evolutions which can easily be prepared, but it is extremely difficult (we say "impossible") to prepare the time-reversed evolutions. For example, it is easy to get a cup of lukewarm water by placing, one hour earlier, a cube of ice in a cup of boiling water.

It is impossible to prepare a cup of lukewarm water in such a way that, one hour later, it will turn into an ice cube floating in boiling water.[35]

The familiar *classical* explanation of irreversibility is based on the notions of *mixing* and *coarse graining*. Mixing is a property of chaotic systems which can be proved rigorously, under suitable assumptions, in the limit $t \to \infty$.[36] It essentially means the following: Consider two finite (and *fixed*) subsets of phase space, say V_1 and V_2, whose measures are fractions μ_1 and μ_2 of the total phase space. Suppose that the distribution $f_1(\mathbf{p},\mathbf{q})$ is uniform in V_1 at time t_1, with $\int f_1 dV_1 = 1$. The dynamical evolution causes this distribution to become $f_2(\mathbf{p},\mathbf{q})$ at time t_2. Then, for any t_2 *sufficiently remote* from t_1 (in the future *or* the past) and for sufficiently large μ_1 and μ_2, we have $|\int f_2 dV_2 - \mu_2| < \delta$, with arbitrarily small δ, irrespective of where V_1 is. The system thus appears to have no memory of its origin. This is the property called *mixing*. Notice that it is *time symmetric*. By itself it cannot explain irreversibility. Notice also that the smaller δ, or μ_1, or μ_2, the larger the time $|t_1 - t_2|$ needed for mixing. (In this paper I consider only *finite* times, not the unattainable mathematical limit of infinite time.)

In the ice-cube paradigm we have μ_1(ice cube plus boiling water)$\ll \mu_2$(lukewarm water) and μ_2 is of order 1. Therefore, with a suitable value of the total energy, *almost* every evolution will lead to a cup of lukewarm water, with only extremely small inhomogeneities. Indeed, the probabilities of finding the final state in V_1 and in V_2 are, respectively, $\mu_1/(\mu_1 + \mu_2) \ll 1$ and $\mu_2/(\mu_1 + \mu_2) \simeq 1$. Nevertheless, we *can*, conceptually, prepare the lukewarm water at time time t_2 so that, at a *later* time t_1 it will separate into an ice cube and boiling water. This "only" requires a *very* special preparation (not just any cup of lukewarm water, but one with delicate correlations between all the molecules) and this preparation has an *extremely* small μ_2, so small indeed that the mixing property, as defined above, will not yet be valid after the given *finite* time $t_1 - t_2$.

Now comes *coarse graining*. The standard textbook claims that we are unable, because of the coarseness of our instruments, to achieve such a preparation. We cannot coerce the initial state to reside within such a tiny V_2 (i.e., one with μ_2 so small that mixing has not yet occurred after a finite $t_1 - t_2$). In summary, we are incapable of preparing the system at time t_2 so that, after a finite time $t_1 - t_2$, it will be located with certainty in the small desired region V_1 of phase space. There are evolutions (e.g., from lukewarm water to an ice cube in boiling water) which cannot be made to proceed.

This argument deserves close scrutiny. The claim that "we are unable to achieve such a preparation" ought to be *proved*. What is the degree of difficulty which is called "impossible?" (How, indeed, do we *measure* difficulty?) One should never underestimate the skill of experimental physicists! The well know spin echo[37,38] phenomenon shows that an apparently irreversible behavior can, under appropriate circumstances, be reversed. In that experiment, a large number of spins, initially aligned, acquire random phases under the influence of random, unknown, but *fixed* internal fields, acting during a *known* time t. A suitable rf pulse can then invert all these phases and, after the *same* time t, all the spins are again aligned. This is possible because all the spin precessions are *decoupled* from each other. Their equations of motion are integrable (the system is random, but not chaotic). The phases *appear* random to the untrained eye, but actually they are *correlated*. They are similar to a text which has been encrypted by a deterministic key, looks incomprehensible, but *can* be deciphered by anyone knowing the key. Information is not destroyed by encryption, it is only made less accessible. Likewise, in the spin echo case, the "key" is the strength of the rf pulse which inverts the phases and realigns the spins.

An even simpler case is Kac's ring model.[39,40] Consider a circle with n equidistant points on the circumference, and n balls, one at each point, which can be either black or white. Moreover, m markers are placed on the circumference, in a random way, at *fixed* positions. The dynamical law is the following: During each time interval, all the balls move simultaneously one step clockwise, with the additional stipulation that when a ball *leaves* a marker it will change color. The initial color distribution is given. The problem is to find this distribution after a large number of steps (after a long time). It is assumed that $n \gg m \gg 1$.

Intuitively, we expect a monotonic approach to an equilibrium state, with equal numbers of white and black balls, except for small fluctuations. This is indeed what happens at the beginning of the evolution. Yet, this cannot continue forever.The system is manifestly periodic with period $2n$ (for after $2n$ steps each ball is certainly back to its original state after having passed each marker twice). Here again, the colors of the balls may look random to the untrained eye, but they are strictly correlated: The encryption key is the fixed distribution of markers.

These two examples show that we must proceed with extreme caution before claiming that it is impossible to achieve this, that, or the other preparation. Consider again the problem of motion in phase space which was discussed at the beginning of this Section. The argument is graphically represented in Fig. 3: Our task is to prepare initial data for a dynamical evolution *ending* at time $t = 0$ in a *designated* area of phase space (drawn as a circle). The experimental error box, due to the coarseness of our instruments, is represented by the small black rectangle (note that $\Delta\mathbf{p}$ and $\Delta\mathbf{q}$ are *uncorrelated*). Now, if we compute where the boundaries of the tar-

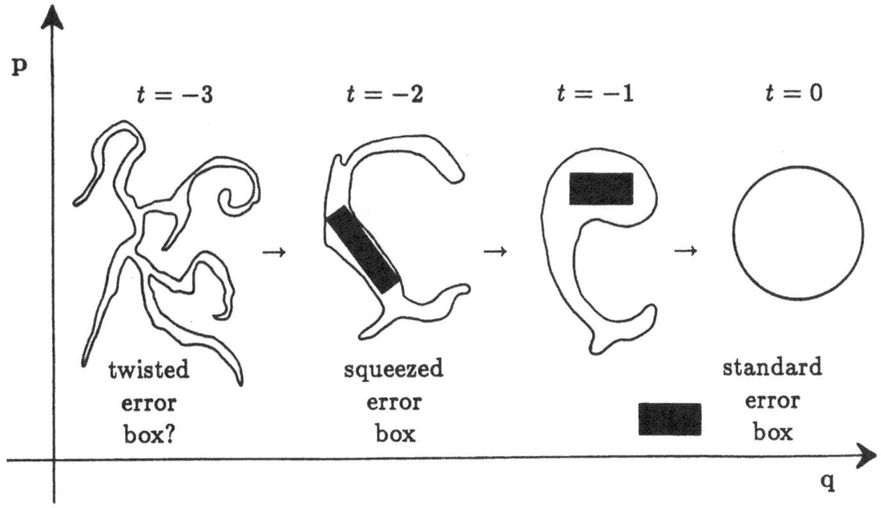

Figure 3: Dynamical evolution of a phase space domain, compared to the size of an experimental error box. Here $\mathbf{p}$ and $\mathbf{q}$ denote collectively all the coordinates and momenta.

get area were located at $t = -1$ (in some arbitrary time units) we see that the error box *could* enter within these boundaries. This means that the required preparation was *possible*, if we performed it one unit of time in advance. However, if we want to perform this preparation two or more units of time in advance, we see that the shape of the bounday is so distorted that it is impossible to fit the error box inside it. We may conclude that we need a higher precision (a smaller error box). For a chaotic system—that is, for nearly all nonlinear systems—the demand in precision increases exponentially with the time separation (the coefficient is called the Lyapunov exponent[41]). Obviously, we may not be able to satisfy these more stringent precision requirements: They may not only be too costly in labor and parts, but also finally run afoul of the uncertainty principle (by this, I mean that the classical equations of motion would cease to be valid).

There is nevertheless a way to circumvent this difficulty, for some time at least. While the shape of the target area is distorted by the time evolution, its volume in phase space remains constant—this is Liouville's theorem. We may therefore try to likewise *adapt* the shape of the error box, without changing its size, by *correlating* Δp and Δq. This can be done by preparing a *squeezed state*[42] of the measuring apparatus, as shown in Fig. 3 for time $t = -2$. Going to still earlier preparation times, we find that no similar squeezed state could fit into the required domain at $t = -3$ but perhaps, if we are able to pay the price, some very skilled experimental physicist could prepare a "twisted state" still fitting in it. After all, these different shapes of the target area are nothing but canonical transforms of each other. What appears complicated in one canonical coordinate system may be simple in another; all these systems are *mathematically* equivalent.

They are not, however, *physically* equivalent. The relevant question is: In a given *fixed* system, for example the one in which the observer appears to have a relatively simple shape, is there an end to this process of higher and higher refinement? It seems that there must be one, even if we ignore the uncertainty principle: The higher precision involves handling a larger amount of *information*, at the very least supplying more and more digits in the initial data. Getting this information is possible only at the cost of a corresponding increase of entropy in the rest of the world.[43] Now, if we assume that the Universe is finite, there must be an upper limit to the amount of entropy which may be generated in it.[34] Our wisdom may be large, but it is not infinite.

It thus appears that some of the laws of physics are subjective and may depend on the intellectual and experimental skills of physicists.[44] The spin echo example, discussed above, shows that it is indeed so: Some people may see only disorder (high entropy) where others may recognize a perfectly ordered—although encrypted— state, with zero entropy. This subjectivity should not be a matter of concern, as long as we admit that there is *some* limit to our skills. The laws of physics are almost indifferent to where this limit is set,[44] because the Lyapunov exponent of a *macroscopic system* is so huge that it will overcome, within an exceedingly brief time, even the most extravagant claims of reversibility.

This situation is also referred to as *chaos*, because the long term evolution of a physical system is *unpredictable*, except statistically, even when the basic dynamical law is deterministic. The final outcome appears to be *random*. The encryption key which we would need to decipher it is too long to be handled.

The above argument thus explains classical irreversibility. The next question naturally is: What would be the corresponding argument in quantum theory? This is not at all a straightforward question because a *quantum state is not the analogue of a point in the classical phase space.*[45] If we consider two classical trajectories starting at neighboring points, their "distance" will usually increase, in a roughly exponential manner. Here, I put the word "distance" between quotes, because phase space has no metric structure (it has a symplectic structure) and the metric distance between two points is not canonically invariant. However, in any arbitrary but *fixed* metric, such as $ds^2 = d\mathbf{p}^2 + d\mathbf{q}^2$ say, an infinitesimal "distance" will ultimately grow exponentially if the system is chaotic.[46]

On the other hand, if we consider two quantum states which are initially almost identical (their scalar product is very close to 1) they will remain almost identical forever, because the Hamiltonian evolution is a unitary mapping which preserves scalar products. In fact, the same property appears in classical mechanics too: The analogue of a quantum state is a *distribution* in phase space[45] and, if we consider two such distributions initially centered around neighboring points, with some overlap, we likewise find that *their overlap remains constant* in time. Each one of these distributions may get distorted and grow "whorls and tendrils"[47] until all of phase space appears thoroughly mixed, when seen on a coarse scale. Yet, the overlap of the two distributions remains constant, by virtue of Koopman's theorem.[48]

This can easily be seen as follows. Let $f(\mathbf{p}, \mathbf{q})$ be a solution of the Liouville equation

$$\frac{\partial f}{\partial t} + \frac{\partial H}{\partial \mathbf{p}} \cdot \frac{\partial f}{\partial \mathbf{q}} - \frac{\partial H}{\partial \mathbf{q}} \cdot \frac{\partial f}{\partial \mathbf{p}} = 0. \tag{2}$$

This can be written as

$$i\frac{\partial f}{\partial t} = Lf, \tag{3}$$

where

$$L = (\frac{\partial H}{\partial \mathbf{p}}) \cdot (-i\frac{\partial}{\partial \mathbf{q}}) - (\frac{\partial H}{\partial \mathbf{q}}) \cdot (-i\frac{\partial}{\partial \mathbf{p}}) \tag{4}$$

is the Liouville operator,[49,50] which is defined over continuous functions in phase space. This operator is "Hermitian." (Whether it is truly self-adjoint or only symmetric[51] depends on the explicit form of H. I shall later return to this delicate point.) The time evolution of f can therefore be described as a unitary mapping in phase space. Now let $g(\mathbf{p}, \mathbf{q})$ be another classical distribution satisfying Eq. (2). It follows that the scalar product $\int fg\,d\mathbf{p}d\mathbf{q}$ is invariant in time. (We could also define the scalar product, if we wished, as $\int \sqrt{(fg)}\,d\mathbf{p}d\mathbf{q}$, because $\sqrt{f}$ —or in fact any function of f—satisfies the Liouville equation if f does.)

The lesson taught to us by this is that *we must carefully screen the questions which can legitimately be asked.* No one should be interested in how some distribution, which was initially given, will later overlap with the tortuous domain covered by the time evolution of another, initially given distribution. The legitimate question (the only one which is amenable to an experimental test) is how each one of these time evolving domains overlaps with a *fixed* domain of phase space, such as the error box drawn in Fig. 3. The answer was given at the beginning of this Section: On a *fixed* coarse scale, the final distribution is homogeneous and roughly independent of the initial conditions. This is the property called *mixing*.[36]

How would we expect *quantum* chaos to manifest its existence? A natural question could be: Does Wigner's distribution[52,53]

$$W(\mathbf{p},\mathbf{q}) = (\pi\hbar)^{-n} \int d\mathbf{y}\, \psi^*(\mathbf{q}+\mathbf{y})\, \psi(\mathbf{q}-\mathbf{y})\, e^{2i\mathbf{p}\cdot\mathbf{y}/\hbar}, \tag{5}$$

which is the quantum analogue of the Liouville distribution, also lead to mixing? Now, it appears in some models[54] that $W(\mathbf{p},\mathbf{q})$ has a much smoother evolution in time than the Liouville distribution. In particular, it cannot develop "whorls and tendrils" on scales smaller than $\hbar$. Therefore, $W(\mathbf{p},\mathbf{q})$ would not have the mixing property, as defined above. (This does not conflict with the fact that, as $\hbar \to 0$, $W(\mathbf{p},\mathbf{q})$ does not tend to a limit, but rather has infinitely rapid oscillations.[55,56])

Indeed, it has been known empirically for a long time that quantum mechanics tends to suppress the appearance of chaos.[57] Quantum wave packets may remain localized, even though classical orbits are strongly chaotic, because cantori, associated with the breakup of invariant KAM surfaces, may effectively act as barriers to quantum wave packet evolution while permitting extensive classical flow. A similar phenomenon appears in models where the Hamiltonian includes a time-dependent external force.[54,58–60] In these models, which usually have a *single* degree of freedom, the physical system is prepared in a state involving only one, or at most a few energy levels of the unperturbed Hamiltonian. One then finds that, in its time evolution, the quantum system visits only a few more neighboring energy levels, so that the energy remains "localized" in a narrow domain (while in the corresponding classical evolution the energy would increase without bounds, in a diffusive way). As a consequence, the quantum motion is *almost periodic* and the initial state recurs repeatedly.[61] The same if of course true for any time-independent Hamiltonian with a discrete spectrum.[62] The problem here is not the recurrence itself, which has the same character as a Poincaré cycle[63] and is completed after an inordinately long time,[64] but the fact that the quantum state after an arbitrarily long time t can be computed accurately with a finite amount of work: The solution of the equations of motion has *null complexity*.[65] Everything is predictable. Nothing is left to chance.

The foregoing argument represents one possible approach to the quantum chaos problem and its conclusion is that there is *no* chaos in quantum theory. On the other hand, one also finds claims that quantum theory is the paradigm of chaos. In the closing lecture of the 1983 Como conference, Lamb exclaimed: "If it is *quantum*, there is more than enough *chaos*!"[66] Indeed, what could be more chaotic (i.e., unpredictable) than the result of the measurement of J_x of a spin-$\frac{1}{2}$ particle, following its preparation in a J_y eigenstate? In his book, Landé[67] elevates this randomness to the status of a fundamental physical principle: "The random distribution is a physical reality, and the determinism which only looks random is a purely academic construction."

In view of these conflicting opinions, it is necessary to clearly formulate what is the *real* problem of quantum chaos: If we prepare a spin-$\frac{1}{2}$ particle in an eigenstate of J_z, and then *immediately* measure J_x or J_y, the result is unpredictable, for sure. However, the outcome of a measurement of J_z can still be predicted with 100% certainty. On the other hand, the way we understand the meaning of *classical* chaos is completely different: If we prepare a classically chaotic system in a small region of phase space and then let it evolve for a *long time*, the uncertainty in its position increases, first exponentially and then at a slower rate, until it finally reaches saturation when the entire accessible phase space is filled. (All this is true provided that the experimental error box is kept constant.) Thus, the corresponding problem for a quantum system would be[65] whether or not there is randomness in its behavior *over and above the intrinsic randomness contained in the wavefunction* ψ. For example, one could conjecture that *no* nontrivial dynamical variable exists which, if measured after a long time t, would yield a definite and predictable outcome. This could perhaps be due to some unpredictability in the time evolution of the wavefunction, or in the quantum eigenvalues or eigenfunctions. I shall show in the final Section of this paper that this conjecture is indeed very plausible.

The other ingredient needed for classical chaos is *coarse graining*, as symbolized by the "experimental error box" of Fig. 3. Now, coarse graining seems to make no sense in the quantum language, nor indeed in any language where dynamical variables occur with *discrete* values.[35] It was even suggested by Lamb[68] that it is in principle possible to prepare arbitrary, pure quantum states. In other words, there is no limit to our resolving power. This is a very bold conjecture. However, the opposite conjecture[69] would be equally bold and should be accompanied by some (possibly vague) specification of which states can—or cannot—be prepared (even the apparently trivial case of the 3-dimensional Hilbert space of a spin-1 particle has been a subject of controversy[70,71]). In this paper, I shall take the (admitedly optimistic) attitude that *any state which can be specified by means of a finite and explicit algorithm, can also be prepared experimentally.* The remaining problem thus is: Are there states which are *algorithmically* inaccessible?

4. THE PHYSICAL LIMITS OF COMPUTATION

In the realm of pure mathematics, a "proof of existence" is considered satisfactory, even if it does not show how to construct explicitly the "existing" object. Physics is more pragmatic: We want *operational* definitions of the concepts which we use. The millionth digit of π may "exist" and even be computable today, but it will *never* be possible to write 10^{100} digits of π, simply because there is not enough matter in the Universe to do the writing. It is therefore time to turn our attention to the "computational errox box." What can, or cannot, be computed?

This is a very frustrating problem. If you ask a computer scientist what is a computation, he will answer, unflinchingly, that a computation is what may be done by a computer. And then, if you try to obtain the definition of a computer, it will come as "a computer is anything which can mimic another computer," except possibly for limitations due to size (memory) and speed of execution.[72] This is not a joke: It was Turing's fundamental discovery[73,74] that *any* type of computation carried out by a human (then, in 1936, or today carried out by a machine) could be formalized and reduced to a *finite* set of standard instructions. Therefore any physical system, living or inanimate, capable of executing these instructions, can be considered as a *universal computer*, and can simulate any other universal computer.

The crucial point here is that computers are *physical systems* and therefore are subject to limitations imposed by the laws of physics.[75–82] In particular, any real computer is *finite* and therefore it is incapable of representing *most* numbers, namely all those numbers which cannot be generated by a program small enough to fit in the computer itself.[83–85]

A computer does not create information. It only *processes* information so that the latter will conform to some desired rules. For example, the words "the millionth digit of π" have an unambiguous meaning, from the point of view of pure mathematics. That digit undoubtedly exists and is unique. However, we may want its *explicit* value—one out of the ten possibilities which appear equally likely in our present state of ignorance. Fulfilling this aim requires an algorithm (a program of finite size) and then a long, but finite sequence of computational steps. The computational process itself will require some additional working space (a scratchpad) and there may be a trade-off between the size of the program, that of the scratchpad, and the number of required computational steps.

Other examples of computation are encryption and deciphering. A *plaintext* consists of words taken from a finite vocabulary and assembled according to some grammatical and syntactical rules. Encryption is a deterministic mapping, yielding a *ciphertext* which conforms none of these rules and appears random and meaningless. Therefore, encryption seems analogous to the *mixing* process in phase space,[86] which was discussed in Section 3. There is, however, a fundamental difference: Mixing occurs in the classical phase space, which is *continuous*. The amount of computational work needed to overcome mixing increases exponentially with the duration of the process and will therefore outpace every finite computer. On the other hand, cryptography deals with a *discrete* set and a *finite* vocabulary. Given a long enough ciphertext, it is only a matter of sheer computing power to decipher it in a finite time.

One would like to have some objective measure for the amount of computational work needed to accomplish a given task. The notion of "logical depth" was introduced by Bennett.[87] It is, roughly speaking, the number of elementary logical steps required to compute a message from its minimal algorithmic description. Unfortunately, neither algorithmic information nor depth are effectively computable properties.[87] This limitation follows from the most basic result of computability theory, the unsolvability of the halting problem.[73,74] Moreover, the length of a computation can decrease dramatically if the minimal program which generates it is replaced by a slightly longer, but more efficient program.[88] Nevertheless, if we impose some arbitrary rate of exchange between run time and program size, the resulting logical depth can be shown to be reasonably machine-independent.[87]

As a concrete example, consider again the time evolution of a nonintegrable dynamical system. As already stated, a system having *deterministic* dynamics can be called chaotic if, in order to predict the outcome of its motion, the *demand in precision of the initial data increases exponentially with the time separation.* For example, looking again at Fig. 3, if we want to be *sure* that the state at time $t = 0$ will be located in the given $2n$-dimensional ball, we must specify the initial data with a number of digits which increases exponentially with the time separation. This is because neighboring trajectories diverge exponentially[89] when we numerically integrate the Hamilton equations of motion

$$\frac{d\mathbf{q}}{dt} = \frac{\partial H}{\partial \mathbf{p}} \quad \text{and} \quad \frac{d\mathbf{p}}{dt} = -\frac{\partial H}{\partial \mathbf{q}}. \tag{6}$$

If the complexity of that computation is beyond our means, we cannot predict what the final state will be (except statistically). Chance is supreme. The final state is said "random" and the evolution is "chaotic."

Let us now examine the corresponding problem in *quantum theory* and integrate the Schrödinger equation, starting with a given initial (or final) wavefunction $\psi_0(\mathbf{q})$. This could be, for instance, a wave packet located in the same region of phase space as before, if that region is much larger than $\hbar$. As well known, a possible procedure is to compute the eigenvalues and eigenfunctions of the Hamiltonian by solving

$$H\, u_n(\mathbf{q}) = \hbar\omega_n\, u_n(\mathbf{q}), \tag{7}$$

where it was assumed that the Hamiltonian has a discrete spectrum (see however below). The $u_n(\mathbf{q})$ are then normalized, and we can expand the initial state as

$$\psi_0(\mathbf{q}) = \sum c_n\, u_n(\mathbf{q}), \tag{8}$$

with

$$c_n = \int u_n^*(\mathbf{q})\, \psi_0(\mathbf{q})\, d\mathbf{q}. \tag{9}$$

Finally we obtain

$$\psi(\mathbf{q};t) = \sum c_n\, u_n(\mathbf{q})\, \exp(-i\omega_n t). \tag{10}$$

This result is essentially different from the classical one, because the computational complexity does *not* increase exponentially with time. It must increase somewhat, because in order to have meaningful phases in Eq. (10), the errors $\Delta\omega_n$ must be $\ll |t|^{-1}$. The higher precision required in ω_n means more numerical work, but the latter increases only as a polynomial[88] in t, not exponentially. Actually, there is a swindle here, which is not difficult to find: The sum in Eq. (10) is *infinite*, so that this equation, taken literally, represents an infinite amount of numerical work, even if $t = 0$. This is already true for Eq. (7), unless the latter can be solved analytically, which would happen only for an integrable system. What is done in practice is to *truncate* the sums (8) and (10). Since these sums must converge (because the Hamiltonian is self-adjoint[51]) they can be truncated with an arbitrarily small, albeit finite, error. We thus replace $\psi_0(\mathbf{q})$ by

$$\psi_0'(\mathbf{q}) = \sum{}' \, c_n\, u_n(\mathbf{q}), \tag{11}$$

where $\sum'$ denotes the summation of a finite number of terms. Likewise we define

$$\psi'(\mathbf{q};t) = \sum{}' \, c_n\, u_n(\mathbf{q})\, \exp(-i\omega_n t). \tag{12}$$

Note that if $\psi_0'(\mathbf{q})$ is very close to the true initial distribution $\psi_0(\mathbf{q})$, then $\psi'(\mathbf{q};t)$ is very close to the true final distribution $\psi(\mathbf{q};t)$, by unitarity.

We thus reach the following rather curious conclusion: If we ask what are the initial conditions for a classical orbit so that, after a long time t, it will land with 100% certainty in a small region of phase space, the computational complexity of the answer increases exponentially with t. On the other hand, if we want to specify a *quantum state* such that, a time t later, it will become a small wave packet localized in the *same* small region of phase space, then the computational complexity, although formidable, will increase only as a polynomial in t. (Note however that in the quantum problem, the initial state from which we wish to obtain a small wave packet is *not* itself a small wave packet, but is spread throughout all accessible phase space.[90])

It follows that if we attach to the word "chaos" the meaning given above (the computational complexity increases exponentially with time) then *there can be no quantum chaos*, unless the Hamiltonian has a continuous spectrum. This could possibly happen in scattering problems. However, scattering by a short range potential cannot normally lead to chaos, since the scattered particle spends only a finite time under the influence of the potential. A potential of *infinite* extent, such as in an almost periodic crystal, is known to lead to chaos, although of a different kind. In that case, it is the band structure which is a pathological function of the energy and other parameters.[91] It still remains to be seen whether a potential of *finite* extent can generate arbitrarily narrow *resonances*, causing arbitrarily long time delays. If such a potential exists, the integration of the Schrödinger equation $i\hbar\partial\psi/\partial t = H\psi$ would perhaps have a computational complexity increasing exponentially with time.

This also shows, incidentally, that the spectrum of the Liouville operator (4), for a nonintegrable system, cannot be discrete. If it were, we could solve the Liouville equation (3) with arbitrary accuracy, just as we solve the Schrödinger equation, with an amount of work increasing only as a polynomial in time. This is obviously impossible because, as we already know, the final result *must* be specified with a number of digits which increases exponentially with the time elapsed (the whorls and tendrils become exponentially thin).

Here is an example: The spectrum of L for two *uncoupled* harmonic oscillators with incommensurable frequencies is dense over the reals: $\omega_{mn} = m\omega_1 + n\omega_2$, where m and n are integers. Each ω_{mn} is degenerate: Eigenfunctions of L can be multiplied by arbitrary functions of the partial Hamiltonians H_1 and H_2. Now, introducing any *nonlinear* coupling between the oscillators leaves a *single* constant of motion $H_1 + H_2$, and thus is likely to convert the dense point spectrum into a continuous one.

5. WHY IT IS DIFFICULT TO RESURRECT SCHRODINGER'S CAT

We are still faced with our original problem, namely to explain how irreversibility occurs in quantum systems (in particular, during a measurement). The answer which was proposed by Peierls[92] is: "Events are irreversible when *we choose* not to allow reversal." In this final Section, I shall show that this choice may be *imposed* on us by the computational complexity of the problem.

As a trivial exercise, let us coerce a physical system to evolve from an arbitrary state $|D\rangle$ to another arbitrary state $|L\rangle$. This can be done as follows.[93,94] Define a real parameter α and a sequence of states

$$|\alpha\rangle = \cos\alpha|D\rangle + \sin\alpha|L\rangle, \tag{13}$$

so that $|0\rangle = |D\rangle$ and $|\frac{\pi}{2}\rangle = |L\rangle$. Note that

$$\langle\alpha'|\alpha''\rangle = \cos(\alpha' - \alpha''). \tag{14}$$

Now, define a sequence of projection operators $P_k = |\alpha_k\rangle\langle\alpha_k|$, where $\alpha_k = k\pi/2N$, and $k = 1,\ldots,N$. It then follows from (14) that if we measure consecutively the operators P_1, P_2, ..., P_N, the probability to get every time a result 1 ("yes, the system is in the state $|\alpha_k\rangle$") is

$$(\cos\frac{\pi}{2N})^{2N} \to 1 - \frac{\pi^2}{4N}, \qquad \text{for large } N. \tag{15}$$

We can therefore transform $|D\rangle$ into $|L\rangle$ with a probability of success arbitrarily close to 1.

Why is it so difficult to achieve this result in the real world? To rotate $|D\rangle$ into $|L\rangle$ as suggested, we must build the set of apparatuses which test the presence of each state $|\alpha_k\rangle$. For this, we must specify *exactly* what are the states $|L\rangle$ and $|D\rangle$. Now, in the Schrödinger's cat paradigm, there are many "equivalent" $|L\rangle$ states, and still many more $|D\rangle$ states. If we make the slightest mistake and replace $|D\rangle$ by another state $|D'\rangle$ (say, the same cat with just one atom removed) then, as $|D\rangle$ and $|D'\rangle$ are *orthogonal* (even though they are macroscopically similar) the resurrection process described above will fail.

From now on, I shall no longer mention cats, but rather consider inanimate apparatuses, so that the argument won't appear frivolous. Recall that I generously assumed, at the end of Section 3, that any quantum state which can be specified by means of an explicit algorithm can also be prepared experimentally. I shall now show that, even though formally there is no quantum chaos (as defined above) there are nevertheless quantum states which *cannot* be prepared reproducibly, because the algorithm which describes them is too long to fit in any computer in the Universe where we live.

Consider an apparatus with N degrees of freedom, which are coupled to each other in a nonlinear way. This apparatus is assumed perfectly isolated, but it is so only *after* it has been prepared in a laboratory which released it at a finite temperature T, say 1μK. It follows that the energy of this apparatus has an uncertainty of order kT, in our case 10^{-10}eV. By "uncertainty" I mean $\sqrt{(\langle H^2\rangle - \langle H\rangle^2)}$, where H is the Hamiltonian, as usual. Now, the crucial point is that in order to give a complete quantum description of this closed system and to compute its evolution explicitly, we must specify the amplitude of each one of its energy levels. The problem thus is how many of these energy levels have nonvanishing amplitudes.

It is well known[95] that the coarse grained density of *quantum* states can be approximated by *classical* statistical mechanics as:

$$\rho(E) = h^{-N} \int \delta[E - H(\mathbf{p},\mathbf{q})]\, d\mathbf{p}d\mathbf{q}. \tag{16}$$

Here, it may be objected that this formula is not valid near the ground state of a quantum system. The usual situation, for a typical system, is that most energy levels lie much higher than 10^{-10}eV. For example, in a large molecule, only the rotational degrees of freedom would be excited. However, a measuring apparatus is *not* a typical system.[96] It must have *macroscopically distinguishable states with very nearly the same energy*, as otherwise it would not be able to react to the presence of a microscopic quantum system. As a crude example, think of a pointer moving along a dial: The potential energy of the pointer must be almost independent of its position[34] since otherwise the pointer would keep moving under the influence of forces due to other parts of the apparatus, and thus it would not indicate the correct result of a measurement.

In summary, an apparatus must, by construction, have a nearly degenerate ground state, or else be prepared in a highly excited, metastable state. In both cases, its state is far enough in the semi-classical range to make Eq. (16) approximately valid, give or take a few orders of magnitude. Let us make a rough estimate for $\rho(E)$ in a "typical" case. By dimensional analysis,

$$\rho(E) \sim (A/\hbar)^N/E, \tag{17}$$

where A is a typical action and E a typical energy. A very, crude, but realistic estimate could be $A \sim N\hbar$, which is valid both for small N (low lying atomic states) and very large N (A is of the order of cgs units when N equals Avogadro's number). Likewise, take E=1eV, a typical atomic energy. We thus obtain

$$\rho \sim N^N \,\text{eV}^{-1}, \tag{18}$$

so that some $10^{-10}N^N$ energy levels are involved, if the temperature is 1μK, as assumed (provided of course that $N^N \gg 10^{10}$, as otherwise we cannot use these statistical arguments). For example, if we take N=100, still a microscopic system, we obtain 10^{190} levels, more than we can write down by using all the matter in the Universe!

Thus, finally, we are in a position to solve the "measurement problem" and to show that neither an inconsistency, nor new results, can arise if we quantize the observer.[14] Recall that in the standard presentation of a measurement, the observer is left outside the dynamical description, and the act of observation is a "collapse" of the wavefunction,

$$\psi = \sum c_n u_n \to u_k, \tag{19}$$

occuring with probability $|c_k|^2$. Here, the u_n are the orthonormal eigenstates of the operator being measured, and $c_n = \langle u_n|\psi\rangle$. This "collapse" is *not* a dynamical process, but simply the replacement of obsolete information by fresh information, thanks to the act of observation.[11] (Recall that the *only* meaning of ψ is the information which it represents.)

Now, let us include in the dynamical description of this phenomenon the measuring apparatus itself, which initially had wavefunction ϕ. Let v_n be a set of *macroscopically distinguishable*[11] states of that apparatus. Then, instead of (19), we have

$$\psi\phi \to \sum c_n u_n v_n, \tag{20}$$

which is a *linear superposition* (it cannot be anything else, if we believe in quantum theory). For example, Eq. (1) is such a superposition. There is nothing strange or unnatural in it. In the general case, the nth state of the system is *correlated* with the nth state (or set of states) of the observer.[14] The evolution (20) is unitary and it is not difficult to concoct a Hamiltonian which generates it.

When we describe the situation in this way, there is *no* definite outcome: *No observation has been carried out* as yet. The act of observation, which yields fresh information, must *then* be performed by a *new* observer who, in turn, remains outside the dynamical description. From *his* point of view, Eq. (19) is replaced by

$$\sum c_n u_n v_n \to u_k v_k. \tag{21}$$

The outcome must be the *same* in both cases: If the first observer says that he found result "k", then the second observer will also say that he found result "k" and, moreover, that he sees the first observer aware that the result is "k". Recall that u_k and v_k are *correlated*.

To get a *new* result (that is, a *contradiction*) the second observer should have the ability to reverse the arrow in Eq. (20), so as to decorrelate the observed system from the first observer. To do that, he ought to specify in (20) not only a few "macroscopically distinguishable states," but every single *microscopic* energy eigenstate with the correct amplitude and phase. We have seen that the latter are far too numerous, if the first observer has many nonlinearly coupled degrees of freedom.

There is thus a very close resemblance between the situation in quantum theory and that known to us from classical physics. The meaning of classical chaos is that, once enough time has elapsed, it is impossible to predict the value of *any* nontrivial dynamical variable. Only statistical predictions are possible. Likewise, in quantum physics, no nontrivial operator can be constructed such that its measurement would yield a predictable outcome. Indeed, if we want a dynamical variable A to have a well defined value a at time t, it must satisfy $A\psi(t) = a\psi(t)$. If we know $\psi(t)$ explicitly, it is trivial to construct an operator A satisfying this equation. But, if the system is sufficiently complex, we cannot specify $\psi(t)$ explicitly, and therefore it is also impossible to define A.

To get this result, namely an unmanageable computational complexity which is the same as we would have expected in a true "quantum chaos," one needs many—perhaps a dozen—nonlinearly coupled degrees of freedom. However, it is not enough for that to consider just any macroscopic body. The fundamental vibration mode of a 10 ton Weber bar[97] may be so well decoupled from the other degrees of freedom that it behaves as a quantum harmonic oscillator, at least over a time scale of a few seconds. The macroscopic phase of a Josephson junction[96] may also be decoupled from its myriad of other degrees of freedom, to a very good approximation. Any macroscopic system, whether classical or quantal, may have a few of these "robust" variables which are well decoupled from the chaotic ones. In principle, these macroscopic degrees of freedom can even be correlated to microscopic ones, as in Eq. (20), but they *cannot be considered as measuring apparatuses*, precisely because they are so well isolated and their behavior is reversible. A correlation like (20) is *not* by itself a measurement. The two electrons in the ground state of the helium atom are correlated, but no one in his right mind would say that each electron "measures" its partner.

Note that in the absence of *nonlinear* coupling, irreversibility may be illusory, even if very many degrees of freedom are involved. An example is the spin echo phenomenon[37,38] which was mentioned earlier. Another one is the correction of optical distortions by phase conjugated mirrors[98,99] which also yield a kind of sharp echo, with restoration of apparently lost information.

Another problem worth mentioning is the parasitic but unavoidable coupling with the environment.[100–102] It is equivalent to a lack of precise knowledge of the Hamiltonian, making it even more difficult to decorrelate the right hand side of (20) and reverse the natural evolution.

Finally, it must be recognized that by invoking cosmological limitations on the maximum admissible computational complexity of a problem, I have introduced a foreign element in quantum theory. If you demand an analytically pure quantum formalism, unadulterated by extraneous "practical" considerations, then there can be no irreversibility (at least for finite t) and then there also are *no* quantum measurements![103]

ACKNOWLEDGMENTS

Part of this research was done during visits to the University of Texas at Austin (Center for Theoretical Physics), the International Center for Theoretical Physics in Trieste, and Massachusetts Institute of Technology (Laboratory for Computer Science). I am grateful to all these institutions for their hospitality and support.

Research done at Technion was supported by the Gerard Swope Fund, the New York Metropolitan Research Fund, and the Fund for Encouragement of Research at Technion.

REFERENCES

1. J. A. Wheeler and W. H. Zurek, "Quantum Theory and Measurement," Princeton Univ. Press, Princeton (1983).
2. G. Casati, ed., "Chaotic Behavior in Quantum Systems," Plenum, New York (1985).
3. A. Peres and W. H. Zurek, Is Quantum Theory Universally Valid? *Am. J. Phys.* 50:807 (1982).
4. A. Peres, The Classic Paradoxes of Quantum Theory, *Found. Phys.* 14:1131 (1984).
5. E. Schrödinger, Die gegenwärtige Situation in der Quantenmechanik, *Naturwiss.* 23:807, 823, 844 (1935) [transl. in ref. 1, pp. 152-167].
6. H. P. Stapp, The Copenhagen Interpretation, *Am. J. Phys.* 40:1098 (1972).
7. J. Rothstein, Information, Measurement, and Quantum Mechanics, *Science* 114:171 (1951).
8. J. B. Hartle, Quantum Mechanics of Individual Systems, *Am. J. Phys.* 36:704 (1968).
9. L. E. Ballentine, The Statistical Interpretation of Quantum Mechanics, *Rev. Mod. Phys.* 42:358 (1970).
10. A. Peres, What is a State Vector? *Am. J. Phys.* 52:644 (1984).
11. A. Peres, When is a Quantum Measurement? *Am. J. Phys.* 54:688 (1986).
12. P. A. M. Dirac, "The Principles of Quantum Mechanics," Oxford Univ. Press, Oxford (1947) p. 36.
13. D. Bohm, "Quantum Theory," Prentice-Hall, Englewood Cliffs (1951) p. 120.
14. H. Everett, III, Relative State Formulation of Quantum Mechanics, *Rev. Mod. Phys.* 29:454 (1957).
15. M. A. B. Whitaker, The Relative States and Many-Worlds Interpretations of Quantum Mechanics and the EPR Problem , *J. Phys. A* 18:253 (1985).
16. I. Bloch, Some Relativistic Oddities in the Quantum Theory of Observation, *Phys. Rev.* 156:1377 (1967).

17. Y. Aharonov and D. Z. Albert, States and Observables in Relativistic Quantum Field Theories, *Phys. Rev. D* 21:3316 (1980).
18. Y. Aharonov and D. Z. Albert, Can we Make Sense out of the Measurement Process in Relativistic Quantum Mechanics? *Phys. Rev. D* 24:359 (1981).
19. J. S. Bell, On the Einstein Podolsky Rosen Paradox, *Physics* 1:195 (1964).
20. A. Einstein, B. Podolsky and N. Rosen, Can Quantum Mechanical Description of Reality be Considered Complete? *Phys. Rev.* 47:777 (1935).
21. J. F. Clauser and A. Shimony, Bell's Theorem. Experimental Tests and Implications, *Rep. Prog. Phys.* 41:1881 (1978).
22. A. Aspect, J. Dalibard and G. Roger, Experimental Test of Bell's Inequalities Using Time-Varying Analyzers, *Phys. Rev. Lett.* 49:1804 (1982).
23. S. J. Feingold and A. Peres, How Often Are Bell's Inequality Premises Violated? *J. Phys. A* 13:3187 (1980).
24. S. Kochen and E. P. Specker, The Problem of Hidden Variables in Quantum Mechanics, *J. Math. Mech.* 17:59 (1967).
25. J. S. Bell, On the Problem of Hidden Variables in Quantum Mechanics, *Rev. Mod. Phys.* 38:447 (1966).
26. A. Peres and A. Ron, Cryptodeterminism and Quantum Theory, *in*: "Microphysical Reality and Quantum Formalism," G. Tarozzi and A. van der Merwe, ed., Reidel, Dordrecht (in press).
27. N. D. Mermin, Is the Moon There when Nobody Looks? Reality and the Quantum Theory, *Physics Today* 38:38 (April 1985).
28. S. J. Feingold and A. Peres, Reality and the Quantum Theory, *Physics Today* 38:15 (Nov. 1985).
29. A. Peres, Existence of "Free Will" as a Problem of Physics, *Found. Phys.* 16:573 (1986).
30. A. Peres, Quantum Measurements are Reversible, *Am. J. Phys.* 42:886 (1974).
31. D. Deutsch, Quantum Theory as a Universal Physical Theory, *Int. J. Theor. Phys.* 24:1 (1985).
32. J. von Neumann, "Mathematical Foundations of Quantum Mechanics," Princeton Univ. Press, Princeton (1955) [transl. from "Mathematische Grundlagen der Quantenmechanik, Springer, Berlin (1932)].
33. F. W. London and E. Bauer, "La Théorie de l'Observation en Mécanique Quantique," Hermann, Paris (1939) [transl. in ref. 1, pp. 217-259].
34. A. Peres, Can we Undo Quantum Measurements? *Phys. Rev. D* 22:879 (1980).
35. A. Peres, Stability of Quantum Motion in Chaotic and Regular Systems, *Phys. Rev. A* 30:1610 (1984).
36. V. I. Arnold and A. Avez, "Ergodic Problems in Classical Mechanics," Benjamin, New York (1968).
37. A. Abragam, "The Principles of Nuclear Magnetism," Clarendon Press, Oxford (1961).
38. J. Rothstein, Nuclear Spin Echo Experiments and the Foundations of Statistical Mechanics, *Am. J. Phys.* 25:510 (1957).
39. M. Kac, "Probability and Related Topics in Physical Sciences," Interscience, New York (1959).

40. M. Dresden, The Kac Ring Model (Mark I), *in*: "Studies in Statistical Mechanics. Vol. I," J. De Boer and G. E. Uhlenbeck, ed., North Holland, Amsterdam (1962) p. 316.
41. L. E. Reichl, "A Modern Course in Statistical Physics," Univ. of Texas Press, Austin (1980).
42. L. Mandel, Squeezed States and Sub-Poissonian Photon Statistics, *Phys. Rev. Lett.* 49:136 (1982).
43. L. Szilard, Über die Entropieverminderung in Einen Thermodynamischen System bei Eingriffen Intelligenter Wesen, *Z. Phys.* 53:840 (1929) [transl. in ref. 1, pp. 539-548].
44. A. Peres, The Physicist's Role in Physical Laws, *Found. Phys.* 10:631 (1980).
45. I. R. Senitzky, Classical Statistics Inherent in Pure Quantum States, *Phys. Rev. Lett.* 47:1503 (1981).
46. S. J. Feingold and A. Peres, Linear Stability Test for Hamiltonian Systems, *Phys. Rev. A* 26:2368 (1982).
47. M. V. Berry, N. L. Balazs, M. Tabor, and A. Voros, Quantum Maps, *Ann. Phys. (N. Y.)* 122:26 (1979).
48. B. O. Koopman, Hamiltonian Systems and Transformations in Hilbert Spaces, *Proc. Nat. Acad. Sc.* 17:315 (1931).
49. A. Isihara, "Statistical Physics," Academic Press, New York (1971) p. 196.
50. P. Résibois and M. De Leener, "Classical Kinetic Theory of Fluids," Wiley, New York (1977) p. 205.
51. F. Riesz and B. Sz.-Nagy, "Functional Analysis," Ungar, New York (1955) p. 228.
52. E. Wigner, Quantum Correction for Thermodynamic Equilibrium, *Phys. Rev.* 40:749 (1932).
53. M. Hillery, R. F. O'Connell, M. O. Scully, and E. P. Wigner, Distribution Functions in Physics: Fundamentals, *Phys. Reports* 106:121 (1984).
54. H. J. Korsch and M. V. Berry, Evolution of Wigner's Phase-Space Density under a Nonintegrable Quantum Map, *Physica D* 3:627 (1981).
55. A. M. Ozorio de Almeida and J. H. Hannay, Geometry of Two Dimensional Tori in Phase Space: Projections, Sections and the Wigner Function, *Ann. Phys. (N. Y.)* 138:115 (1982).
56. A. M. Ozorio de Almeida, The Wigner Function for Two Dimensional Tori: Uniform Approximation and Projections, *Ann. Phys. (N. Y.)* 145:100 (1983).
57. R. C. Brown and R. E. Wyatt, Quantum Mechanical Manifestation of Cantori: Wave-Packet Localization in Stochastic Regions, *Phys. Rev. Lett.* 57:1 (1986).
58. S. Fishman, D. R. Grempel and R. E. Prange, Chaos, Quantum Recurrences, and Anderson Localization, *Phys. Rev. Lett.* 49:509 (1982).
59. T. Hogg and B. A. Huberman, Quantum Dynamics and Nonintegrability, *Phys. Rev. A* 28:22 (1983).
60. G. Casati, B. V. Chirikov, I. Guarneri and D. L. Shepelyansky, Dynamical Stability of Quantum "Chaotic" Motion in a Hydrogen Atom, *Phys. Rev. Lett.* 56:2437 (1986).
61. T. Hogg and B. A. Huberman, Recurrence Phenomena in Quantum Dynamics, *Phys. Rev. Lett.* 48:711 (1982).

62. I. C. Percival, Almost Periodicity and the Quantal *H* Theorem, *J. Math. Phys.* 2:235 (1961).
63. L. S. Schulman, Note on the Quantum Recurrence Theorem, *Phys. Rev. A* 18:2379 (1978).
64. A. Peres, Recurrence Phenomena in Quantum Dynamics, *Phys. Rev. Lett.* 49:1118 (1982).
65. J. Ford, Chaos: Solving the Unsolvable, Predicting the Unpredictable! *in*: "Chaotic Dynamics and Fractals," M. F. Barnsley and S. G. Demko, ed., Academic Press, New York (1986) pp. 1-52.
66. W. Lamb, Quantum Chaos and the Theory of Measurement, *in* ref. 2, pp. 353-361.
67. A. Landé, "Foundations of Quantum Theory, A Study in Continuity and Symmetry," Yale Univ. Press, New Haven (1955) p. 4.
68. W. Lamb, An Operational Interpretation of Nonrelativistic Quantum Mechanics, *Phys. Today* 22:23 (April 1969).
69. E. P. Wigner, The Problem of Measurement, *Am. J. Phys.* 31:6 (1963).
70. B. O. Hultgren, III, and A. Shimony, The Lattice of Verifiable Propositions of the Spin-1 System, *J. Math. Phys.* 18:381 (1977).
71. A. R. Swift and R. Wright, Generalized Stern-Gerlach Experiments and the Observability of Arbitrary Spin Operators, *J. Math. Phys.* 21:77 (1980).
72. M. Davis, What is a Computation? *in*: "Mathematics Today: Twelve Informal Essays," L. A. Stern, ed., Springer-Verlag, New York (1978) pp. 241-267 [Published in softcover by Vintage Books, New York (1980)].
73. A. M. Turing, On Computable Numbers, with an Application to the Entscheidungsproblem, *Proc. London Math. Soc.* 42:230 (1936).
74. J. E. Hopcroft, Turing Machines, *Sci. Am.* 250:86 (May 1984).
75. C. H. Bennett, The Thermodynamics of Computation—a Review, *Int. J. Theor. Phys.* 21:905 (1982).
76. H. J. Bremermann, Minimum Energy Requirements of Information Transfer and Computing, *Int. J. Theor. Phys.* 21:203 (1982).
77. D. W. Hillis, New Computer Architectures and their Relation to Physics, or Why Computer Science is No Good, *Int. J. Theor. Phys.* 21:255 (1982).
78. L. B. Levitin, Physical Limitations of Rate, Depth and Minimum Energy in Information Processing, *Int. J. Theor. Phys.* 21:299 (1982).
79. T. Toffoli, Physics and Computation, *Int. J. Theor. Phys.* 21:165 (1982).
80. J. D. Bekenstein, Entropy Content and Information Flow in Systems with Limited Energy, *Phys. Rev. D* 30:1669 (1984).
81. C. H. Bennett and R. Landauer, The Fundamental Physical Limits of Computation, *Sci. Am.* 253:48 (July 1985).
82. R. Landauer, Fundamental Physical Limitations of the Computational Process, *Ann. New York Acad. Sc.* 426:161 (1985).
83. G. Chaitin, Randomness and Mathematical Proof, *Sci. Am.* 232:47 (May 1975).
84. G. Chaitin, Algorithmic Information Theory, *IBM J. Res. Develop.* 21:350 (1977).
85. J. Ford, How Random is a Coin Toss? *Phys. Today* 36:40 (April 1983).

86. G. Y. Vichniac, Cellular Automata Models of Disorder and Organisation, *in*: "Disordered Systems and Biological Organization," E. Bienenstock, F. Fogelman Soulié and G. Weisbuch, ed., Springer-Verlag, Berlin (1986) pp. 3-20.

87. C. H. Bennett, Dissipation, Information, Computational Complexity and the Definition of Organization, *in*: "Emerging Syntheses in Science," D. Pines, ed., Santa Fe Institute, Santa Fe (1985) pp. 297-313.

88. G. Dahlquist and Å. Björk, "Numerical Methods," Prentice-Hall, Englewood Cliffs (1974).

89. J. Ford and G. H. Lunsford, Stochastic Behavior of Resonant Nearly Linear Oscillator Systems in the Limit of Zero Nonlinear Coupling, *Phys. Rev. A* 1:59 (1970).

90. M. Feingold, N. Moiseyev and A. Peres, Ergodicity and Mixing in Quantum Theory, *Phys. Rev. A* 30:509 (1984).

91. A. Peres, Chaotic Band Structure of Almost-Periodic Potentials, *Phys. Rev. B* 27:6493 (1983).

92. R. Peierls, private conversation (1984).

93. A. Peres, A Single System has No State, *Am. J. Phys.* 43:1015 (1975).

94. Y. Aharonov and M. Vardi, Meaning of an Individual "Feynamn Path," *Phys. Rev. D* 21:2235 (1980).

95. M. V. Berry, Semi-Classical Mechanics of Regular and Irregular Motion, *in*: "Chaotic Behavior of Deterministic Systems," G. Iooss, R. H. G. Helleman and R. Stora, ed., North-Holland, Amsterdam (1983) pp. 171-271.

96. A. J. Leggett, Macroscopic Quantum Systems and the Quantum Theory of Measurement, *Suppl. Prog. Theor. Phys.* 69:80 (1980).

97. C. M. Caves, K. S. Thorne, R. W. P. Drever, V. D. Sandberg and M. Zimmermann, On the Measurement of a Weak Classical Force Coupled to a Quantum-Mechanical Oscillator, *Rev. Mod. Phys.* 52:341 (1980).

98. V. V. Shkunov and B. Ya. Zel'dovich, Optical Phase Conjugation, *Sci. Am.* 253:40 (Dec. 1985).

99. D. M. Pepper, Applications of Optical Phase Conjugation, *Sci. Am.* 254:74 (Jan. 1986).

100. H. D. Zeh, On the Interpretation of Measurement in Quantum Theory, *Found. Phys.* 1:69 (1970).

101. W. H. Zurek, Pointer Basis of Quantum Apparatus: Into what Mixture does the Wave Packet Collapse? *Phys. Rev. D* 24:1516 (1981).

102. W. H. Zurek, Environment-Induced Superselection Rules, *Phys. Rev. D* 26:1862 (1982).

103. J. S. Bell, On Wave Packet Reduction in the Coleman-Hepp Model, *Helv. Phys. Acta* 48:93 (1975).

SEMICLASSICAL CHAOLOGY

Michael Berry

H.H.Wills Physics Laboratory, Tyndall Avenue

Bristol BS8 1TL, U.K.

The connection between classical and quantum mechanics (i.e. the semiclassical limiting asymptotics as $\hbar \to 0$) must be subtle and complicated, because classical mechanics itself (i.e. the classical limit $\hbar=0$) is subtle and complicated: the orbits of systems governed by Hamilton's equations of motion may be predictable (regular) or unpredictable (irregular) depending on subtle details of the form of the Hamiltonian $H(\{q_i\},\{p_i\})$[1-3]. A natural question is: how does the 'chaology' of classical orbits reflect itself in the correspnding quantum system? Sometimes this question is put in the form: what is quantum chaos?

There are many approaches to this question. One is to study the dynamics of quantum systems which are classically chaotic, that is, to study non-stationary states. There have been many studies of mathematical models of such quantum evolution[4-10], which have found important recent application in interpreting experiments on the microwave ionization of Hydrogen atoms[11-13]. Another approach is to look at stationary states and concentrate on the form of the wave functions: these are remarkably different for eigenstates corresponding to regular and chaotic systems[14-19].

Here however I will concentrate on the energies of stationary states, and ask how the distribution of eigenvalues $\{E_n\}=E_1,E_2\ldots$ of a quantum Hamiltonian $\hat{H} = H(\{\hat{q}_i\},\{\hat{p}_i\})$ reflects the chaology of the classical trajectories generated by the classical H, in which $\{q_i\}$ and $\{p_i\}$are variables rather than operators. Of course the energies $\{E_n\}$depend on $\hbar$. I will consider only the nontrivial case where the number of freedoms N exceeds unity, and confine myself to a brief review of current ideas.

Ideally one would like an explicit asymptotic formula giving $\{E_n(\hbar)\}$ with an error that decreases as $\hbar\to 0$ faster than the mean level spacing. Such a formula has been found only for classically integrable (i.e. nonchaotic) systems[20,21,22] and is a generalization of the familiar WKB theory for one dimension. For integrable systems with N freedoms, there are N constants of motion (including the energy) which confine motion to N-dimensional tori in the 2N-dimensional phase space[23]. In lowest order, quantization selects the energies E of those tori whose N actions are separated by multiples of $\hbar$, i.e.

$$E_{\{n_i\}} = H(\{I_i = (n_i + \tfrac{1}{4}\alpha_i)\hbar\}) \qquad (1)$$

where $\{n_i\} = n_1 \text{---} n_N$ are the quantum numbers, $\{I_i\}$ are the actions

$$I_i \equiv \oint_{\gamma_i} p_j dq_j \qquad (2)$$

round the irreducible cycles γ_i of the torus, and the α_i are constants (Maslov indices[20]). Obviously (1) works only when tori exist. In the chaotic extreme, motion is ergodic and there are no constants of motion apart from E; therefore there are no tori and the semiclassical rule (1) cannot be applied: so far nobody has found a semiclassical quantization rule for chaotic systems.

In these circumstances one must seek less precise information, in the form of average properties of the distribution of energies. These spectral averages can be defined semiclassically, because as $\hbar \to 0$ infinitely many levels crowd into any fixed energy interval however small. The simplest spectral average is the mean spectral density $\langle d(E) \rangle$. This is the average of

$$d(E) \equiv \Sigma\, \delta(E-E_n) = \text{Tr}\, \delta(E-H), \qquad (3)$$

and is given semiclassically by the 'Weyl rule'[21]

$$\langle d(E) \rangle = \frac{d\Omega(E)/dE}{h^N} \qquad (4)$$

where Ω is the phase volume

$$\Omega(E) = \int_{(H<E)} d^N q \int d^N p, \text{ i.e. } d\Omega/dE = \int d^N q \int d^N p\, \delta(E-H(\{q_i\},\{p_i\})). \qquad (5)$$

The Weyl rule formalises the old idea of 'one quantum state per volume h^N of phase space'. When applied to quantum billiards (vibrating drums[49]), the Weyl rule plus corrections[50] can be made the basis of a method[51] for accurately reconstructing aspects of billiard geometry (e.g. area and length) from sequences of eigenvalues.

The result (4) tells us nothing about quantum chaos, because the classical volume $\Omega(E)$ is insensitive to the regularity or chaos of the orbits. This is disappointing but nevertheless two useful pieces of information can be obtained. First, the mean level spacing $\langle d \rangle^{-1}$ is of order $\hbar^N$; thus for example in a classically small energy range of size $\hbar$ there are many levels (of order $\hbar^{-(N-1)}$); this will be important later. And second, a rough quantization rule can be found by realizing that the integral of d(E) is the spectral staircase

$$\mathcal{N}(E) \equiv \Sigma\Theta(E-E_n) = \int_{-\infty}^{E} dE' d(E') \qquad (6)$$

where Θ is the unit step; the rule, expressing the idea that the smooth curve of the average staircase might intersect the steps halfway, on average, is then

$$(n+\tfrac{1}{2}) = \langle \mathcal{N}(E_n) \rangle = \Omega(E_n)/h^N. \qquad (7)$$

The above rule is rough because it fails to describe the fine-scale fluctuations in the levels (in graphs of the E_n as a function of a parameter on which $\hat{H}$ depends, these fluctuations appear as avoided crossings[24-26]). To describe these fluctuations it is necessary to employ statistics which (unlike $\langle d \rangle$ and $\langle \mathcal{N} \rangle$) involve correlations between nearby levels, that is on scales $\hbar^N$. Such fluctuation measures have been devised in random-matrix theory[27] and applied to sequences of excited resonance levels of atomic nuclei. The fluctuation statistics depend not on the raw spectrum levels $\{E_n\}$ but on the 'unfolded' spectrum of levels $\{x_n\}$ which have been scaled so as to have unit mean spacing. Thus

$$x_n + \frac{1}{2} \equiv \langle \mathcal{N}(E_n)\rangle. \tag{8}$$

(without the fluctuations, (7) shows that x_n would be simply n). Two particular useful statistics are the probability distribution P(S) of the level spacings $\{S_n \equiv x_{n+1} - x_n\}$, and the spectral rigidity[28-30]

$$\Delta(L) \equiv \min \frac{1}{L} \int_{x-L/2}^{x+L/2} d\zeta [\mathcal{N}(\zeta) - A - B\zeta]^2 > \tag{9}$$

(This is the least squares deviation of the staircase from a straight line, over a range of L mean spacings, averaged over an interval of energies x that includes many levels). P(S) is useful in describing spectral correlations on the finest scales - i.e. between neighbouring levels - and $\Delta(L)$ is useful for describing how spectral correlations depend on range - i.e. large or small L.

When spectral statistics are computed for sequences of levels of Hamiltonian systems with classical limits, a remarkable 'experimental' fact emerges: the statistics display universality, and the spectral universality class depends on the chaology of the classical orbits. The universality classes are

(a) Classically integrable systems. Here the spectral statistics are those of a Poisson - i.e. uncorrelated random - distribution of levels[30,31]. At first sight it is surprising that the quantum conditions (1) can give rise to a random sequence, but the surprise dissipates with the realisation that neighbouring levels E_n, E_{n+1} can have very different sets of quantum numbers $\{n_i\}$. For Poisson statistics,

$$P(S) = \exp(-S), \text{ and } \Delta(L) = L/15 \tag{10}$$

(b) Classically chaotic spinless (or integral-spin) systems with time-reversal symmetry. Here the spectral statistics are those of the Gaussian orthogonal ensemble (GOE), which consists[27] of real symmetric matrices whose elements are Gauss-distributed so as to make the statistics of the ensemble invariant under orthogonal rotations. Only real symmetric matrices are involved because time-reversal symmetry implies that the wavefunctions are real. For the GOE, to a close approximation,

$$P(S) \approx \frac{\pi}{2} S \exp(-\pi S^2/4) \tag{11}$$

and

$$\begin{aligned} \Delta(L) &\to L/15 \quad (L \ll 1) \\ &\to \frac{\ln L}{\pi^2} - .000695 \quad (L \gg 1) \end{aligned} \tag{12}$$

(c) Classically chaotic systems without time-reversal symmetry. This is in a sense the generic case which best justifies the label 'quantum chaos'. One way to break time-reversal symmetry (T) is with magnetic fields. These may be smoothly-varying[32] or may consist of a single (Aharonov-Bohm) flux line[33] (which has the advantage of breaking T without altering the classical chaology). (An even simpler way[52], not involving magnetic fields, is through the massless Dirac equation with boundary conditions corresponding to a 4-scalar potential ('neutrino billiards').) Here the spectral statistics are those of the Gaussian unitary ensemble (GUE)[27], of complex Hermitian matrices whose elements are Gauss-distributed so as to make the statistics of the ensemble invariant under unitary transformations. For the GUE, to a close

$$P(S) \approx \frac{32S^2}{\pi^2} \exp(-4S^2/\pi) \tag{13}$$

and

$$\begin{aligned} \Delta(L) &\to L/15 \quad (L \ll 1) \\ &\to \frac{\ln L}{2\pi^2} + 0.05902 \quad (L \gg 1) \end{aligned} \tag{14}$$

Before anticipating that a system without T will have GUE statistics, care must be taken to determine whether it has any geometric symmetries, because these can act so as to mimic T-symmetry and generate levels with GOE statistics; in the theory of this 'false time-reversal symmetry-breaking'[34] it is shown that for GUE the system must possess no antiunitary symmetry operator $\hat{A}$[27,35] (commuting with $\hat{H}$) and satisfying $\hat{A}^2=1$ or $\hat{A}\hat{A}^* = 1$ (T is represented in position representation by the operator $\hat{A}$=complex conjugation). A set of numbers recently discovered to have GUE statistics is the imaginary parts of zeros of Riemann's zeta function[36]; this is surprising and suggestive[37]

(d) Classically chaotic systems with half-integer spin and with T (or more generally, chaotic systems with an antiunitary symmetry satisfying $\hat{A}^2=-1$). Here there are so far no numerical experiments (I am planning one now) but the spectral statistics are expected to be those of the Gaussian symplectic ensemble (GSE)[27], of quaternion real Hermitian matrices whose elements are Gauss-distributed so as to make the ensemble invariant under symplectic transformations. For the GSE, to a close approximation,

$$P(S) \approx \frac{2^{18}S^4}{3^6\pi^3} \exp(-64S^2/9\pi) \tag{15}$$

and

$$\begin{aligned} \Delta(L) &\to L/15 \quad (L \ll 1) \\ &\to \frac{\ln L}{4\pi^2} + .006076 \quad L \gg 1) \end{aligned} \tag{16}$$

That completes the list of universality classes. But I now reveal that life is really not so simple, and describe two ways in which universality is compromised. First, most classical systems are neither purely regular nor purely chaotic[1], but exhibit mixed (or, in the jargon 'KAM') behaviour in which some orbits are regular and some unpredictable, depending on initial conditions. Such cases are important in quantum mechanics because they correspond to the anharmonically coupled oscillators describing vibrating molecules and to atoms in strong magnetic fields occurring astrophysically. It is natural to expect that in lowest approximation the spectral statistics will interpolate between those of the Poisson and the appropriate random-matrix universality classes, to a degree which depends on the relative phase-space volumes of regions of regular and chaotic motion; a theory along these lines[38] is supported by numerical experiments[39]. Second, even for purely regular or purely chaotic systems the domain of universal behaviour is limited to energy ranges not exceeding a quantity of order $\hbar$; for the rigidity $\Delta(L)$, this means that universality holds when $L<L_{max} \sim \hbar^{-(N-1)}$, so that in the semiclassical limit the domain of universality shrinks to zero in energy but nevertheless extends over infinitely many levels. When $L>L_{max}$, numerical experiments[40] (so far restricted to integrable systems) show, and a theory[30] (for both integrable and chaotic systems) explains, that $\Delta(L)$ does not continue to increase as in (10), (12), (14) and (16), but saturates at nonuniversal values characteristic of the particular system.

In spite of these caveats, the universality of semiclassical spectra is a remarkable phenomenon that demands explanation. One class of theories[41,42] considers the energies to depend on a parameter t which is regarded as akin to a time variable, and the 'motion' of the eigenvalues $\{E_n(T)\}$ on the E axis is put into correspondence with the statistical mechanics of particles on a line. These theories can be made to generate random-matrix behaviour but the derivations rest on statistical assymptions about the matrix elements of the t derivatives of $\hat{H}$ between different eigenstates. It is desirable to understand spectral statistics directly, without introducing parameters or extra statistical assymptions. Some progress has been made as I now describe.

The behaviour of P(S) as $S\to 0$ can be related to the codimensiona K of degeneracies when the system is embedded in an ensemble of similar ones[43,21,24]. K is the number of parameters that must be varied to produce a degeneracy; for separable systems K=1, for real symmetric matrices K=2, for complex Hermitian matrices K=3 and for quaternion real matrices K=5. The result is

$$P(S) \sim S^{K-1} \text{ as } S\to 0 \qquad (17)$$

and this agrees with (10), (11), (13) and (15). But K is only roughly related to the classical symmetries (subtleties arise from barrier penetration[44]) and a semiclassical understanding of P(S) is still lacking.

The behaviour of $\Delta(L)$ on the other hand is rather well understood[30] in terms of a semiclassical theory. According to (9), $\Delta(L)$ is a quadratic functional of the spectral staircase (6). This can be expressed as its average $\langle\mathcal{N}\rangle$ (equation 7) plus a series of correction terms which are oscillatory functions of E. Each such correction comes from a <u>closed orbit</u> of the classical system[45,47], and gives an oscillation with energy period $\hbar/T$ where T is the time period of the orbit. So values of $L<<L_{max}$, corresponding to energy scales $<<\hbar$, correspond to <u>very long orbits</u>. In particular, any fixed L corresponds as $\hbar\to 0$ to periods of order $\hbar^{-(N-1)}$. But for these very long orbits there exist <u>universal sum rules</u>[48] which depend on the classical chaology, and it is these that enable the theory[30] to reproduce the random-matrix results (10), (12), (14) (but not-yet-(16)). When $L>>L_{max}$, the previously-mentioned breakdown of universality occurs and is explained by $\Delta(L)$ depending only on short closed orbits which of course differ from system to system.

Much work remains to be done in understanding semiclassical spectra. The most pressing and also fundamental problem is to discover whether the semiclassical sum over the closed orbits of a chaotic system can be extended (or interpreted, or analytically continued[37]) so as to describe the finest spectral scales such as those embodied in P(S) (or even - we can at least hope - a complete quantization formula). Then there are 'crossover' phenomena associated with the breakdown of universality when $L \sim L_{max}$. Finally, higher spectral statistics, depending more than quadratically on $\mathcal{N}(E)$, should be studied semiclassically. It is likely that from the program outlined in the last two sentences there might emerge a statistic which for chaotic systems depends on the Kolmogorov-Sinai[1] entropy which is so important in classical chaology.

REFERENCES

1. A.J.Lichtenberg and M.A.Lieberman, "Regular and Stochastic Motion," Springer: New York (1983).
2. M.V.Berry, Regular and Irregular Motion in "Topics in Nonlinear Mechanics" S.Jorna, ed., Am.Inst.Phys.Conf.Proc. 46: 16-120 (1978).
3. H.G.Schuster, "Deterministic Chaos", Physik-Verlag GMBH: Mannheim (1984)
4. G.Casati, B.V.Chirikov, J.Ford and F.M.Izraelev in "Stochastic Behaviour in Classical and Quantum Hamiltonian Systems," G.Casati,J.Ford, ed., Springer Lecture Notes in Physics 93: 334-352 (1979).
5. M.V.Berry, N.L.Balazs, M.Tabor andA.Voros, Ann.Phys. (N.Y) 122:26-63 (1979).
6. H.J.Korsch and M.V.Berry, Physica 3D: 627-636 (1981).
7. M.V.Berry, Physica 10D: 369-378 (1984).
8. B.V.Chirikov, F.M.Izraelev and D.L.Shepelyansky, Sov.Sci.Revs. 2C: 209-267 (1981).
9. D.L.Shepelyansky, Physica 8D: 208-222 (1983).
10. Shmuel Fishman, D.R.Grempel and R.E.Prange, Phys.Rev.Lett. 49: 509-512 (1982).
11. R.V.Jensen, in "Chaotic Behavior in Quantum Systems", G.Casati,ed., Plenum, 171-186 (1985).
12. J.E.Bayfield, L.D.Gardner and P.M.Koch, Phys.Rev.Lett. 39: 76-79 (1977).
13. G.Casati, B.V.Chirikov and D.L.Shepelyansky, Phys.Rev.Lett. 53: 2525-2528 (1984).
14. M.V.Berry, Phil.Trans.Roy.Soc. A287: 237-271 (1977).
15. S.W.McDonald and A.N.Kaufman, Phys.Rev.Lett. 42: 1189-1191 (1979)
16. M.V.Berry, J.Phys.A 10: 2083-2091 (1977).
17. M.V.Berry, J.H.Hannay and A.M.Ozorio de Almeida, Physica 8D: 229-242 (1983).
18. E.J.Heller, Phys.Rev.Lett. 16: 1515-1518 (1984).
19. M.V.Berry and M.Robnik, J.Phys.A 1365-1372 (1986).
20. V.P. Maslov and M.V.Fedoriuk, "Semiclassical Approximation in Quantum Mechanics," D.Reidel: Dordrecht, (1981).
21. M.V.Berry Semiclassical Mechanics of Regular and Irregular Motion in "Chaotic Behavior of Deterministic Systems, " Les Houches Lectures XXXVI, G.Iooss, R.H.G.Helleman and R.Stora, eds., North-Holland, Amsterdam, pp 171-271 (1983).
22. I.C.Percival, Adv.Chem.Phys. 36: 1-61 (1977).
23. V.I.Arnold, "Mathematical Methods of Classical Mechanics," Springer, New York, (1978).
24. M.V.Berry, Ann.Phys. (N.Y.) 131: 163-216 (1981).
25. R.Ramaswamy and R.A.Marcus, J.Chem.Phys. 74: 1379-1384, 1385-1393 (1981).
26. M.V.Berry and M.Wilkinson, Proc.Roy.Soc.Lond. A392: 15-43 (1984).
27. C.E.Porter, "Statistical Theories of Spectra: Fluctuations," Academic Press, New York (1965).
28. F.J.Dyson and M.L.Mehta, J.Math.Phys. 4: 701-712 (1963).
29. O.Bohigas and M.J.Giannoni, Chaotic motion and random-matrix theories, in "Mathematical and Computational Methods in Nuclear Physics," J.S.Dehesa, J.M.G.Gomez and A.Polls,eds., Lecture Notes in Physics 209, Springer-Verlag, N.Y. pp 1-99 (1984).
30. M.V.Berry, Proc.Roy.Soc.Lond. A400: 229-251 (1985).
31. M.V.Berry and M.Tabor, Proc.Roy.Soc.Lond. A356: 375-394 (1977).
32. T.H.Seligman and J.J.M.Verbaarschot, Phys.Lett. 108A: 183-187 (1985).
33. M.V.Berry and M.Robnik, J.Phys.A 19: 649-668 (1986).
34. M.Robnik and M.V.Berry, J.Phys.A 19: 669-682 (1986).

35. J.J.Sakurai, "Modern Quantum Mechanics," Benjamin, USA (1985).
36. A.M.Odlyzko, On the distribution of spacings between zeros of the zeta function, Bell. Lab. Preprint (1986).
37. M.V.Berry, Riemann's zeta function - a model for quantum chaos? in "Proc. 2nd Int. Conf. Quantum Chaos," T.Seligman ed., Springer (1987).
38. M.V.Berry and M.Robnik J.Phys.A 17: 2413-2421 (1984).
39. H-D.Meyer, E.Haller, H.Köppel and L.S.Cederbaum, J.Phys.A 17: L831-836 (1984).
40. G.Casati, B.V.Chirikov and I.Guarneri, Phys.Rev.Lett. 54: 1350-1353 (1985).
41. P.Pechukas, Phys.Rev.Lett. 51: 943-946 (1983).
42. T.Yukawa, Phys.Rev.Lett. 54: 1883-1886 (1985).
43. M.V.Berry, Aspects of degeneracy in "Proc. Como Conference on Quantum Chaos," G.Casati, ed., Plenum, London 123-140 (1985).
44. M.Wilkinson, J.Phys.A. In press.
45. M.C.Gutzwiller, J.Math.Phys. 12: 343-358 (1971).
46. M.C.Gutzwiller, in "Path Integrals and Their Applications in Quantum, Statistical and Solid-State Physics," G.J.Papadopoulos and J.T. Devreese, eds., Plenum, N.Y. 163-200 (1978).
47. R.Balian and C.Bloch, Ann.Phys.(N.Y.) 69: 76-160 (1972).
48. J.H.Hannay and A.M.Ozorio de Almeida, J.Phys.A 17: 3429-3440 (1984).
49. M.Kac, Amer.Math.Monthly 73 no. 4: 1-23 (1966).
50. H.P.Baltes and E.R.Hilf, "Spectra of Finite Systems," B.I.Wissenschaftsverlag: Mannheim (1976).
51. M.V.Berry, to be published 1987.
52. M.V.Berry, to be published 1987.

CHAOS IN DISSIPATIVE QUANTUM MAPS

Robert Graham

Fachbereich Physik
Universität Essen GHS
Essen, W. Germany

ABSTRACT

In this lecture a review is given of recent work on chaos and the onset of chaos in dissipative quantum systems. For simplicity only dynamical systems described by discrete maps are considered which correspond to systems periodically driven by short kicks. An exactly solvable model (Kaplan Yorke map) is presented and its density matrix in the steady state is constructed analytically. The dissipative Hénon map corresponding to a nonlinearly kicked damped harmonic oscillator and the dissipative standard map corresponding to a nonlinearly kicked damped free rotator are also studied. Their density matrices in the steady state are obtained numerically and represented as Wigner quasi-probability densities over the classical phase space. The case of very weak dissipation is discussed concentrating on two aspects - the influence of weak dissipation and the associated quantum noise in the period doubling route to chaos in the Hénon map, and the modification due to weak dissipation of quantum mechanical localization phenomena in the standard map. The material of this lecture is based on results published in refs. [1-8] and makes use, in part, of results obtained in joint work with T. Dittrich, T. Tél, and S. Isermann, whose collaboration on this problem is gratefully acknowledged.

1. INTRODUCTION

The study of chaos in dynamical systems with only a few degrees of freedom has been a very active area of research in the last decade and many properties of such systems seem by now reasonably well understood. The influence of quantum effects in systems which are chaotic in the classical limit has also attracted much interest [9], with less definite conclusions, so far. However, it has emerged that quantum systems are much less susceptible to chaotic behavior than their classical counterparts, due to the possible presence of very subtle and nontrivial quantum mechanical interference effects. These interference effects may lead to delocalization phenomena via quantum tunnelling in cases where the classical system is localized by impenetrable barriers such as KAM manifolds. They may also lead to localization phenomena in the space of action variables, in cases where the classical system shows stochastic spreading by diffusion. This localization effect (which does not occur under all circumstances [10]) has much in common with the phenomenon of Anderson localization [11]. So far the study of quantum effects in chaotic systems has mainly been concentrated on conservative systems. However, for several reasons the study of chaos in dissipative quantum systems also deserves interest. Chaos in classical conservative and dissipative systems has quite different properties, e.g. strange attractors can appear only in dissipative systems. Therefore, the properties of the classical limit in the two types of systems must also be quite different. Indeed the limits $t \to \infty, \hbar \to 0$, which are well known not to commute in conservative systems, do commute in dissipative systems [1,2]. Furthermore, the suppression of chaos in quantum systems depends on localizing interference effects, which may be partially or totally destroyed in dissipative systems, due to the randomizing interaction of the system with the reservoirs, responsible for the dissipation. It may therefore be surmised that dissipative quantum systems may in some sense behave in a more chaotic fashion than conservative quantum systems. Also from a more practical point of view there could be interesting applications, where chaotic dynamics, dissipation and quantization need to be taken into account at the same time. E.g. if energy is fed into a sufficiently complicated molecule by driving one of its degrees of freedom by an external field, inducing classically chaotic dynamics, other degrees of freedom may act as a reservoir into which energy is dissipated. Yet the intramolecular energy transfer is governed by quantum mechanics. Quite generally, systems which are dissipative but behave quantum mechanically, are of a mesoscopic nature, i.e. they are large and complex on the atomic scale, but they are extremely small on the macroscopic scale. Examples are quantum well structures in semiconductors or minaturized Josephson junctions with extremely small capacitances, or lasers (or masers) with only a small number of active atoms. Chaotic behavior in such systems is known to occur, dissipative effects are not negligible, and as the drive towards smaller systems continues, quantum effects will become increasingly important.

In the present lecture I give an overview over some recent work [1-6] on chaos and the onset of chaos via period doubling [7,8] in dissipative quantized systems described by discrete maps. The method of quantization of such maps in terms of a master equation is briefly described in section 2. In section 3 the Kaplan Yorke map is quantized and its steady state is analyzed in some detail. Section 4 is devoted to the quantization and numerical analysis of the Hénon map. In section 5 the scaling of quantum noise under period doubling is discussed. Section 6 reviews some results on the quantized dissipative standard map. Finally, in section 7 we discuss the obstruction of localization in the quantized standard map in the presence of weak dissipation.

2. REMARKS ON THE QUANTIZATION OF DISSIPATIVE MAPS

Dissipation in quantum theory is described by the coupling of a system to a reservoir. The elimination of the degrees of freedom of the reservoir leads to a reduced dynamical description of the system, in which energy is no longer conserved, and in which the system can no longer be described by a pure state. The mixed state which occurs instead is described by a statistical operator, which satisfies a master equation. This method, which is well known from the theory of spin resonance and from quantum optics, can also be applied to dynamical systems described by dissipative maps. A convenient way for the quantization of area preserving (i.e. conservative) maps is to interpret them as stroboscopic maps of periodically kicked Hamiltonian systems and to apply the canonical quantization rule to find a corresponding quantum system with the desired classical limit. Like all quantization rules also this procedure is not unique, but just a recipe for model building. In addition to the usual ambiguities of quantization, an additional ambiguity arises here from the fact that in physics a discrete map is not a complete dynamical description and can represent quite different continuous dynamical systems.

Having chosen a periodically kicked system to represent an area preserving map, dissipation can be introduced by coupling the system to a heat reservoir in between the kicks. This has been done for a number of examples [2,3,5], namely the Kaplan Yorke map [2], the Hénon map with dissipation [3], and the standard map with dissipation [5]. These maps are discussed in the following sections.

3. KAPLAN YORKE MAP

The classical form of this 2-dimensional map [12,13] is given by

$$\begin{aligned} q_{n+1} &= 2q_n \pmod 1 \\ p_{n+1} &= \frac{\lambda}{2} p_n - g'(q_n) \end{aligned} \tag{3.1}$$

where $0 \le q < 1$, $-\infty < p < \infty$, $g(q+\frac{1}{2}) = g(q)$. In the following only the special case $\lambda = 1$ will be considered (for the more general case cf. [2]) where the classical map is locally area preserving, even though it is globally contracting since the two areas $0 \le q < \frac{1}{2}$, $|p| < 2\max|g'(q)|$ and $\frac{1}{2} \le q < 1$, $|p| < 2\max|g'(q)|$ are mapped into overlapping areas. For $n \to \infty$ the classial system approaches a strange attractor in the (p,q)-plane and is described by the stationary phase-space density [13,2]

$$W(p,q) = \lim_{n\to\infty} \frac{1}{2^n} \sum_{m=0}^{2^n-1} \delta(p - F_m^{(n)}(q)) \tag{3.2}$$

with

$$F_m^{(n)}(q) = -\sum_{l=0}^{n} 2^{-l} g'\left(\frac{q+m}{2^{l+1}}\right) \tag{3.3}$$

The classical strange attractor for the case $\lambda = 1$, $g(q) = (4\pi)^{-1}\cos 4\pi q$ generated numerically by iterating the map is shown in fig. 1. The reduced probability density of p is obtained from $W(p,q)$ by integration over q and turns out to be a rather singular function analytically. A numerical example again for $g(q) = (4\pi)^{-1}\cos 4\pi q$ is shown in fig. 2.

The quantization of the map [1,2] can be achieved by splitting the quantum map into two subsequent steps. The first step describes dissipation. It is non-unitary and consists in projecting the state $|\psi_n\rangle$ of the system on the two stripes $0 \le q < 1/2$, and $1/2 \le q < 1$. Non-unitary projections of quantum mechanical states on eigenstates of an observable A are a familiar and logically necessary feature in the quantum theory of measurement and have long been understood, in principle, to arise from the interaction of a quantum system with a macroscopic system serving as a measuring device of A. In precisely the same manner, the projection of the state required in the first step of the Kaplan Yorke map can be considered as the result of the interaction of the system with a macroscopic reservoir serving as measuring device of the observable

$$\Omega = \int_0^{1/2} dq\, |q\rangle\langle q| \tag{3.4}$$

As a result of the interaction with the reservoir the state vector $|\psi_n\rangle$ at the discrete time n is collapsed into two mutually incoherent components

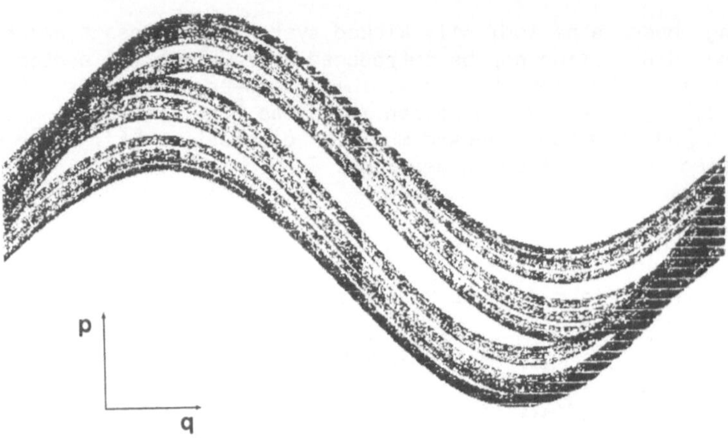

Fig. 1. Attractor of the classical Kaplan Yorke map with λ = 1, $g(q)=(4\pi)^{-1}\cos 4\pi q$ (after [2])

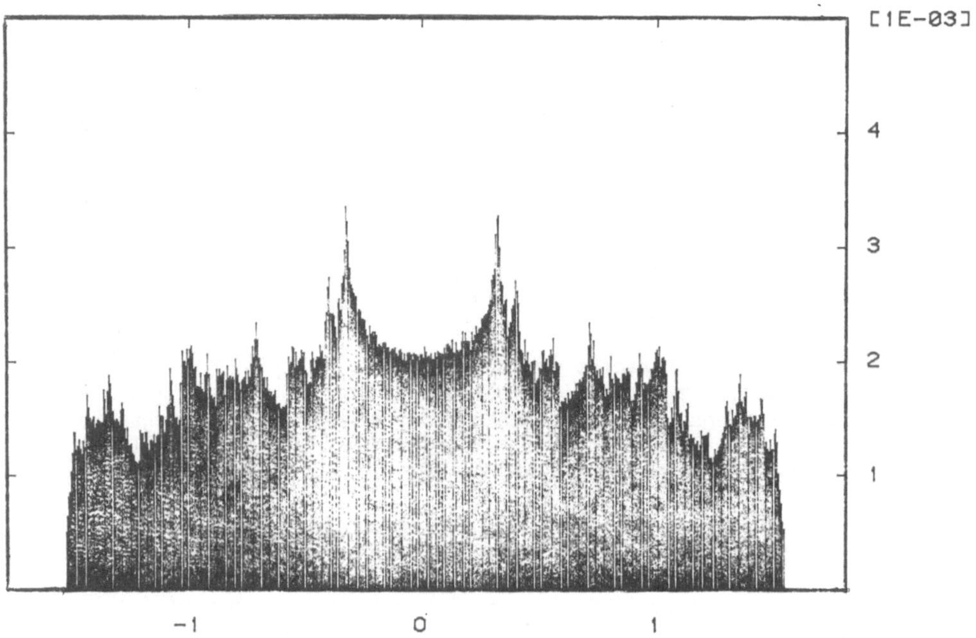

Fig. 2. Classical probability density $W(p)$ on the strange attractor of Fig. 1 (after [2])

$$|\tilde{\psi}_n\rangle = \left(\Omega + e^{i\delta_n}(1-\Omega)\right)|\psi_n\rangle \tag{3.5}$$

where δ_n is a random phase. The next step of the quantized Kaplan Yorke map now consists in a unitary propagation of each of the mutually incoherent components of the state $|\tilde{\psi}_n\rangle$ to yield the final state $|\psi_{n+1}\rangle$. A unitary propagation corresponding to the classical map (3.1) in the classical limit is

$$\langle q|U|q'\rangle = \begin{cases} \sqrt{2}\,\delta(q-2q')\exp\left(-\frac{2i}{\hbar}g(q')\right) & 0 \le q' < \frac{1}{2} \\ \sqrt{2}\,\delta(q-1-2q')\exp\left(-\frac{2i}{\hbar}g(q')\right) & \frac{1}{2} \le q' < 1 \end{cases} \tag{3.6}$$

Introducing the statistical operator

$$\rho_n = \overline{|\Psi_n\rangle\langle\Psi_n|}^{(\delta_m,\, m<n)} \tag{3.7}$$

by averaging over the random phases δ_m prior to n, the quantized map can be written succinctly as the master equation [1,2]

$$\langle q|\rho_{n+1}|q'\rangle = \frac{1}{2}\exp\left(-\frac{2i}{\hbar}\left(g\left(\frac{q}{2}\right)-g\left(\frac{q'}{2}\right)\right)\right)\cdot \\ \cdot\left(\left\langle\frac{q}{2}\right|\rho_n\left|\frac{q'}{2}\right\rangle + \left\langle\frac{q+1}{2}\right|\rho_n\left|\frac{q'+1}{2}\right\rangle\right) \tag{3.8}$$

where we introduced the bras and kets of the coordinate representation with the understanding that $|q\rangle = |q \text{ (mod 1)}\rangle$. An exact solution of this quantum map in the steady state can be obtained by analytically iterating the map to $n\to\infty$ [1,2]. The result is

$$\langle q|\rho_\infty|q'\rangle = \lim_{n\to\infty}\frac{1}{2^n}\sum_{m=0}^{2^n-1} e^{-\frac{2i}{\hbar}\sum_{\ell=1}^{n}\left(g\left(\frac{q+m}{2^\ell}\right)-g\left(\frac{q'+m}{2^\ell}\right)\right)} \tag{3.9}$$

It is useful to transform the density matrix to the Wigner quasi-probability density, which can be done without any loss of information by [1,2]

$$W_n(p,q) = \sum_{\ell=-\infty}^{+\infty} W_n^{(\ell)}(q)\,\delta(p-\pi\hbar\ell)$$
$$W_n^{(\ell)}(q) = \int_0^1 dq'\, e^{-2\pi i\ell q'}\langle q+q'|\rho_n|q-q'\rangle \tag{3.10}$$

We draw here attention to the fact [1,2,5,6] (cf. also section 6) that the Wigner function on a cylindrical phase space has support not only on the physically quantized integer values of angular momentum $p_\ell = 2\pi\hbar\ell$ but also on half-integers $p_{\ell/2} = 2\pi\hbar(\ell/2)$. The steady state expressed in terms of the Wigner function bears a remarkable formal resemblance to the classical phase space distribution (3.2). Indeed, the result obtained in [1] may be expressed most simply by the correspondence rule that each branch $\delta(p-F_m^{(n)}(q))$ of the classical distribution is replaced in the Wigner distribution by

$$\delta(p-F_m^{(n)}(q)) \rightarrow \sum_{\ell=-\infty}^{+\infty}\delta(p-\pi\hbar\ell)\,W_n^{(\ell,m)}(q) \tag{3.11}$$

with

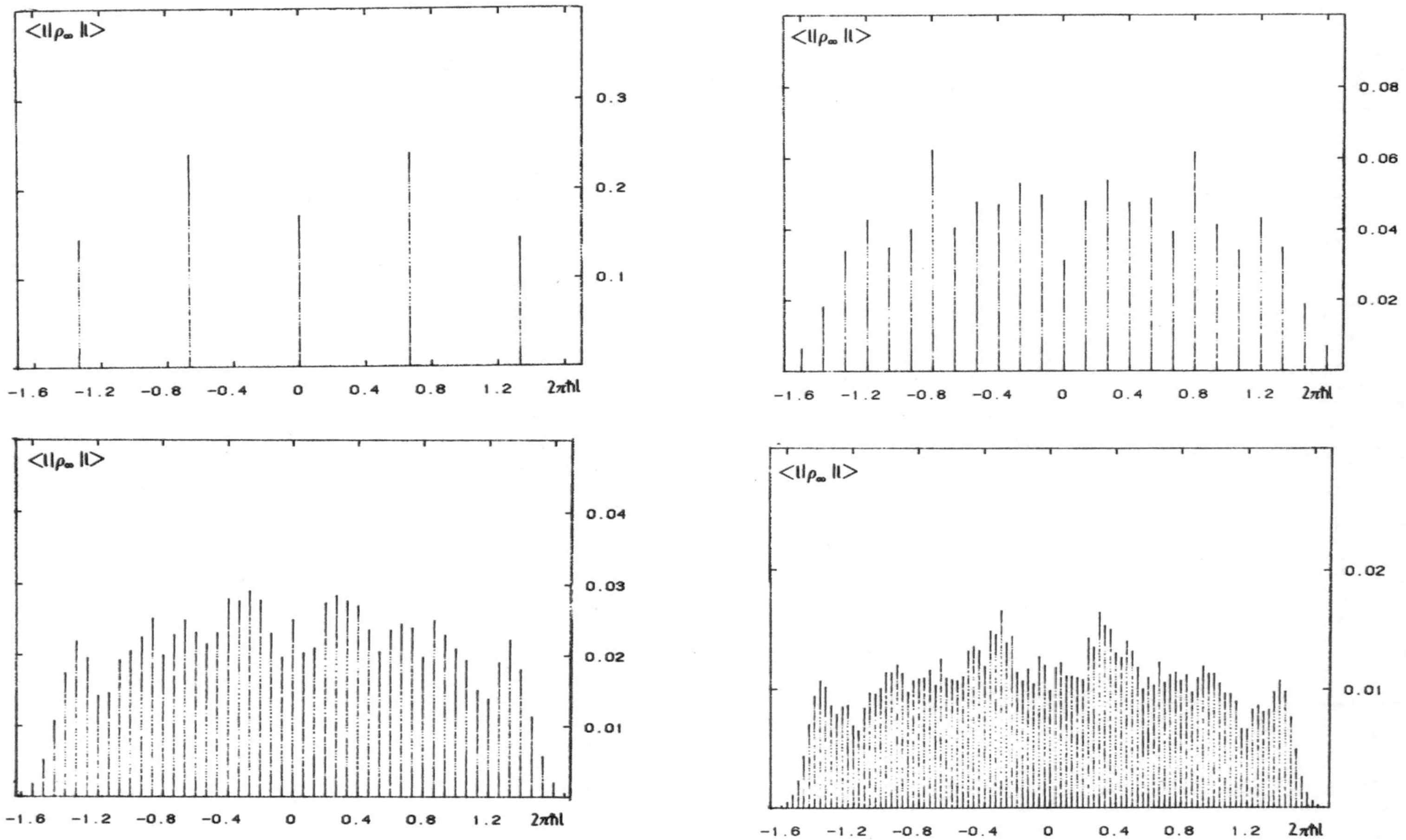

Fig. 3. Probability $\langle \ell | \rho_\infty | \ell \rangle$ of the quantized Kaplan Yorke map for the same case as in Figs. 1,2 for (a) $\hbar$ = 0.1, (b) $\hbar$ = 0.02, (c) $\hbar$ = 0.01, (d) $\hbar$ = 0.005.

$$W_n^{(\ell,m)}(q) = \int_0^1 dx\, \exp\left\{-2\pi i x\ell - \frac{2i}{\hbar}\sum_{k=1}^{n}\left[g\left(\frac{(q+x)(\mathrm{mod}\,1)+m}{2^k}\right) - (x \to -x)\right]\right\} \tag{3.12}$$

In other words all branches $F_m^{(\infty)}(q)$ of the strange attractor in phase space are delocalized into angular momentum packets consisting of many integer and half-integer angular momentum quantum numbers, whose positive and negative weights in the Wigner function are given by the Fourier transform of an infinite $(n\to\infty)$ product of phase factors. For $\hbar \to 0$ these factors interfere destructively, except on the strange attractor, and the classical distribution (3.2) is recovered. Hence, $\hbar \to 0$ and $t \to \infty$ commute. For $\hbar$ not negligible compared to the amplitude of $g(q)$ the individual branches of the strange attractor are completely delocalized in angular momentum. However, each classical branch of the strange attractor still leaves its trace in the Wigner distribution in that it contributes incoherently with respect to all other branches, as is made explicit by the existence of a correspondence rule (3.11) for each individual branch.

Taking the integral over q in the Wigner distribution the reduced probability density of angular momentum is obtained in the form

$$W(p) = \sum_{\ell=-\infty}^{+\infty} P_\ell\, \delta(p - 2\pi\hbar\ell) \tag{3.13}$$

with

$$P_\ell = \lim_{n\to\infty} 2^{-n} \sum_{m=0}^{2^n-1} \left| \int_0^1 dq\, \exp\left[-2\pi i q\ell - \frac{2i}{\hbar}\sum_{k=1}^{n} g\left(\frac{q+m}{2^k}\right)\right]\right|^2 \tag{3.14}$$

Again the incoherent contribution from the different classical branches of the attractor can be discerned in the sum over m. In eq. (3.14) angular momentum is, of course, correctly quantized in the reduced distribution. Figs. 3a-d give plots of expression (3.14) for decreasing values of $\hbar$ and show the approach of the exact quantum result towards the classical singular distribution function $W(p)$ shown in Fig. 2.

4. HENON MAP

The Hénon map [14] can be considered as the stroboscopic map describing the dynamics of a kicked oscillator [15]. Neglecting dissipation, to begin with, the oscillator is described by the Hamiltonian

$$H_0 = \hbar\omega a^+ a - \sum_{n=-\infty}^{+\infty} \delta(t - n\tau)\, \bar{g}\left(\sqrt{\frac{\hbar}{2\omega}}(a + a^+)\right) \tag{4.1}$$

Here ω is the frequency of the oscillator in between the kicks with period τ. We shall make the choice $\omega\tau = 2\pi(k+\frac{1}{4})$ later on, with k integer. The operators a^+, a are the usual raising and lowering operators of the oscillator satisfying

$$[a, a^+] = 1 \tag{4.2}$$

The function $-\bar{g}(x)$ is the potential of the kicking force $\bar{f}(x) = -\bar{g}'(x)$. Dissipation is introduced in the model by coupling the oscillator to a reservoir, which is most conveniently represented by the ensemble of harmonic oscillators in thermal equilibrium,

$$H = H_0 + H_{res} + H_{int} \tag{4.3}$$

with

$$H_{res} = \hbar \sum_i \omega_i a_i^+ a_i$$
$$H_{int} = \hbar \sum_i g(\omega_i)(a a_i^+ + a^+ a_i) \tag{4.4}$$

Here ω_i and $g(\omega_i)$ are the frequencies and coupling constants of the reservoir oscillators, which are also represented by their raising and lowering operators a_i^+, a_i. The coupling $g(\omega_i)$ is assumed to be weak, the density of states $\rho(\omega_i)$ of the reservoir oscillators is assumed to be dense and flat in a broad neighbourhood of the frequency ω so that the dissipation rate 2γ of energy is described by the golden rule

$$2\gamma = 2\pi\, g^2(\omega)\, \rho(\omega) \tag{4.5}$$

and the temperature T of the reservoir is assumed to be not too low [16]

$$k_B T \gg \hbar\gamma \tag{4.6}$$

Then, the elimination of the reservoir oscillators is known [17,18] to lead to the master equation

$$\frac{\partial \rho}{\partial t} = -\frac{i}{\hbar}[H_0, \rho] + \Lambda(a,a^+)[\rho] \tag{4.7}$$

with

$$\Lambda(a,a^+)[\rho] = \gamma\{(\bar{n}+1)[a\rho, a^+] + \bar{n}[a^+\rho, a] + h.c.\} \tag{4.8}$$

where $h.c.$ means the hermitian conjugate and $\bar{n}$ is the thermal quantum number of the oscillator

$$\bar{n} = \left(\exp\left(\frac{\hbar\omega}{k_B T}\right) - 1\right)^{-1} \tag{4.9}$$

The great virtue of this description is the fact that it can be exactly converted to a map, in the classical and also in the quantized formulation.

The classical map corresponding to the oscillator described by eqs. (4.1), (4.7) (with $\omega\tau = 2\pi(k+\frac{1}{4})$) is obtained by solving the classical equations over the time interval τ starting with the state after the nth kick. The solution reads

$$x_{n+1} = \frac{E p_n}{\omega} \quad ; \quad p_{n+1} = -\omega E x_n + \bar{f}(x_{n+1}) \tag{4.10}$$

where x_n, p_n is the coordinate and momentum after the nth kick, and

$$E = \exp(-\gamma\tau) \tag{4.11}$$

is a factor which measures dissipation ($0 \leq E \leq 1$). Hénon's form of the map is obtained by introducing

$$y = \frac{1}{\omega}(-p + \bar{f}(x)) \quad ; \quad \bar{f}(x) = \frac{\omega}{E} f(x) \tag{4.12}$$

It reads

$$
\begin{aligned}
x_{n+1} &= -E y_n + f(x_n) \equiv r(x_n, y_n) \\
y_{n+1} &= E x_n \equiv s(x_n, y_n)
\end{aligned}
\tag{4.13}
$$

The quantized map is obtained by carrying out the corresponding steps. Denoting by $\tilde{\rho}_n^{(-)}$, $\tilde{\rho}_n^{(+)}$ the statistical operator in interaction representation with respect to the free oscillator Hamiltonian $\hbar\omega a^+ a$ before and after the nth kick, respectively, we can write the solution of the master equation between two kicks as

$$
\tilde{\rho}_{n+1}^{(-)} = \exp\left(\Lambda(\tilde{a}, \tilde{a}^+)\tau\right)\left[\tilde{\rho}_n^{(+)}\right]
\tag{4.14}
$$

and for a single kick as

$$
\tilde{\rho}_n^{(+)} = U(\tilde{a}, \tilde{a}^+)\,\tilde{\rho}_n^{(-)}\,U^+(\tilde{a}, \tilde{a}^+)
\tag{4.15}
$$

with

$$
\begin{aligned}
U(a, a^+) &= \exp\left(\frac{i}{\hbar}\,\bar{g}\left(\sqrt{\frac{\hbar}{2\omega}}\,(a + a^+)\right)\right) \\
\tilde{a} &= a \exp(i\omega\tau) \,, \quad \tilde{a}^+ = a^+ \exp(-i\omega\tau)
\end{aligned}
\tag{4.16}
$$

The total quantum map therefore reads

$$
\rho_{n+1}^{(+)} = U(a, a^+)\, e^{\Lambda(a, a^+)\tau}\left[\rho_n^{(+)}\right] U^+(a, a^+)
\tag{4.17}
$$

where we transformed back to the Schrödinger representation.

A convenient c-number representation of the map is obtained by introducing the Wigner quasi-probability density

$$
W_n(x, p) = \int \frac{dq}{2\pi\hbar}\, e^{-\frac{ipq}{\hbar}} \left\langle x + \frac{q}{2} \right| \rho_n^{(+)} \left| x - \frac{q}{2} \right\rangle
\tag{4.18}
$$

and introducing the variable y instead of p via eq. (4.12). As shown in [13] the map can then be written as

$$
W_{n+1}(x, y) = \int dx'\,dy'\; K(x, y; x', y')\, W_n(x', y')
\tag{4.19}
$$

with the kernel

$$
\begin{aligned}
K(x, y; x', y') = {} & \left(\frac{\omega}{2\pi\hbar Q}\right)^{1/2} \exp\left[-\frac{\omega}{2\hbar Q}\left(x - r(x', y')\right)^2\right] \cdot \\
& \cdot \int \frac{d\eta}{2\pi} \exp\left[i\eta\left(y - s(x', y')\right) - \frac{\hbar Q}{2\omega}\eta^2 + i\,G(x, \eta, \hbar)\right]
\end{aligned}
\tag{4.20}
$$

where

$$
\begin{aligned}
G(x, \eta, \hbar) &= \hbar^{-1}\left(\bar{g}\left(x + \frac{\eta\hbar}{2\omega}\right) - \bar{g}\left(x - \frac{\eta\hbar}{2\omega}\right) - \frac{\eta\hbar}{\omega}\, g'(x)\right) \\
Q &= (1 - E^2)\left(\bar{n} + \tfrac{1}{2}\right) = \frac{1 - E^2}{2} \coth\left(\frac{\hbar\omega}{2k_B T}\right)
\end{aligned}
\tag{4.21}
$$

Eq. (4.19) formally looks like a map of a probability density, but, of course, the Wigner function $W_n(x,y)$ is not everywhere positive and therefore is only a quasi-probability density. As is well known stochastic maps can be used as alternative representations for maps of probability densities. Recently, we introduced the idea to generalize this for maps of quasi-probability densities at the cost of working with 'quasi-stochastic' maps, in which non-deterministic c-number quantities appear which are distributed according to some quasi-probability density [17]. The quasi-stochastic map corresponding to eqs. (4.19), (4.20) reads

$$
\begin{aligned}
x_{n+1} &= f(x_n, y_n) + \xi_n \\
y_{n+1} &= S(x_n, y_n) + \eta_n
\end{aligned}
\tag{4.22}
$$

where the non-deterministic c-number forces ξ_n , η_n are neither autocorrelated nor crosscorrelated for different values of n and have correlation coefficients

$$
M_{lm} = \langle \xi_n^l \eta_n^m \rangle \tag{4.23}
$$

which follow from eq. (4.20).

An important special case to which we restrict our discussion in the following is obtained for

$$
\bar{f}(x) = \frac{\omega}{E} c\,(1 - ax^2) \tag{4.24}
$$

which, classically, yields the original form of Hénon's map [14]. Then $G(x,\eta,\hbar)$ reduces to

$$
G(x,\eta,\hbar) = -\frac{\hbar^2 ac}{12 E\omega^2}\,\eta^3 \tag{4.25}
$$

independent of x . It is straight forward to read off all the non-vanishing cumulants from eq. (4.20)

$$
\begin{aligned}
\langle \xi_n^2 \rangle_c &= \frac{\hbar Q}{\omega} = \langle \eta_n^2 \rangle_c \\
\langle \eta_n^3 \rangle_c &= -\frac{\hbar^2}{2} \frac{ac}{\omega^2 E}
\end{aligned}
\tag{4.26}
$$

We emphasize again the important point that the quantum map (4.19) can be restated exactly as a quasi-stochastic map (4.22) with non-deterministic c-number forces ξ_n , η_n with non-classical correlations (4.26). The non-vanishing cubic correlation coefficients are not compatible with classical positive probability distributions and require the use of quasi-probability densities which are capable of negative values, also. In analytical calculations eqs. (4.22) may be handled just like stochastic maps as long as the positivity of probability densities need not be invoked. Numerical simulations can, of course, not be done with eqs. (4.22) since, as we emphasize again, the correlations (4.26) cannot be generated from a positive probability distribution.

Eqs. (4.22), (4.26) are very convenient to discuss the semi-classical limit $\hbar \to 0$ of the map. In this limit the leading moments are $\langle \xi_n^2 \rangle = \langle \eta_n^2 \rangle = \hbar Q/\omega$ and $\langle \eta_n^3 \rangle = -\hbar^2 ac/2\omega^2 E$. The latter moment becomes negligible when $\hbar \ll 4E^2 Q^3 \omega / a^2 c^2$ i.e. in this limit only the second order moment survives and the quasi-stochastic process becomes a true stochastic process. This result has also been derived in [3] without the use of the quasi-stochastic representation of the map.

The steady state quasi-probability distribution function of the map (4.19) can now be obtained numerically by iterating an arbitrary initial distribution under the map, provided the function $\bar{f}(x)$ is chosen in such a way that a unique steady state does exist. This has been done in joint work with S. Isermann and T. Tél [4]. The function $\bar{f}(x)$ chosen in this work is given by

$$\bar{f}(x) = \frac{\omega}{E} f(x) = \frac{\omega}{E}\left(1 - \frac{a x^2}{1+x^4}\right) \tag{4.27}$$

Fig. 4 shows the classical strange attractor in the (x,y)-plane

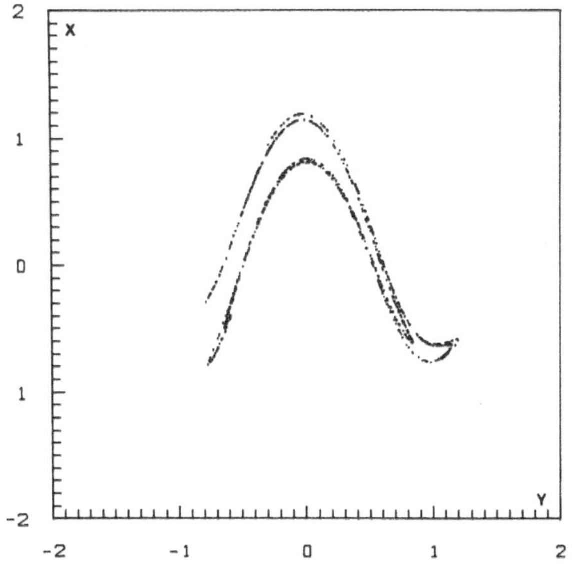

Fig. 4. Attractor in the (y,x)-plane of the classical map (4.13) with $f(x) = 1 - ax^2/(1+x^4)$ for a = 3.4, E^2 = 0.3, ω = 1.

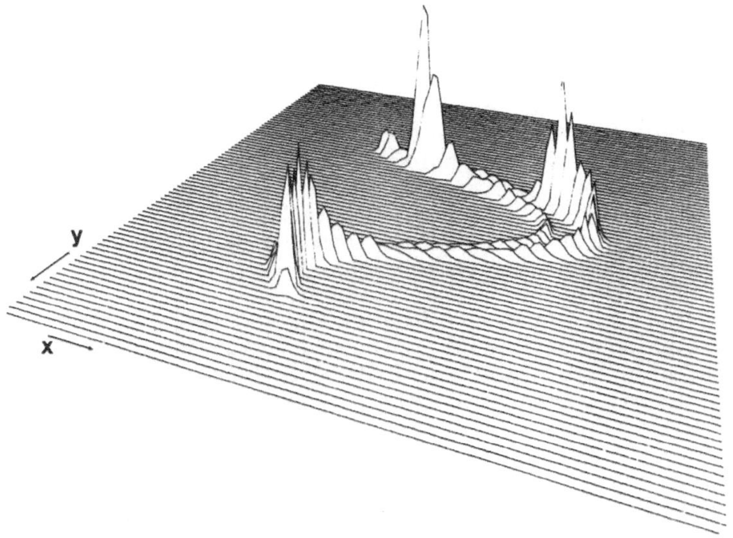

Fig. 5. Probability density of the map (4.13) for the case of Fig. 4 with resolution $\Delta x = \Delta y = 0.04$.

which is obtained for $a = 3.4$, $B = E^2 = 0.3$, $\omega = 1$. Parts of the Cantor substructure of the strange attractor is clearly visible.

Fig. 5 shows the classical probability distribution on the strange attractor, obtained for the same parameter values as in Fig. 4, by dividing the square $-2 < (x,y) < 2$ into 10^4 boxes of size 0.04 x 0.04 and iterating under the classical map an initial distribution over the boxes of total mass unity until convergence is obtained. It is seen that the limited resolution due to the finite box size is still sufficient to resolve the main branches of the attractor. Of course, making the box size smaller, more and more of the full Cantor set structure of the classical attractor must be resolved. The effective value of $\hbar$, which would lead to a minimum uncertainty $\Delta x \Delta p_3 = \omega \Delta x \Delta y = \hbar/2$ equal to the area of the boxes would be $\hbar = 3.2 \times 10^{-3}$.

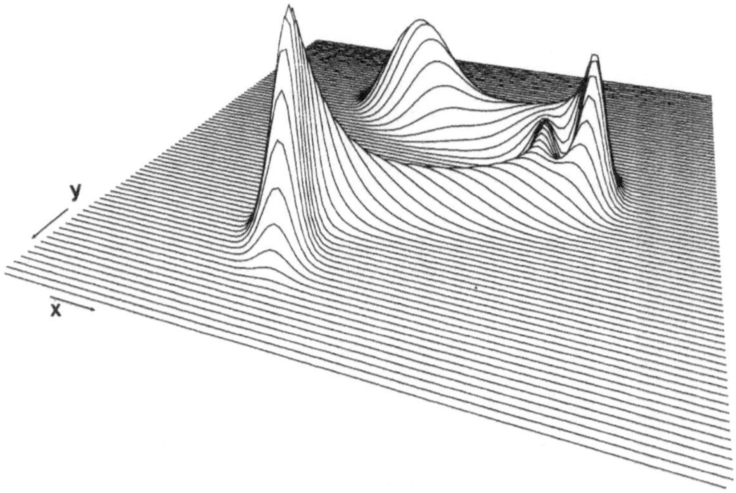

Fig. 6. Steady state Wigner distribution of the quantum map (3.19) for $\hbar = 0.01$ keeping only second order cumulants of the quantum noise (same parameters as Fig. 4)

Figs. 6,7 show the Wigner function in the steady state obtained by numerically iterating two different approximations of the map (4.19) until convergence is found (10 iterations were found to be more than sufficient for the parameters chosen). The choice of parameters and the partition of the (x,y) -plane into boxes was made in the same way as in Fig. 2. In addition, the effective value of $\hbar$ in eqs. (4.20), (4.21) was chosen as $\hbar = 0.01$, which is significantly larger than the value corresponding to the numerical box size.

In Fig. 6 the semi-classical limit of the map (4.19) is used, in which all correlation coefficients in the quasi-stochastic map of higher than the second order are neglected. In this limit the quasi-stochastic map becomes a stochastic map, and the Wigner distribution reduces to a positive probability density. It can be seen in Fig. 5 that this probability density is smooth and concentrated around the classical attractor, but the different branches of the attractor are hardly resolved, due to the quantum noise.

Taking $\hbar$ still smaller, the distribution must approach the classical distribution of Fig. 5. The limits $\hbar \to 0$ and $n \to \infty$ commute for this dissipative system, unlike for conservative systems, where these limits

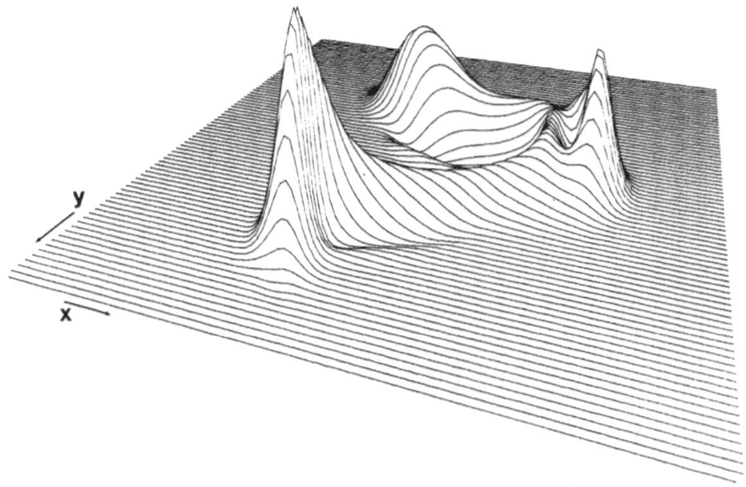

Fig. 7. Steady state Wigner distribution of the quantum map (3.19) keeping third order cumulants of the quantum noise (same parameters as Fig. 4)

cannot be interchanged. A comparison of Fig. 6 with Fig. 5 shows that the fluctuations not only broaden the distribution transverse to the attractor but also tend to smoothen the probability density parallel to the attractor, where the classical motion is unstable.

In Fig. 7 the third order correlation coefficient of the quasi-stochastic map is kept, but the higher order correlation coefficients are still neglected. The kernel (4.20) for this case is derived in [3]. Due to the presence of the third order cumulant the Wigner function must become negative in some regions of the (x,y) -plane. Fig. 7 shows that this indeed occurs, but it can also be seen by comparison with Fig. 5 that these corrections are not drastic. The oscillations of the Wigner function to negative values occur transverse to the attractor. This behavior is in good agreement with the analytical semi-classical analysis which was given in [3].

5. PERIOD DOUBLING IN DISSIPATIVE QUANTUM SYSTEMS

Now we make use of the quasi-stochastic representation of the Hénon map of the preceding section and determine the scaling properties of quantum noise in period doubling [7]. We choose the map defined by eqs. (4.13), (4.22) with $f(x) = (E/\omega)\tilde{f}(x)$ given by eq. (4.24). Combined to a single two-step recursive map in standard form (cf. [19]) it reads

$$z_{n+1} + B z_{n-1} = 2C z_n + 2 z_n^2 + \zeta_n \tag{5.1}$$

with

$$\begin{aligned} z_n &= -\frac{ac}{2} x_n - \frac{1}{2} C \\ B &= E^2 \\ C &= \frac{1+B}{2} - \sqrt{\frac{(1+B)^2}{4} + ac^2} \end{aligned} \tag{5.2}$$

The quasi-stochastic force is defined by

$$\Gamma_n = -\frac{ac}{2}(\xi_n - E\,\eta_{n-1}) \tag{5.3}$$

where ξ_n and η_n have the cumulants given in eq. (4.26). We obtain as the non-vanishing cumulants of Γ_n for all n

$$\langle \Gamma_n^2 \rangle \equiv \hbar Q_0 = \frac{\hbar a^2 c^2}{\omega} \cdot \frac{(1+B)Q}{4}$$

$$\langle \Gamma_n^3 \rangle \equiv \hbar^2 R_0 = \left(\frac{\hbar a^2 c^2}{\omega}\right)^2 \cdot \frac{B}{16} \tag{5.4}$$

The second order cumulant would also occur if was a classical noise source, and, in this sense, describes classical noise properties of the quantum fluctuation arising from dissipation, i.e. from the coupling to the reservoir. The third order cumulant would also occur in a conservative quantum system and therefore describes intrinsical quantum mechanical nois properties.

Renormalization transformations will later generate also non-vanishin cumulants of the form

$$\langle \Gamma_n \Gamma_{n\pm 1} \rangle_c \equiv \hbar Q_1$$

$$\langle \Gamma_n^2 \Gamma_{n\pm 1} \rangle_c \equiv \hbar^2 R_{\pm 1} \tag{5.5}$$

even though these cumulants vanish for the map (5.1) according to eqs. (4.26).

For $\Gamma_n = 0$ the map (5.1) is well known to undergo an infinite sequence of period doublings as C is varied, and an approximate analytical renormalization scheme was developed by Helleman and McKay [20] to derive the scaling behavior. After period doubling a new variable z_n' and new parameters B', C' are introduced in terms of which the map can be recast, at least approximately, into the original form

$$z_{n+1}' + B' z_{n-1}' = 2C' z_n' + 2 z_n'^2 + \Gamma_n' \tag{5.6}$$

Here we have extended this scheme to the quasi-stochastic map (5.1) by taking along the force Γ_n. The recursion relations are [19,20]

$$z_n' = \alpha\,(z_{2n} - z_{2+})$$

$$\Gamma_n' = \alpha\,(B\,\Gamma_{2n-1} + e\,\Gamma_{2n} + \Gamma_{2n+1})$$

$$\alpha = e + (1+B)^{-1}(2C + 4z_{2+})^2 \tag{5.7}$$

$$z_{2\pm} = a \pm b$$

$$e = 2C + 4z_{2-}$$

$$a = -\frac{1+B}{4} - \frac{C}{2}$$

$$b = \frac{1}{2}\left(C + \frac{1+B}{2}\right)^{\frac{1}{2}}\left(C - \frac{3(1+B)}{2}\right)^{\frac{1}{2}}$$

$$B' = B^2$$

$$C' = -2C^2 + 2(1+B)C + 2B^2 + 3B + 2$$

The recursion relations for B and C are well known. The recursion formula for the quasi-stochastic force yields the following additional recursion relations for the cumulants

$$\begin{pmatrix} Q_0' \\ Q_1' \end{pmatrix} = \alpha^2 \begin{pmatrix} (1+e^2+B^2) & 2e(1+B) \\ B & e(1+B) \end{pmatrix} \begin{pmatrix} Q_0 \\ Q_1 \end{pmatrix} \tag{5.8}$$

and

$$\begin{pmatrix} R_0' \\ R_1' \\ R_{-1}' \end{pmatrix} = \alpha^3 \begin{pmatrix} (1+e^3+B^3) & 3e(e+B^2) & 3e(1+Be) \\ B & e(1+Be) & 2Be \\ B^2 & 2Be & e(e+B^2) \end{pmatrix} \begin{pmatrix} R_0 \\ R_1 \\ R_{-1} \end{pmatrix} \tag{5.9}$$

These recursion relations determine the scaling behavior of the quantum fluctuations. Let us first consider the case of very weak dissipation $(1-B) \ll 1$, where the system remains close to the fixed point [19,20] $B = 1$, $C_\infty(1) = (3 - \sqrt{65})/4$ for many iterations before crossing over to the fixed point $B = 0$. The matrix renormalizing the cubic cumulants for $B = 1$ has the largest eigenvalue 7

$$\lambda_2 = \alpha^3\left[\frac{e^3+e^2+3e+2}{2} - \sqrt{\frac{(e^3-e^2-3e+2)^2}{4} + 6e(e+1)}\right] \tag{5.10}$$
$$\simeq 4856.3$$

which implies a rescaling of the effective value of $\hbar$ by

$$\hbar' = \sqrt{|\lambda_2|}\,\hbar \simeq 69.69\,\hbar \tag{5.11}$$

as long as B is close to 1. Grempel et.al. [21] find a numerically very similar rescaling factor of $\hbar$ by a different method, which is only applicable to conservative maps, however. For $B \to 1$ the largest eigenvalue of the matrix which renormalizes second order cumulants is found as

$$\lambda_3 = \left[\frac{e^2+2e+2}{2} + \sqrt{\frac{(e^2+2-2e)^2}{4} + 4e} \right] \alpha^2 \tag{5.12}$$

$$\simeq 312.8$$

which implies the rescaling

$$\coth\left(\frac{\hbar\omega}{2k_BT}\right)' = \frac{\lambda_3}{2\sqrt{\lambda_2}} \coth\left(\frac{\hbar\omega}{2k_BT}\right) \tag{5.13}$$

with $\lambda_3/2\sqrt{\lambda_2} \simeq 2.244$. Therefore, even though $\hbar$ effectively increases by renormalization, the ratio $\hbar\omega/k_BT$ effectively decreases. In other words, the quantum fluctuations associated with dissipation are driven, by renormalization, towards higher noise temperature, even if initially one deals with quantum fluctuations near T = 0, and with a second order cumulant $Q_0 \sim (1-B^2)$ which is very small for $B^2 \to 1$. Another way to discuss how renormalization affects the quantum fluctuations is to consider how the dimensionless ratio $\langle \xi_n^3 \rangle^2 / \langle \xi_n^2 \rangle^3 = \hbar R_0^2/Q_0^3$, measuring the relative importance of the cubic cumulant, is renormalized under period doubling. We find

$$\left(\hbar \frac{R_0^2}{Q_0^3}\right)' = \frac{\lambda_2^2}{\lambda_3^2}\left(\hbar \frac{R_0^2}{Q_0^3}\right) \simeq 0.879 \left(\hbar \frac{R_0^2}{Q_0^3}\right) \tag{5.14}$$

which shows that this ratio decreases in each period doubling quite slowly. For the more strongly dissipative case described by the fixed point B = 0, $C_\infty(0) = (1-\sqrt{17})/4$ the second order cumulant dominates over the cubic cumulant even before period doubling occurs, and the cubic cumulant is further reduced in relative size by renormalization.

Our results for the renormalization of second and third order cumulants imply scaling behavior in the usual way. E.g. the rescaling of $\hbar$ according to eq. (5.11) and Feigenbaum's scaling law

$$|C_\infty - C_n| = \frac{\text{const}}{\delta^n} \tag{5.15}$$

(with $\delta \simeq 9.06$ for B = 1 in the present approximation) imply that for $C \to C_\infty$, $\hbar$ enters any physical property only via the scaling variable $\hbar/|C_\infty - C|^{\ln\lambda_2/\ln\delta}$. Similarly, the noise intensity Q defined in eq. (4.21) is renormalized by the factor λ_3 and therefore enters only via the scaling variable $Q/|C_\infty - C|^{\ln\lambda_3/\ln\delta}$.

The fact that the classical noise properties gradually outgrow the quantum mechanical ones in the renormalization process has important physical consequences. It seems to be clear that quantum mechanical interference effects as those giving rise to the localization phenomena mentioned in the introduction (cf. also sections 6,7) are impossible under conditions where the classical noise dominates. In section 7 the destruction of localization even by very weak dissipation is discussed in a qualitative manner from a somewhat different point of view, again emphasizing the essential role of noise associated with dissipation.

6. DISSIPATIVE STANDARD MAP

The conservative standard map

$$p_{n+1} = p_n - \frac{K}{2\pi} \sin 2\pi q_n$$
$$q_{n+1} = q_n + p_{n+1} \tag{6.1}$$

has been intensely studied, both classically (cf. [19,22]) and quantum mechanically [9-11]. Classically, for $K \gtrsim 1$ chaos on large scales appears and the angular momentum variable undergoes a stochastic diffusive motion with

$$\langle p_n^2 \rangle \simeq \frac{1}{2} \left(\frac{K}{2\pi}\right)^2 n \tag{6.2}$$

Quantum mechanically, if in the dimensionless units for p and q employed in eq. (6.1) $2\pi\hbar$ is irrational, it seems that the quasi-energy eigenstates

$$U|\varphi_\lambda\rangle = e^{-i\omega_\lambda}|\varphi_\lambda\rangle \tag{6.3}$$

are localized. Here the unitary operator

$$U = \exp\left[-\frac{i}{\hbar}\frac{p^2}{2}\right] \exp\left[i\frac{K}{4\pi^2\hbar}\cos 2\pi q\right] \tag{6.4}$$

represents the quantum map

$$|\psi_{n+1}\rangle = U|\psi_n\rangle \tag{6.5}$$

corresponding to (6.1). As a consequence of the localization of quasi-energy states the classical diffusion (6.2) occurs in the quantum map only on short time scales $n < n^*$ where n^* is related to the average spacing $\delta\omega_\lambda$ of the eigenphases ω_λ of quasi-energy states overlapping with a given angular momentum eigenstate, to the localization length $2\pi\hbar L$ of angular momentum, and to the kicking strength K by [23]

$$n^* \simeq \frac{2\pi}{\delta\omega_\lambda} \simeq 2L \simeq 2K^2/(4\pi^2\hbar)^2 \tag{6.6}$$

It is interesting to consider the effects of dissipation on this scenario [5,6]. Classically, the dissipative standard map reads

$$p_{n+1} = \lambda p_n - \frac{K}{2\pi} \sin 2\pi q_n$$
$$q_{n+1} = q_n + p_{n+1} \tag{6.7}$$

In order to quantize, we proceed as in sections 3 and 4, split the classical map into the dissipative part

$$p'_{n+1} = \lambda p_n \;;\; q'_n = q_n \tag{6.8}$$

and the conservative part (6.1) for the mapping $(q'_n, p'_n) \to (q_{n+1}, p_{n+1})$, quantize the first part by introducing a reservoir which absorbs angular momentum [2,5], and quantize the second part by the unitary operator (6.4). The result is a master equation for the density matrix, which, in angular momentum representation

$$p|l\rangle = 2\pi\hbar|l\rangle \tag{6.9}$$

reads

$$\langle l' | \rho_{n+1} | m' \rangle = \sum_{l,m} G(l',m' | l,m) \langle l | \rho_n | m \rangle \qquad (6.10)$$

with

$$G(l',m'|l,m) = \lambda^{\frac{1}{2}(|l|+|m|)} \Big(G^{(0)}(l',m'|l,m) + + \theta_{l\cdot m,0} \sum_{j=1}^{\min(|l|,|m|)} \left(\frac{1-\lambda}{\lambda}\right)^j \sqrt{\binom{|l|}{j}\binom{|m|}{j}} \cdot \cdot G^{(0)}\left(l',m' | l-\frac{l j}{|l|}, m-\frac{m j}{|m|}\right)\Big) \qquad (6.11)$$

where

$$G^{(0)}(l',m'|l,m) = \langle l' | U | l \rangle \langle m | U^{+} | m' \rangle \qquad (6.12)$$

The map (6.11) has been analyzed in some detail in [5,6]. We first discuss some numerical results obtained by iterating the map (6.10) until the steady state is reached. Fig. 8 shows the result for the angular momentum distribution in the steady state compared to the classical result (fine-dashed line) and the semi-classical result obtained by keeping only the second order correlation coefficients in the corresponding quasi-stochastic map. The classical distribution has support only for $|p| \leq 0.9$ which,

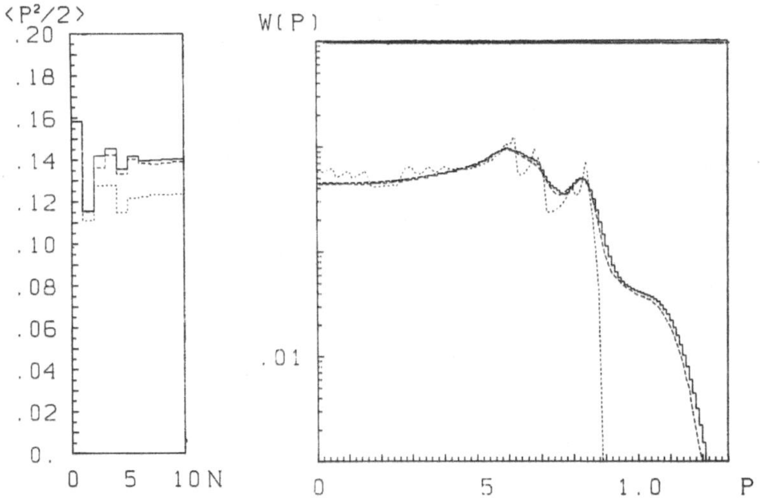

Fig. 8. Average rotation energy as function of discrete time N and angular momentum distribution for the dissipative standard map in the steady state (full line) compared to classical result (fine-dashed line) and semi-classical result (coarse-dashed line) ($K = 5$, $\lambda = 0.3$, $\hbar = 0.01/2\pi$)

for the parameter values $K = 5$, $\lambda = .3$ of Fig. 8 does not reach the classical upper bound

$$|p_{max}| = \frac{K}{2\pi(1-\lambda)} \approx 1.137 \qquad (6.13)$$

following from eq. (6.7). Due to quantum noise this classical upper bound is slightly exceeded by the quantum distribution and the semi-classical distribution, which are very close to each other.

In Fig. 9 the classical attractor is shown for the same parameter values as in Fig. 8. Only the positive-p part is shown because the negative-p part follows from the point symmetry with respect to the origin. The 'dead end' of the attractor on its leftmost arm shows that the attractor does not exhaust the large-p region of the corresponding invariant manifold, which explains why the upper bound $|p_{max}|$ of eq. (6.13) is not reached in the classical distribution on Fig. 9. In Fig. 10 the Wigner distribution corresponding to Fig. 9 is shown. In addition to structure corresponding to the classical strange attractor it has additional non-classical structure. E.g. there is a 'shadow' of the attractor shifted in phase by a half period and oscillating between nearly equal positive and negative values between $p = \pi\hbar l, \pi\hbar(l+1)$. Furthermore, the Wigner distribution shows coherent oscillations between positive and negative values in some regions of phase space far outside the classical strange attractor. It can therefore be concluded that some quantum mechanical coherence effects are still present in the steady state, despite of the rather large dissipation rate corresponding to Figs. 8 - 10. The simple physical reason for this result is the periodic external force acting on the system: Coherence effects in the quantum mechanical response to this periodic force in the stationary periodic state of the system would have to be totally destroyed within a single oscillation period in order to be absent from the steady state distribution.

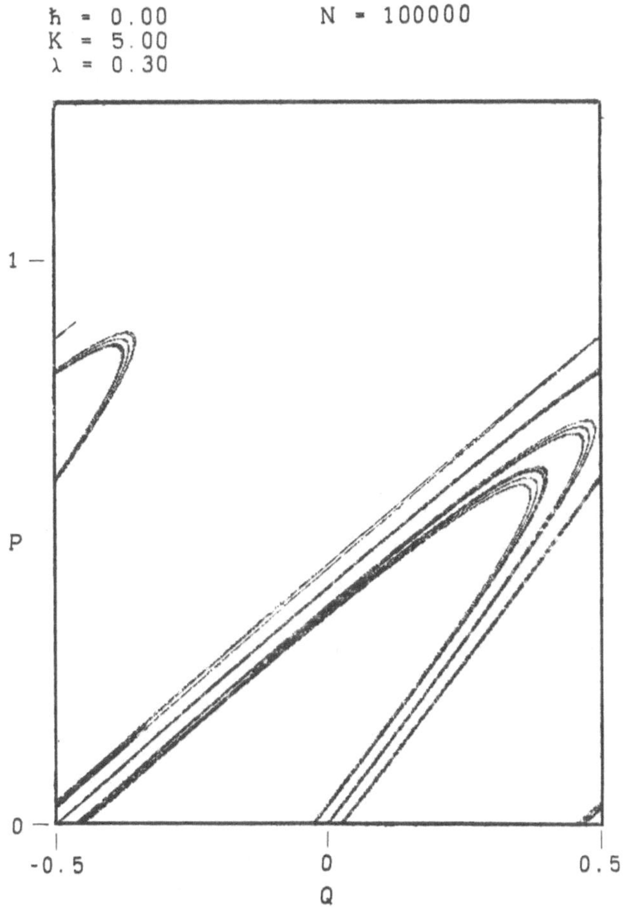

Fig. 9. Attractor in the (q,p) plane of the classical standard map with dissipation ($K = 5$, $\lambda = 0.3$)

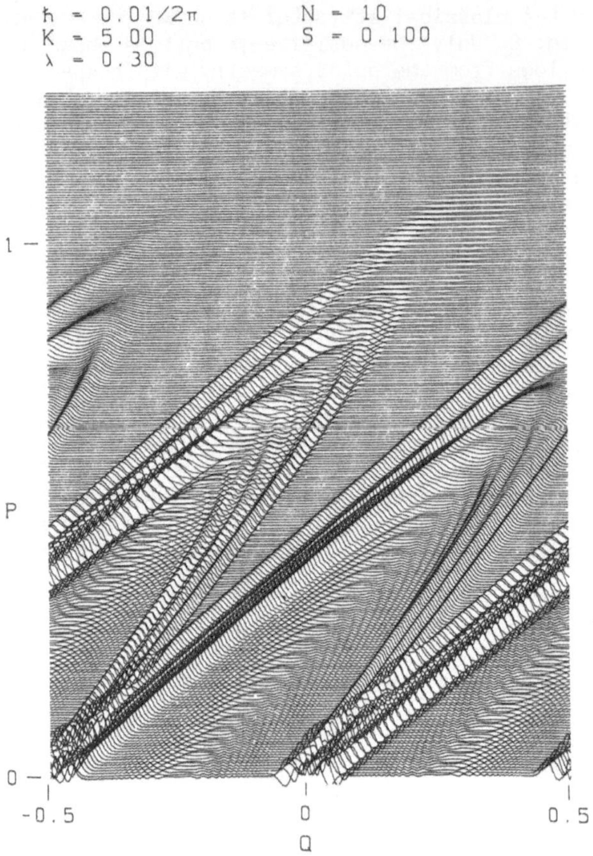

Fig. 10. Wigner distribution for the case shown in Fig. 8

7. LOCALIZATION AND VERY WEAK DISSIPATION

Localization of quasi-energy states with respect to the action variable has been extensively studied for the quantized version (6.5) of the conservative standard map (cf. [21,24]). In [17] the influence of external classical noise on localization in the quantized standard map was considered with the conclusion that even for small external noise localization is destroyed on a long time scale and a diffusion of the action variable p reappears. The influence of very weak dissipation on localization in the standard map is qualititatively discussed in [5]. Qualitative arguments of the kind given here have been remarkably successful in the past in cases where they could be checked by numerical simulations and have provided a good intuitive basis for a physical understanding of quantized conservative chaotic systems (cf. e.g. [23 - 25]).

The physical mechanism which influences localization is the infrequent absorption or emission of a quantum of action (angular momentum p in the case of the standard map) by the reservoir which is responsible for dissipation. This process is dissipative and incoherent, in contrast to the coherent transfer of action (angular momentum p) to or from the system described by the conservative part of the map. As was already discussed, the fully conservative quantum map, in the classically chaotic regime, gives rise to localization of quasi-energy states with a localization length $2\pi\hbar L \simeq K^2/8\pi^3\hbar$ (with dimension of action). This localization is a coherence phenomenon and brought about by constructive and destructive interference of the quasi-energy states contained in the initial state, whose mutual coherence is introduced by the initial state.

Introducing now some weak dissipation and denoting λ in eq. (6.7) as $\lambda = 1-\nu$, $\nu \ll 1$, there is a small exponential decay rate

$$\gamma_\varkappa = (\nu/2\pi\hbar)\langle |p| \rangle_\varkappa$$

of quasi-energy states $|\varphi_\varkappa\rangle$. This can be derived from the master equation (6.10) - (6.12) (cf. appendix). If the system is started at $p=0$ the decay probability of the quasi-energy states contained in this packet in n^* time steps is roughly $n^*\nu L$, where L is the localization length, n^* is given by eq. (6.6), and we assume that the dissipation is so small that $n^*\nu L \ll 1$, i.e.

$$\nu < \frac{1}{n^* L} \simeq \frac{(4\pi^2\hbar)^4}{2K^4} \tag{7.1}$$

This defines the case of weak dissipation. In this case, by definition, the localizing quantum effects are still operative on a time scale of the order n^*. The initial time scale of incoherent transitions n_c can be expressed in terms of ν as

$$n_c = \frac{1}{\nu L} \simeq \frac{(4\pi^2\hbar)^2}{\nu K^2} \tag{7.2}$$

These incoherent transitions disrupt the coherence of the original wave packet on the time scale n_c and start a new wave packet which, subsequently, on a short time scale $< n^*$, can again undergo the classical diffusion process over a distance, in the action variable, of order L. Thus, the action can again spread by diffusion over an additional localization length L.

After N such incoherent processes we therefore have a diffusive growth of order

$$\frac{\langle \Delta p^2 \rangle}{(2\pi\hbar)^2} \simeq L^2 N \tag{7.3}$$

The time scale $\partial N/\partial n$ for the incoherent processes itself does not remain constant, however, but changes according to

$$\frac{\partial N}{\partial n} \simeq \frac{\nu}{2\pi\hbar}\sqrt{\langle \Delta p^2 \rangle} \simeq \nu L \sqrt{N} \tag{7.4}$$

As a result N incoherent processes occur in about

$$n(N) \simeq 2\sqrt{N}/\nu L \simeq 2n_c\sqrt{N} \tag{7.5}$$

time steps and we therefore obtain by this mechanism

$$\frac{\langle \Delta p^2 \rangle}{(2\pi\hbar)^2} \simeq \frac{L^2}{4n_c^2} n^2 \simeq \left(\frac{K}{4\pi^2\hbar}\right)^8 \frac{\nu^2}{4} n^2 \tag{7.6}$$

A derivation of this result from the master equation is given in the appendix. However eq. (7.6) is true only as long as the time scale of the incoherent processes $\partial n(N)/\partial N$ is still sufficiently long for the localization to be operative on the time scale n^*, i.e. if

$$n^* < \frac{\partial n}{\partial N} \simeq \frac{n_c}{\sqrt{N}} \simeq \frac{2n_c^2}{n} \tag{7.7}$$

Obviously, this inequality can be invalidated for sufficiently large n. However, due to the presence of dissipation, the system will always reach a steady state on a time scale $n_s \simeq \nu^{-1}$ [6]. This happens before (7.7) is violated if $n^* n_s < 2n_c^2$ or

$$\nu < (4\pi^2\hbar/K)^6 \tag{7.8}$$

which is a stronger condition than (7.1) and defines the case of very weak dissipation. In this extreme case the estimate (7.6) applies until the steady state is reached and we find for the action scale in the steady state

$$\frac{\langle \Delta p^2 \rangle}{(2\pi\hbar)^2} \simeq \frac{1}{4}\left(\frac{K}{4\pi^2\hbar}\right)^8 \tag{7.9}$$

which is independent of ν and much larger than the corresponding scale $\langle \Delta p^2 \rangle/(2\pi\hbar)^2 \simeq K^4/4.(4\pi^2\hbar)^4$ of the conservative system. Infrequent but random incoherent absorption of action by the reservoir therefore drastically increases the accessible scale of the action variable and in this sense destroys localization. It should be noted, however, that this obstruction of localization by dissipation only occurs on the very long time scale $n_s \simeq \nu^{-1} > (K/4\pi^2\hbar)^6$.

It should also be noted that in some sense a trace of localization is still left in eq. (7.9) since the scale (7.9) is still smaller than the corresponding classical scale (cf. (7.10)), because (7.8) must hold.

If ν satisfies (7.1) but not (7.8), the inequality (7.7) is violated before the steady state is reached. In this case eq. (7.6), which supposes the localization mechanism to be active on the time scale n^*, looses its validity and the system returns to classical diffusion on the time scale of incoherent processes. The momentum scale in the steady state is then the same as in the classical system

$$\frac{\langle \Delta p^2 \rangle}{(2\pi)^2} \simeq \frac{1}{4}\left(\frac{K}{4\pi^2}\right)^2 \nu^{-1} \tag{7.10}$$

which is again much larger than the localization scale $\langle \Delta p^2 \rangle/(2\pi\hbar)^2 \simeq K^4/(4\pi^2\hbar)^4$ in the conservative system as can be seen from inequality (7.1). (The upper bound (6.13) is, of course, much larger than the typical scale (7.10)).

If ν is not sufficiently small to satisfy (7.1), the coherence of the initial wave packet is destroyed by dissipation before localization can occur. The momentum scale in the steady state is then again given by eq. (7.10). It is larger than the localization scale of the conservative system if $4(4\pi^2\hbar)^2/K^2 > \nu > (4\pi^2\hbar)^4/2K^4$ and smaller if $\nu > (4\pi^2\hbar/K)^2.4$. In the latter case the dissipation is so strong that the steady state is reached before the mechanism of quantum mechanical localization has ever had a chance to be operative. This case is the one realized in Figs. 8-10.

In summary, it is clear that there exists a number of different regimes distinguished by the strength of dissipation in which quite different physical behavior is expected. Numerical work, up to now, had to be restricted to the regime of strong dissipation, where the steady state is reached before interesting dynamical behavior can occur. The difficulty is, of course that a master equation for a density matrix must be numerically solved while it is sufficient to solve the Schrödinger equation for a state vector in the conservative system. A number of rather severe data storage problems must still be overcome before there is hope for a successful numerical study of the various regimes of weaker dissipation.

APPENDIX

The decay rate $\gamma_\varkappa$ of quasi-energy states $|\varphi_\varkappa\rangle$ for weak dissipation can be obtained from the master equation (6.10) - (6.12) as

$$\gamma_\varkappa = \frac{(1-\lambda)}{2\pi\hbar} \langle |p| \rangle_\varkappa \left(1 + O\left((1-\lambda)^2, (4\pi^2\hbar/k)^2\right)\right) \tag{A.1}$$

where $\langle |p|_\varkappa \rangle$ denotes the expectation of $|p|$ in the state $|\varphi_\varkappa\rangle$. To see this, we rewrite eq. (6.11) to first order in $(1-\lambda)$ as

$$G(\ell',m'|\ell,m) = G^{(0)}(\ell',m'|\ell,m) + (1-\lambda)\, G^{(1)}(\ell',m'|\ell,m)$$

with

$$G^{(1)}(\ell',m'|\ell,m) = -\tfrac{1}{2}\left(|\ell| + |m|\right) G^{(0)}(\ell',m'|\ell,m) \tag{A.2}$$
$$+ \,\Theta_{\ell m,0} \sqrt{\ell\cdot m}\; G^{(0)}\!\left(\ell',m'|\ell - \tfrac{\ell}{|\ell|}, m - \tfrac{m}{|m|}\right)$$

and rewrite this equation in the representation provided by the quasi-energy eigenstates $|\varphi_\varkappa\rangle$. We find from eq. (A.2)

$$G(\varkappa'\mu'|\varkappa\mu) = e^{-i(\omega_{\varkappa'} - \omega_{\mu'})} \Big\{ \delta_{\varkappa\varkappa'}\, \delta_{\mu\mu'} -$$
$$- \frac{1-\lambda}{2}\left(|\ell|_{\varkappa'\varkappa}\, \delta_{\mu\mu'} + |\ell|_{\mu\mu'}\, \delta_{\varkappa\varkappa'} \right) \tag{A.3}$$
$$+ (1-\lambda)\left(C^{(+)}_{\varkappa'\varkappa}\, C^{(+)*}_{\mu'\mu} + C^{(-)}_{\varkappa'\varkappa}\, C^{(-)*}_{\mu'\mu} \right) \Big\}$$

with

$$|\ell|_{\varkappa'\varkappa} = \sum_{\ell=-\infty}^{+\infty} \langle \varphi_{\varkappa'} | \ell \rangle |\ell| \langle \ell | \varphi_\varkappa \rangle$$
$$C^{(+)}_{\varkappa'\varkappa} = \sum_{\ell>0} \sqrt{\ell}\; \langle \varphi_{\varkappa'} | \ell - 1 \rangle \langle \ell | \varphi_\varkappa \rangle \tag{A.4}$$
$$C^{(-)}_{\varkappa'\varkappa} = \sum_{\ell<0} \sqrt{|\ell|}\; \langle \varphi_{\varkappa'} | \ell + 1 \rangle \langle \ell | \varphi_\varkappa \rangle$$

The decay rate $\gamma_\varkappa$ of the occupation probability $\rho_{\varkappa\varkappa}$ of a quasi-energy state $|\varphi_\varkappa\rangle$ to first order in $(1-\lambda)$ can be read off from eq. (A.3) as

$$\gamma_\varkappa = (1-\lambda)\left(|\ell|_{\varkappa\varkappa} - |C^{(+)}_{\varkappa\varkappa}|^2 - |C^{(-)}_{\varkappa\varkappa}|^2 \right) \tag{A.5}$$
$$+ O\left((1-\lambda)^2\right)$$

The transition rate $R_{\varkappa\varkappa'}$ from a quasi-energy state $|\psi_\varkappa\rangle$ to a quasi-energy state $|\psi_{\varkappa'}\rangle$ to first order in $(1-\lambda)$ is obtained from eq. (A.3) as

$$R_{\varkappa\varkappa'} = (1-\lambda)\left(|C^{(+)}_{\varkappa'\varkappa}|^2 + |C^{(-)}_{\varkappa'\varkappa}|^2\right) \tag{A.6}$$

We estimate the sums over ℓ in $|C^{(\pm)}_{\varkappa\varkappa'}|$ in eq. (A.4) as

$$|C^{(\pm)}_{\varkappa\varkappa'}| \simeq \left(|\ell|_{\varkappa\varkappa'}/2L\right)^{1/2} \tag{A.7}$$

where $L \simeq (K/4\pi^2\hbar)^2$ is the localization length by noting that about $2L$ terms under the sum contribute and have mutually random phases and are roughly of size $(2L)^{-1}$[25]. Hence, the terms $|C^{(\pm)}_{\varkappa\varkappa'}|^2$ in eq. (A.5) are smaller in order of magnitude by a factor L^{-1} compared to $|\ell|_{\varkappa\varkappa}$ and the result (A.1) follows from eqs. (A.5), (A.7) with $\langle|p|\rangle_\varkappa = 2\pi\hbar\,|\ell|_{\varkappa\varkappa}$. The transition rate (A.6) is estimated via eq. (A.7) as

$$R_{\varkappa\varkappa'} \simeq \frac{(1-\lambda)|\ell|_{\varkappa\varkappa'}}{2L} \tag{A.8}$$

According to this estimate the mean square displacement of the angular momentum in $|\psi_\varkappa\rangle$ changes per time step as

$$\frac{\partial\langle\Delta p^2\rangle}{\partial n} \simeq \frac{(1-\lambda)|\ell|_{\varkappa\varkappa}}{2L}\cdot(4\pi\hbar L)^2\cdot 2L \tag{A.9}$$

assuming that $|\ell|_{\varkappa\varkappa'}$ is appreciable only for states $\varkappa,\varkappa'$ within a localization length L and there $\simeq |\ell|_{\varkappa\varkappa}$. Estimating $|\ell|_{\varkappa\varkappa}$ by $\langle\Delta p^2\rangle^{1/2}$ and integrating (A.9) with respect to n we recover eq. (7.6).

REFERENCES

1. R. Graham, Phys. Lett. 99A,131 (1983).
2. R. Graham, Z. Phys. B59,75 (1985).
3. R. Graham, T. Tél, Z. Phys. B60,127 (1985).
4. R. Graham, S. Isermann, T. Tél, Z. Phys., to be published.
5. T. Dittrich, R. Graham, Z. Phys. B62,515 (1986).
6. T. Dittrich, R. Graham, to be published.
7. R. Graham, Europhys. Lett., to appear.
8. R. Graham, Physica Scripta, to appear.
9. G. Casati, ed., 'Chaotic Behavior in Quantum Systems', New York, Plenum 1985.
10. G. Casati, B.V. Chirikov, D.L. Shepelyansky, Phys. Rev. Lett. 53,2525 (1984).
11. S. Fishman, D.R. Grempel, R.E. Prange, Phys. Rev. Lett. 49,509 (1982).
12. J.L. Kaplan, J.A. Yorke, Lecture Notes in Mathematics, vol. 730, p. 228, Springer, Berlin 1979.
13. D. Mayer, G. Roepstorff, J. Stat. Phys. 31,309 (1983).
14. M. Hénon, Comm. Math. Phys. 50,69 (1976).
15. R.H.G. Helleman, in 'Fundamental problems in statistical mechanics', ed. E.G.D. Cohen, vol. 5, North Holland, Amsterdam 1980.
16. F. Haake, R. Reibold, Phys. Rev. A32,2462 (1985).
17. H. Haken, 'Laser Theory', Encyclopedia of Physics, Vol. XXV/2c, Springer, Berlin 1970.
18. W.H. Louisell, 'Quantum Statistical Properties of Radiation', Wiley, London 1973.

19. A.J. Lichtenberg, M.A. Lieberman, 'Regular and Stochastic Motion', Springer, Berlin 1983.
20. R.H.G. Helleman, in 'Nonequilibrium Problems in Statistical Mechanics', Vol. 2, eds. W. Horton et.al., Wiley, New York 1981.
21. D.R. Grempel, S. Fishman, R.E. Prange, Phys. Rev. Lett. 53,1212 (1984).
22. B.V. Chirikov, Phys. Rep. 52,263 (1979).
23. B.V. Chirikov, F.M. Izrailev, D.L. Shepelyanski, Sov. Sci. Rev. C2,209 (1981).
24. D.L. Shepelyanski, Physica D8,208 (1983).
25. E. Ott, T.M. Antonsen, J.D. Hanson, Phys. Rev. Lett. 53,2187 (1984).

CHAOS AND QUANTUM BEHAVIOR OF KICKED TOPS+

Fritz Haake, Marek Kus* and Rainer Scharf

Fachbereich Physik
Universität-Gesamthochschule Essen
4300 Essen, Deutschland

Evidence is discussed for two different manifestations of quantum mechanical quasiperiodicity, one associated with classically regular motion and the other with classical chaos. Furthermore, concrete dynamics with linear, quadratic, and quartic level repulsion are presented.

We would like to present two results on the conservative quantum dynamics of a kicked spin which is described stroboscopically by powers of the unitary evolution operator [1]

$$U = e^{-iH_1} e^{-iH_0} . \tag{1}$$

Both H_0 and H_1 are buildt up by angular momentum operators $\hbar\vec{J} = \hbar(J_x, J_y, J_z)$, $[J_i, J_j] = i\varepsilon_{ijk} J_k$. The squared angular momentum is conserved,

$$\vec{J}^2 = j(j+1) \ , \quad j \text{ integer or half integer} \ , \tag{2}$$

and the quantum dynamics can therefore be described in a Hilbert space with $(2j+1)$ dimensions. The classical limit is attained as $j \to \infty$. For at least one of the two operators H_0, H_1 being a nonlinear function of the J_i the classical dynamics will be nonintegrable.

The first of the results to be discussed pertains to the special case [1]

$$H_0 = p\, J_y \ , \quad H_1 = k\, J_z^2/2j \ . \tag{3}$$

---

* Permanent address: Institute of Theoretical Physics, University of Warsaw, Hoza 69, 00-681 Warsaw, Poland

+ Talk presented by Fritz Haake.

Obviously, H_0 describes a uniform precession of $\vec{J}$ around the y axis by an angle p. The perturbation H_1 may be interpreted as a nonuniform precession around the z axis such that the precession angle is itself proportional to kJ_z. When choosing $k = 3$, $p \approx \pi/2$ we are facing classical coexistence of regular and chaotic motion on the sphere $\vec{J}^2 = \text{const}$. Quantum mechanically, we also find rather different behavior depending on whether the initial state of the spin is well localized within regions of classically regular or classically chaotic motion [1]. That difference concerns, for instance, the temporal behavior of expectation values like $\langle J_i \rangle$ even for times of the order of or exceeding the quantum quasiperiod (which is of the order j) when the quasiperiodic quantum motion in no case resembles the classical behavior. Classically regular motion corresponds to nearly periodic (period ~j) sequences of collapse and revival while classical chaos is associated with erratic sequences of recurrence events [1]. Complementary to that temporal distinction is a spectral one. We had previously proposed to base a spectral distinction on the number of eigenvectors of U appreciably populated in a coherent state. For a concrete criterion we may use the minimum number of eigenvectors of U, N_{min}, necessary to exhaust the normalization of the coherent state in question to, say, 99.9%. We had conjectured the power law

$$N_{min} \sim j^x \text{ with } x \approx \begin{cases} 1/2 & \text{regular case} \\ 1 & \text{chaotic case} \end{cases} . \quad (4)$$

Fig. 1 presents numerical evidence for that conjecture, based on computations done for j = 200, 400, and 500. We should note that N_{min} varies in

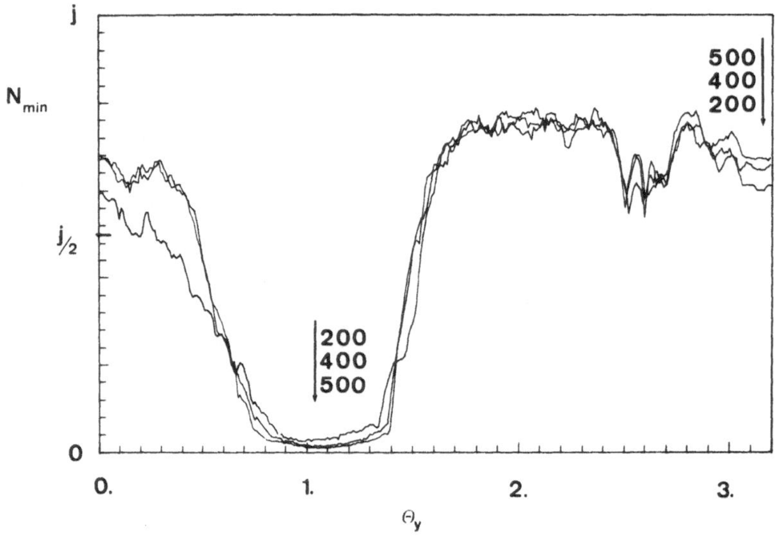

Fig. 1. Number of eigenstates of U excited in an initial coherent state versus polar angle (with respect to y axis) at which the coherent state is located (azimutal angle $\phi_y = \pi/2$ fixed). The three curves pertain to j = 200, 400, and 500. Classically, there is regular motion in $0.6 \lesssim \Theta_y \lesssim 1.4$ and chaos otherwise. The curves follow the classical Lyapunov exponent the better the higher j.

close correlation with the classical Lyapunov exponent as the coherent state is moved on the sphere $\vec{J}^2$ = const.

The second result to be presented concerns the statistics of the eigenphases of U, defined by $U|\phi\rangle = e^{i\phi}|\phi\rangle$. These so called quasienergies tend to repel each other when all external parameters are set such that the classical motion is globally chaotic. There is no level repulsion under conditions of predominantly regular behavior in the classical case. Three universality classes of level repulsion are known. They are characterized by a linear, quadratic, and quartic dependence of the relative frequency of level spacings P(S) on the spacing S for small S,

$$P(S) \sim S^{\beta} \text{ for } S \to 0 \ , \quad \beta = \begin{cases} 1 \\ 2 \\ 4 \end{cases} . \tag{5}$$

Which class a given dynamics belongs to is determined by the full group of symmetries of U [2]. It is well known that linear, quadratic, and quartic level repulsion can also be obtained from ensembles of orthogonal, unitary, and symplectic random matrices [3].

By far the largest number of examples of dynamics (autonomous or driven) reported in the literature belongs to the class of linear level repulsion. These include all measured energy spectra of nuclei. Relatively sparse are instances of quadratic repulsion and we know of no previous mentioning of a model pertaining to the symplectic case. It is quite an attractive feature of kicked spins with conserved $\vec{J}^2$ that they allow realizations of all three universality classes of level repulsion. Fig. 2a,b show level spacing statistics for the model defined by (1,3) with j = 500, Fig. 2a for k = 0.1, p = 1.7 (corresponding to predominantly regular motion in the classical limit) and Fig. 2b for k = 10, p = 1.7. The linear repulsion in the latter case is due to a generalized time reversal invariance of the map (1,3). No such invariance holds for the map [1]

$$U = e^{-i(k'/2j)J_x^2} e^{-i(k/2j)J_z^2} e^{-ipJ_y} \tag{6}$$

the level spacing histogram of which is seen to display quadratic repulsion in Fig. 2c for k' = 0.5, k = 10, p = 1.7, and j = 500.

Level statistics typical of the unitary matrix ensemble are not a priviledge of systems without an antiunitary time reversal symmetry. Fig. 2d pertains to the map of the form (1) with

$$\begin{aligned} H_0 &= (p/2j)\{J_x^2 + \tfrac{1}{2}\left(J_y J_z + J_z J_y\right)\}, \\ H_1 &= (k/2j)J_z^2, \ p = 5, \ k = 10 \end{aligned} \tag{7}$$

which does have such a symmetry [2]. Quadratic level repulsion in this case is made possible by a half integer value of j, j = 499.5 and a discrete unitary symmetry (rotation by π around the x axis).

Finally, quartic level repulsion prevails in Fig. 2e, obtained for j = 499.5 from a map of the form (1) with $H_0 = (k/2j)\ J_z^2$ and V a symplectic matrix with random elements chosen such that an antiunitary time reversal symmetry $TUT^{-1} = U^{-1}$ with $T^2 = -1$ but no discrete rotation symmetry hold [4].

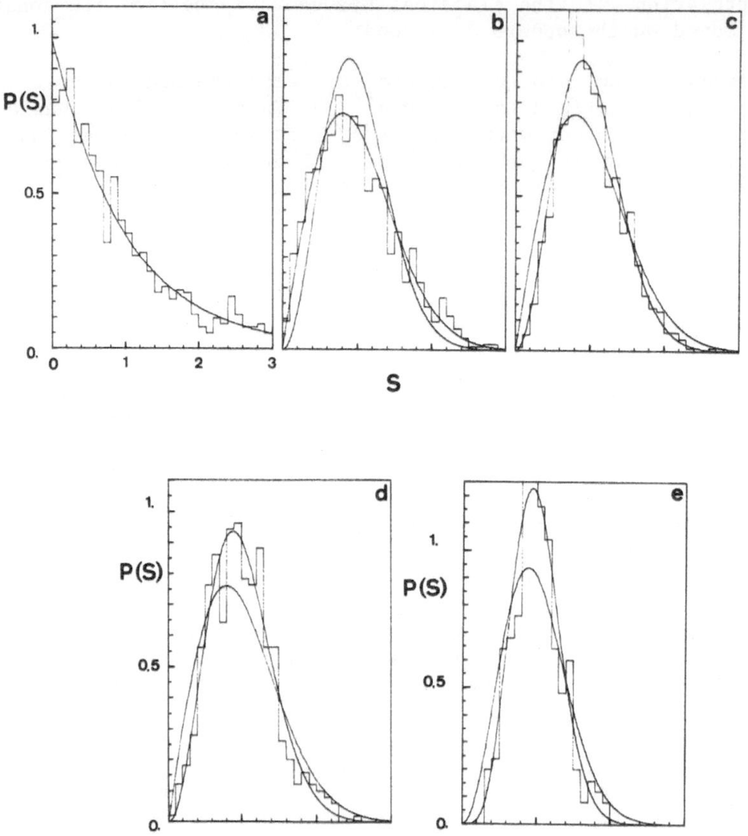

Fig. 2. Level spacing statistics, (a) under conditions of classically regular motion, (b - e) under conditions of classical chaos. (b) integer j, with generalized time reversal invariance T gives linear repulsion. (c) integer j, no T, quadratic repulsion. (d) half integer j, with T and parity, quadratic repulsion. (e) half integer j, with T, no parity, quartic repulsion.

REFERENCES

1. F. Haake, M. Kuś, and R. Scharf, Z. Physik B65, 381 (1987).
2. M. Kus, R. Scharf, and F. Haake, Z. Physik B66, 129 (1987).
3. F. I. Dyson, J. Math. Phys. B140, 1199 (1962).
4. A group theoretical discussion of this case as well as of (3,6,7) is given in [2]. The actual realization of the symplected case was achieved only after ref. [2] was submitted for publication and also after the Como meeting reported on in this volume.

THE DYNAMICS OF THE PURE STATE DENSITY MATRIX

FOR HAMILTONIAN SYSTEMS

E.R. Pike*†, Sarben Sarkar† and J.S. Satchell[x]†

* Dept. of Physics, King's College, London WC2R 2LS, UK

† Centre for Theoretical Studies,
Royal Signals and Radar Establishment,
Gt Malvern, Worcs WR14 3PS, UK

[x] Clarendon Laboratory, Oxford University, Oxford OX1 3PU, UK

Abstract

The dynamics of pure state density matrices are studied for Hamiltonian systems in terms of generic features of the Hamiltonian spectrum. It is shown that it is the localised or extended nature of dressed (quasi) energy eigenstates, the frequency of their occurrence, and the distribution of spacings (not necessarily adjacent) of (quasi) energies, which are important in determining the evolution of density matrix elements.

1. INTRODUCTION

Recent progress in classical non-linear dynamical systems and maps has revealed that stochasticity[1] may arise from deterministic systems with small numbers of degrees of freedom. As a result it has been found useful to describe the attracting sets statistically, e.g., by considering the probability of visiting infinitesimal regions of the attractor, the evolution of average quantities, etc. The quantum analogue[2,3] of such stochasticity is much less well understood. Most effort has been expended on understanding the 'ergodic' properties of energy eigenstates of 'chaotic' Hamiltonians[3]. However, some years ago[4] it was suggested that the quantum version of chaotic motion may allow the formulation of dynamical equations such as the Pauli type master equation (ie, in terms of probabilities or equivalently of diagonal elements of the density matrix) without the assumption of random initial phases or coarse graining[5]. In ref 4 some semi-classical arguments were given which claim to support this. The arguments, however, were not definitive; consequently in this paper we will reconsider the dynamics of density matrix elements. We have restricted ourselves to pure states since the evolution of quantum mechanical mixed states is an unnecessary complication for a first investigation. We will investigate several Hamiltonians, covering different situations. Our choice of Hamiltonians has been influenced by their solubility, at least numerically, without the need for truncations, ie approximating infinite dimensional Hilbert space by finite dimensional vector spaces. Before we proceed to the details we will set up the calculations of the density matrix ρ for a general Hamiltonian, H. It is easy to show that the elements of the pure state density matrix, in an arbitrary 'bare' basis satisfy

$$\rho_{jk}(t) = \sum_{\substack{N,M \\ n,m}} C_N^{(j)} (C_N^{(n)})^* C_M^{(m)} (C_M^{(k)})^* e^{-i(E_N-E_M)t/\hbar} \rho_{nm}(0) \quad (1)$$

where the $C_P^{(q)}$ are overlap integrals between states in an arbitrary or 'bare' representation (in small letters) and dressed (in capital letters) states, and

$$C_P^{(q)} = \langle q \mid P \rangle \quad (2)$$

Moreover the states $|P\rangle$ from a complete set of eigenstates of H, namely

$$H|P\rangle = E_P|P\rangle \quad (3)$$

Formula (1) is of course exact. For kicked systems (1) has to be interpreted somewhat differently. If τ is the unit of time from just after one kick to just after the next then t is a non-negative integer multiple of τ and E_N is a quasi-energy such that

$$e^{-iH\tau/\hbar} |P\rangle = e^{-iE_P\tau/\hbar} |P\rangle \quad (4)$$

We will use (1) to calculate the time dependence of the diagonal occupation probability, $\rho_{nn}(t)$, of single given initial pure states, as a general investigative tool connecting the dynamical properties of a system with its energy spectrum and eigenstates. For $\rho_{nn}(t)$ we have the form (using (1))

$$\rho_{nn}(t) = \int_{-\infty}^{\infty} g(E)\, \rho_{nn}^{(E)}(0)\, e^{-iEt/\hbar}\, dE \quad (5)$$

where g(E), which may be singular, is the probability density of all energy level differences (i.e not just adjacent spacings), and $|n\rangle$ is the initial (bare) pure state; also

$$1 = \int_{-\infty}^{\infty} g(E)\, \rho_{nn}^{(E)}(0)\, dE \qquad (6)$$

This partition of unity reflects both the 'stationary' envelope of occupied dressed states and the spectrum of energy-level differences of the system.

The integral in (5) is similar to one appearing in the theory of Brownian motion[6], with the energy dependent phase differences analogous to the correlated "self-diffusion" phase terms in a two-dimensional random walk. In the Brownian motion calculation one appeals usually to a decoupling of motions about and of the centre of mass to factorise the angle and prefactor averages. The analysis gives diffusive motion. For complicated quantum mechanical problems it is possible to produce heuristic arguments of this sort if necessary. It is easy to see that for the spacings S a Poisson distribution (e^{-S}) will give a Lorentzian like decay whereas a Wigner type distribution ($S\ e^{-S}$) will give a Gaussian like fall off provided $\rho_{nn}^{(E)}$ (0) is fairly independent of E for small E. For g(S) flat, the fall-off is even faster. In general a heavily "bunched" spectrum will give a slower time decay than an "antibunched" one. The insensitivity of $\rho_{nn}^{(E)}$(o) to E is very important, and a distinguishing feature of chaotic systems. When this is not the case, and a certain small set of E's dominate in (5), the motion is likely to be quasi-periodic whatever the form of g(E).

The notions that we have introduced run along similar lines to the thinking of other authors, for example, Einstein, Percival, Berry, Casati, Heller, Peres, etc,[7,3,8] although the usefulness of the general energy spacing distribution does not seem to have been discussed before. The Hamiltonians that we will discuss in order to give quantitative form to these notions are the kicked spin investigated recently by Hakke et al[9], the well-known models of an electron in a 1-D lattice with incommensurate "tan" potential,[10,11] and the Harper model[12] which is the same as the previous one, but with a "cos" potential. This change in potential allows the model to show a transition from localised to extended states at a certain critical value of the potential amplitude. Our interest in the electron models arose from the mapping of a version of the forced rotator model onto a "tan" model due to Fishman et al.[13]. Finally we study the coupled spin model investigated by Feingold and Peres[8].

The paper is organised in four sections. In section two the models that we study are presented in detail. Since the rotator model of Fishman et al[13] is soluble, we use it in section 3 to give some analytic results for the time dependence of density matrix elements. These results serve to motivate the discussion of numerical results obtained in section 4 for the other models.

2. THE MODEL HAMILTONIANS

The kicked spin J model has the calculationally advantageous feature of having the dimension of its state space finite (= 2J+1). In the large J (classical) limit Haake et al[9] have shown that there is chaotic as well as regular dynamics. It is thus an example of a quantum K system. The periodic Hamiltonian H(t) is given by

$$H(t) = \frac{\hbar p}{\tau} J_z + \frac{\hbar k}{2J} J_x^2 \sum_n \delta(t-n\tau) \tag{7}$$

where J_x, J_y and J_z are angular momentum operators satisfying the SU(2) algebra and k and p are constant parameters. Its experimental significance is not clear, and we will use it solely as a convenient model system. A natural basis for describing the quantum theory is in terms of the standard angular momentum states $|J,m\rangle$. We recall that

$$J_z |J,m\rangle = m|J,m\rangle \tag{8}$$

$$J_\pm |J,m\rangle = (J(J+1) - m(m\pm 1))^{\frac{1}{2}} |J,m\pm 1\rangle \tag{9}$$

where

$$J_\pm = (J_x \pm iJ_y)$$

The eigenstates of J_x, $|J,m\rangle_x$, can be expressed in terms $|J,n\rangle$, namely

$$|J,m\rangle_x = \sum_n d_{nm}^{(J)}(\pi/2) |J,n\rangle \tag{10}$$

where $d_{nm}^{(J)}(\pi/2)$ is the rotation matrix for a rotation of $\pi/2$ about the y axis. With the help of eqn 10 it is easy to show that

$$\rho_{nn}(r\tau) = |((G(\tau))^r)_{nn}|^2 \tag{11}$$

where we have abbreviated the label (J,n) by n since J is a good quantum number and

$$G_{nm}(\tau) = e^{-i\varepsilon_m \tau/\hbar} \sum_{m'} e^{-i \frac{\tau k}{2j}(m')^2} d^{(J)}_{nm'}(\pi/2)(d^{(J)}_{nm'}(\pi/2))^* \tag{12}$$

with

$$\varepsilon_m = \frac{\hbar p}{\tau} m \tag{13}$$

Following the work of Heller and Sundberg (in Ref 3), as a heuristic measure of extension for both initial and dressed states, overlap on a set of (2J+1) patches on the sphere describing the wave functions will be useful. We will take this set to be a subset of the atomic coherent sets in such a way that an even cover of the sphere is obtained. The overlap of a coherent state $|\theta,\phi\rangle$ with the states in eqn (8) is given by

$$\langle J,m|\theta,\phi\rangle = \binom{2J}{m+J}^2 \sin^{J+m}(\tfrac{1}{2}\theta) \cos^{J-m}(\tfrac{1}{2}\theta)\, e^{-i(J+m)\phi} \tag{14}$$

and the ranges of θ and ϕ are $0 \leqslant \theta < \pi$, $0 \leqslant \theta < 2\pi$. The atomic coherent states are minimum uncertainty states and so provide a reasonably localized "patch" on the sphere. In terms of a set of patches $\{|\theta_i,\phi_i\rangle \mid i=1,..,(2j+1)\}$ we adopt a simple (but somewhat arbitrary) criterion for a state $|\psi\rangle$ to be extended or localised. We first define ψ_i to be

$$\psi_i = |\langle\psi|\theta_i,\phi_i\rangle|^2 \tag{15}$$

and

$$\Gamma_\psi = \{\psi_i \mid |\psi_i| < \frac{1}{10(2J+1)},\ i=1, \ldots, 2j+1\} \tag{16}$$

If the number of ψ_i in Γ_ψ is greater than J then we define $|\psi\rangle$ to be localised. Otherwise a state is extended.

We can make some general statements[10,13] about the solutions of H(t). From Floquet's theorem it is known that a solution has the form

$$e^{iE_\psi t} |\psi(t)\rangle$$

where E_ψ is a quasi-energy and

$$|\psi(t+\tau)\rangle = |\psi(t)\rangle \tag{17}$$

If $|\psi^+\rangle$ denotes the value of $|\psi(t)\rangle$ just after a kick then we can show that

$$|\psi^+\rangle = \left(\frac{1+iW}{1-iW}\right)\left(\frac{1+iT}{1-iT}\right)|\psi^+\rangle \tag{18}$$

where

$$W = -\tan\frac{k}{4j}J_x^2$$

and

$$T = \tan\tfrac{1}{2}(\omega\tau - pJ_z)$$

On defining $|\tilde{\psi}\rangle$ to be

$$|\tilde{\psi}\rangle = (1+iW)^{-1}|\psi^+\rangle \tag{19}$$

we find that

$$(T+W)|\tilde{\psi}\rangle = 0 \tag{20}$$

This can be cast into the form of a lattice problem on writing

$$|\tilde{\psi}\rangle = \sum_n \tilde{\psi}_n |J,n\rangle \tag{21}$$

Substituting this into Eqn 20 gives

$$T_{n''}\tilde{\psi}_{n''} - \sum_{m,n} W_{n''n}\tilde{\psi}_n = 0 \tag{22}$$

where

$$T_{n''} = \tan\tfrac{1}{2}(\omega\tau - pn) \tag{23}$$

and

$$W_{n''n} = \sum_m d^{(J)}_{n''m}(\pi/2)\,(d_{mn}(\pi/2))\tan\frac{km^2}{4j} \tag{24}$$

This leads naturally to considering hopping models of electron transport in one dimensional lattices where the hopping amplitudes are given by the matrix

W_{nn} and the site energy at site n is T_n. The Hamiltonian H corresponding to such a problem can be given in the following second quantised form

$$H = \sum_n t_n a_n^\dagger a_n - \sum_{\substack{n,m \\ n \neq m}} W_{nm} a_n^\dagger a_m \qquad (25)$$

The operators a_m and $a_m^\dagger$ destroy or create a particle at site m. In eqn (24) we see already the appearance of a 'tan' potential. We will consider a class of models of the form eqn 25 on infinite lattices with

(A) $t_m = \tan \tfrac{1}{2}(\omega - 2\pi\alpha m)$

$W_{nm} = \overline{W}(n-m)$

$\overline{W}_r = -W(\delta_{r1} + \delta_{r,-1})$

(B) $t_m = \tan \tfrac{1}{2}(\omega - 2\pi\alpha m^2)$

but otherwise as in (A)

(C) $t_m = V \cos(2\pi\alpha m)$

but otherwise as in (A)

Fishman et al[13] have shown that models of the form (A) and (B) are closely related to rotator models with Hamiltonians H given by

$$H = 2\pi\alpha\, f\left(-i \frac{\partial}{\partial\theta}\right) + V(\theta)\, \Delta(t) \qquad (26)$$

and in general f is an arbitrary function and

$$V(\theta) = \sum_m V_m e^{im\theta}$$

$$\Delta(t) = \sum_m \Delta_m e^{-2\pi imt} \qquad (27)$$

It is possible to solve for the behaviour of the system when $f(\theta) = \theta$. In particular we consider the cases

$$V(\theta) = k \cos \theta$$

$$\Delta(t) = \cos 2\pi t \qquad (28a)$$

$$\text{or}\quad \Delta(t) = \sum_n \delta(t-n) \tag{28b}$$

for α irrational (< 1). For α rational we take

$$\begin{aligned} V(\theta) &= k(\cos\theta + \cos 2\theta + \cos 3\theta) \\ \text{and}\quad \Delta(t) &= \cos 2\pi t \end{aligned} \tag{29}$$

Other choices are possible but (28) and (29) represent simple cases which adequately illustrate the essentials in the evolution of the diagonal density matrix in the soluble systems.

The coupled spin model of Feingold and Peres(8) has the advantages of being soluble numerically in a finite basis and it also has true eigenstates rather than quasi eigenestates. The model consists of two spins of total angular momentum L^2 and M^2, respectively, each one of which is separately conserved. The interaction is via the x-components. Thus the Hamiltonian is

$$H = H_o + V$$

$$H_o = L_z + M_z, \qquad V = L_x M_x$$

Feingold and Peres have shown that this model has regular and chaotic regions and have carried out extensive numerical experiments of a similar nature to these reported here. We have, in this case, made some preliminary calculations on the form of the complete master equation for the density matrix.

3. SOLUBLE MODELS

It has been shown[10,11] that for α a good irrational (i.e not a Liouville number) the quasi-energy eigenfunctions for H in eqn (26) (with $f(\theta) = \theta$) have the form

$$u^{\nu}(\theta,t) = e^{i\nu\theta}\, e^{-i\phi(\theta,t)}\, e^{-ig(t)} \tag{30}$$

for ν and integer and

$$\phi(\theta,t) = \sum_{\substack{m \\ (\neq 0)}} \sum_n \frac{V_m \Delta_n e^{im\theta} e^{-i2\pi nt}}{2\pi i(\alpha m - n)} \tag{31}$$

and

$$g(t) = V_o \int_0^t ds(\Delta(s) - \Delta_o) \tag{32}$$

For the case (28a)

$$u^{\nu}(\theta,t) = \frac{1}{(2\pi)^{\frac{1}{2}}} e^{i\nu\theta} \exp \frac{-ik}{4\pi} \left(\frac{\sin(\theta-2\pi t)}{\alpha-1} + \frac{\sin(\theta+2\pi t)}{\alpha+1} \right) \tag{33}$$

$$u_m^{\nu} \equiv \int d\theta \, e^{im\theta} \, u^{\nu}(\theta,t) \tag{34}$$

$$= \exp(-i(m+\nu)\phi) \, J_{m+\nu} \frac{k}{4\pi} r(t)$$

where

$$r(t) = \frac{2}{1-\alpha^2} (\alpha^2 + (\sin^2 2\pi t)(1-\alpha^2))^{\frac{1}{2}} \tag{35}$$

$$\cos \phi = - \frac{\alpha \cos 2\pi t}{(\alpha^2 + (1-\alpha^2) \sin^2 2\pi t)^{\frac{1}{2}}} \tag{36}$$

and

$$\sin \phi = \frac{\sin 2\pi t}{(\alpha^2 + (1-\alpha^2) \sin^2 2\pi t)^{\frac{1}{2}}}$$

The quasi-energy ω_{ν} is given by

$$\omega_{\nu} = 2\nu\pi\alpha \tag{37}$$

Hence the distribution of quasi-energy spacings is flat. However the dressed quasi-energy states of (34) are lumpy, in fact localized since the asymptotic expansion of a Bessel function for large order is

$$J_m(z) \sim \frac{1}{\sqrt{2\pi m}} \left(\frac{ez}{2m}\right)^m$$

From the arguments in the introduction since the prefactor of eqn 5 couples only locally into the dressed state system we do not expect the diagonal density matrix elements to decay. If we consider the evolution of an initial density matrix $|m_o\rangle\langle m_o|$ we can show explicitly that

$$\rho_{m_o m_o}(t) = (J_o(W(\alpha,k,t)))^2 \tag{38}$$

and

$$(W(\alpha,k,t))^2 = \frac{k^2}{4\pi^2(1-\alpha^2)^2}\Bigg(2\alpha^2 + (1-\alpha^2)\sin^2 2\pi t + \frac{2\alpha}{k}(\alpha \cos 2\pi t \cos \tau t + \sin 2\pi t \sin t\tau)\Bigg) \tag{39}$$

Clearly there is no decay with time. The classical behaviour of the Hamiltonian of eqn (26) with linear f is

$$\begin{aligned} \dot{p} &= -\left(\frac{d}{d\theta} V(\theta)\right) \Delta(t) \\ \dot{\theta} &= 2\pi\alpha \end{aligned} \tag{40}$$

where p and θ are conjugate momentum and co-ordinate. The solution is

$$\begin{aligned} \theta(t) &= 2\pi\alpha t + \theta(0) \\ p(t) &= \frac{-k}{4\pi} \sum_m \Delta_m \left(\frac{e^{i(2\pi(\alpha+m)t+\theta(0))}}{\alpha + m} - \frac{e^{-i(2\pi(\alpha-m)t+\theta(0))}}{\alpha - m}\right) + \text{constant} \end{aligned} \tag{41}$$

when $V(\theta) = k \cos \theta$.

For (28a) it is clear from (41) that the motion is quasi-periodic and involves a small finite number of independent frequencies. However if $\Delta_m = 1$ for all m the number of independent frequencies in the classical motion is infinite and the motion is aperiodic. (28b) is just such a case. We now find

$$\begin{aligned} u^\nu(\theta,t) &= \frac{1}{(2\pi)^{\frac{1}{2}}} e^{i\nu\theta} \exp\left(\frac{-iW \sin(\theta + \alpha\pi(2n+1-2t))}{\sin \alpha\pi}\right) \qquad \text{for } n < t < n+1 \\ &= \frac{1}{(2\pi)^{\frac{1}{2}}} e^{i\nu\theta} \exp\left(\frac{-iW}{\sin \alpha\pi} \sin \theta \cos \alpha\pi(2n+1-2t)\right) \qquad \text{for } t = n \end{aligned} \tag{42}$$

Moreover

$$\begin{aligned} u_m^\nu &= (2\pi)^{\frac{1}{2}} e^{-i(m+\nu)G(t)} J_{m+\nu}\left(\frac{W}{\sin \alpha\pi}\right) \qquad \text{for } n < t < n+1 \\ &= (2\pi)^{\frac{1}{2}} J_{m+\nu}(W \cot \alpha\pi) \qquad \text{for } t = n \end{aligned} \tag{43}$$

From our general argument we again do not expect a decay in time of $\rho_{m_o m_o}(t)$. Explicitly we find

$$\rho_{m_o m_o}(t) = (J_o(W \cot \alpha\pi - \tilde{W}(t)))^2 \qquad (44)$$

with

$$\begin{aligned} W(t) &= W \cot \alpha\pi && \text{for } t = n \\ &= \frac{W}{\sin \alpha\pi} && \text{for } n < t < n+1 \end{aligned} \qquad (45)$$

which does not show a decay.

For α rational (= p/q with p, q of co-prime integers) Prange et al[10] have shown that

$$u_{\nu,\theta_o}(\theta,t) = e^{i\nu'\theta} e^{-i\phi'(\theta,t)} e^{-if'(t)} \sum_{k=0}^{q-1} \sum_{n} \delta(\theta-\theta_k(t)+2\pi n) \qquad (46)$$

with $\theta_k(t) = 2\pi\alpha(k+t) + \frac{\theta_o}{q}$, and θ_o labels the eigenstate within one of the q bands. ϕ' has the form of eqn 31 except that in the sum over m multiples of q are not allowed. f' is defined by

$$f'(t) = \int^{t} V'(\theta_o,s)\, \Delta(s)\, ds - t \int_{o}^{1} V'(\theta_o,t)\, \Delta(t)\, dt \qquad (47)$$

where

$$V'(\theta_o,t) = \sum_m V_{qm}\, e^{im(\theta_o+2\pi q\alpha t)} \qquad (48)$$

For (29) with $\alpha = 1/3$ we have the quasi-energies $\omega_{\nu'}(\theta_o)$ satisfying

$$\omega_{\nu'}(\theta_o) = \frac{k}{2} \cos \theta_o + 2\pi\alpha\nu' \qquad (49)$$

From our general arguments owing to the lumpiness of the dressed energy states (i.e there is a band index) and the bunched nature of the spacing distribution of quasi-energies we would conclude that there is not a rapid decay. An explicit calculation indeed shows this to be the case.

4 . NUMERICAL EXPERIMENTS

Our general formula (5) for the diagonal density matrix element shows two factors which influence its time dependence and we may therefore

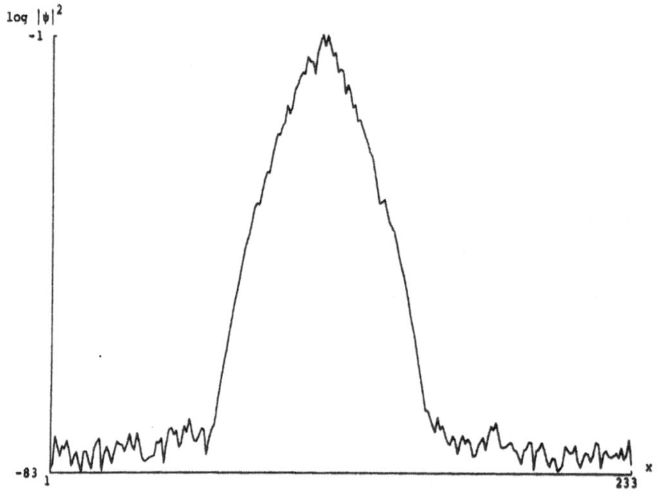

Figure 1a. Evolution of square of modulus of localised state wavefunction for the Harper model with W = 3

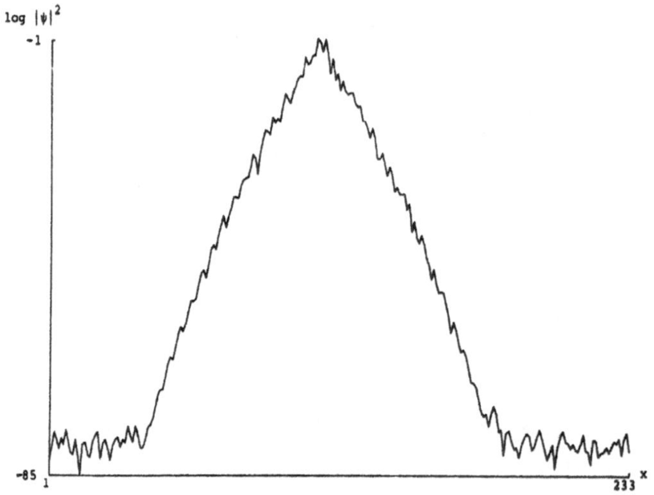

Figure 1b. Evolution of square of modulus of localised state wavefunction for the Harper model with W = 3

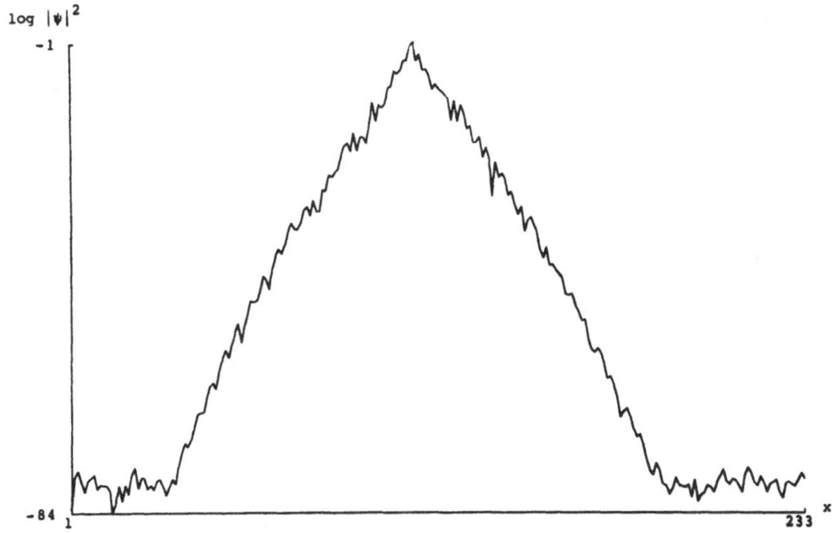

Figure 1c. Evolution of square of modulus of localised state wavefunction for the Harper model with W = 3

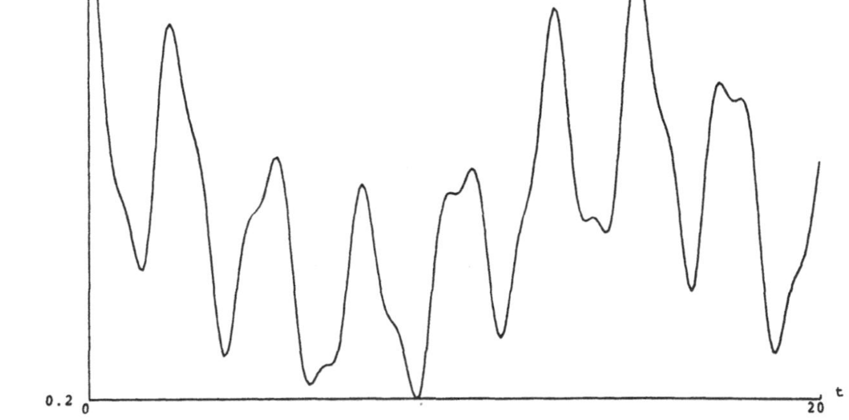

Figure 2. $\rho_{nn}(t)$ for situation of Figure 1

calculate $\rho_{nn}(t)$, in any particular case, directly from equation (1) or indirectly by calculating separately thse two factors, viz: the density g(E) of energy level differences of dressed pairs of states and the projections $\rho_{nn}^{(E)}(0)$ of the initial state on the set of these dressed pairs. The aim of the exercise in the latter case is to find a dominant reason for time behaviour of either decaying or quasi-periodic type. In the previous section where we chose potentials so that the expansion (31) derived by Grempel et al could be summed explicitly, analytic separation of these two terms was possible and the quasi-periodic behaviour seen explained in each case more by the effects of the coupling factor rather than by any property of the energy spacing distribution. Unfortunately, none of these soluble models shows decays. In this section we will be able to explore the effects of the two factors in further "regular" cases, which have both localised and extended solutions, and also in one case of decaying $\rho_{nn}(t)$, that of the kicked spin model in the classically chaotic region. We also show some dressed state wavefunctions and some "snapshots" of evolving wavefunctions in the incommensurate potential linear chain models and finally look at the kinetic equation for the coupled spin model.

4.1 Cases of Localised Dressed State Linear Chain Systems

We may group together a discussion of the "tan n^2", "tan n" and Harper models in the high potential region, since, in all these cases the solutions are localised. The time behaviour of an initial state localised on one point of a chain of 233 sites is exemplified by the sequence of Fig 1, which shows the evolution of the $(\text{modulus})^2$ of the wavefunction until it has essentially settled down under an exponentially decaying envelope. The figure is for the Harper case with W = 3 but the behaviour is similar in the other two cases.

The behaviour of $\rho_{nn}(t)$ for the same initial state is shown in Fig 2. As can be expected from the previous figure, the occupation shows periodic fluctuations without decay.

Since all states are "lumpy" the dominant factor in this time dependence is the coupling term which is limited to very few states. The form of g(E) has little effect on the dynamics but is shown in Fig 3, in this case for the "tan n" model but it is similar in all these cases. It shows a mildly "bunched" behaviour overall but with a definite minimum spacing (antibunching) at the origin.

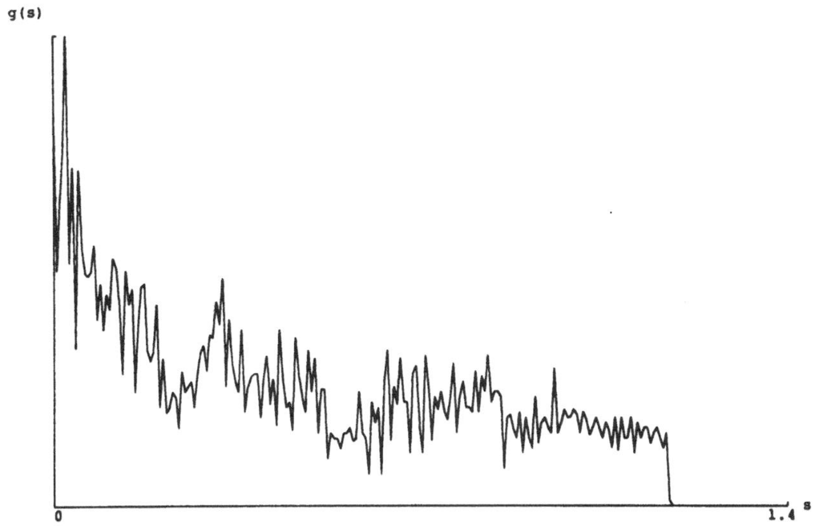

Figure 3. g(E) for the 'tan n' model

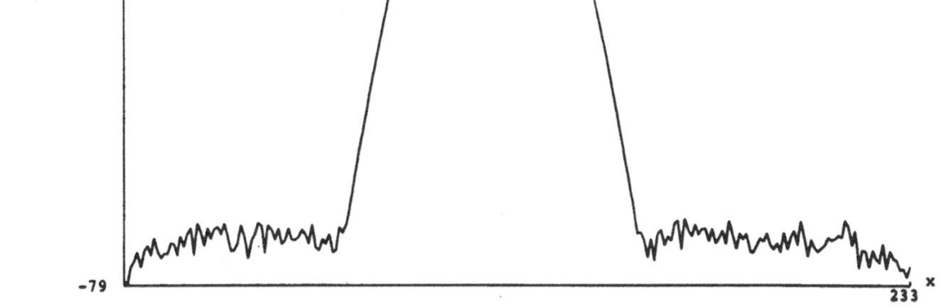

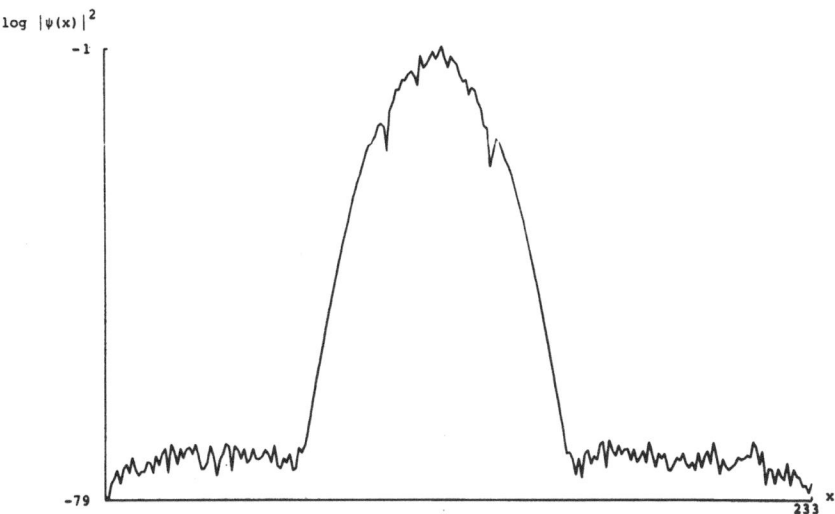

Figure 4a. Evolution of square of modulus of localised state wavefunction for Harper model at criticality (W=2)

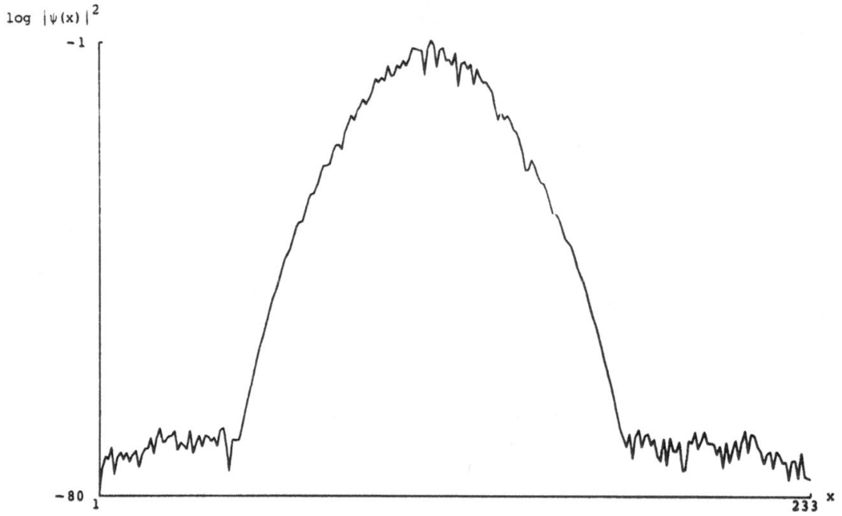

Figure 4b. Evolution of square of modulus of localised state wavefunction for Harper model at criticality (W=2)

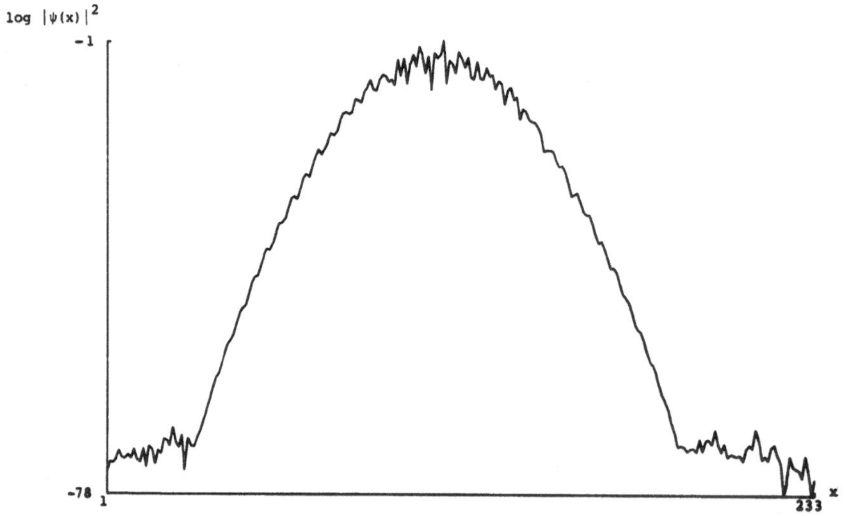

Figure 4c. Evolution of square of modulus of localised state wavefunction for Harper model at criticality (W=2)

4.2 Harper Model in Critical and Extended Regions

The case of the Harper model with W = 2 is the critical case between localised and extended solutions. The temporal behaviour of an initial δ-function state is shown in Fig 4, again on a chain of 233 sites, and a gradual spreading out to an envelope of a Gaussian like shape is observed.

In Figs 5 and 6 we show the same functions for W = 1, where the dressed states are extended. Here the initial state spreads out until it fills the chain uniformly and the occupation of the initial state conducts away until the inevitable recursion begins due to the discrete nature of the spectrum.

4.3 Kicked Spin Model

This model has already been extensively discussed by Haake et al and we used the same parameters. Although in § 2 we mapped this model also on to a one-dimensional chain, we have not utilised this mapping as yet for numerical work. The classical equations of motion implied by the Hamiltonian (7) are

$$y_{n+1} = x_n \cos(\hbar k y_n) + z_n \sin(\hbar k y_n)$$

$$x_{n+1} = -y_n$$

$$z_{n+1} = z_n \cos(\hbar k y_n) - x_n \sin(\hbar k y_n)$$

and these have regions of chaotic dynamics as well as regions of regular dynamics. Using the atomic coherent states (14) we may investigate the evolution of a state localised at (ϕ, θ) on the surface of a sphere starting from both regular and chaotic regions. Similarly, we break up the components of the effects into couplings of the initial state to the dressed state system and to distributions of energy level differences.

We have not made time evolution pictures of the wave functions on the sphere in this case; but we have calculated the time dependence of the diagonal density matrix elements of pure initial states corresponding to atomic coherent states in both regular and chaotic regions of the sphere. Two such examples are shown in Figs 7 and 8. Fig 7 is starting in a regular region (θ = 2.34, ϕ = -0.8) and shows strong oscillatory behaviour. In the chaotic region, on the other hand (θ = 2.8, ϕ = -0.8). Fig 8 shows a very fast decay indeed of the localised state followed by a low occupation for some time before the recurrence begins to appear. The coupling, $\rho_{nn}^{(E)}(0)$, in

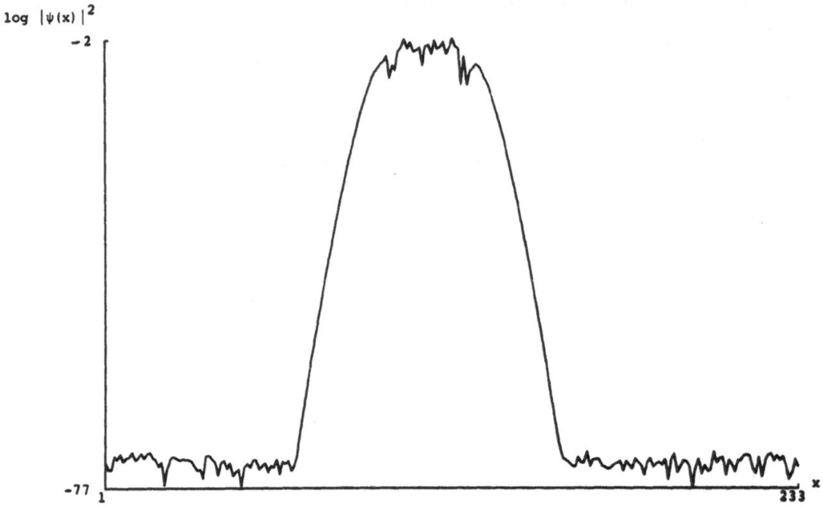

Figure 5a. Evolution of square of modulus of localised state wavefunction for Harper model for W = 1

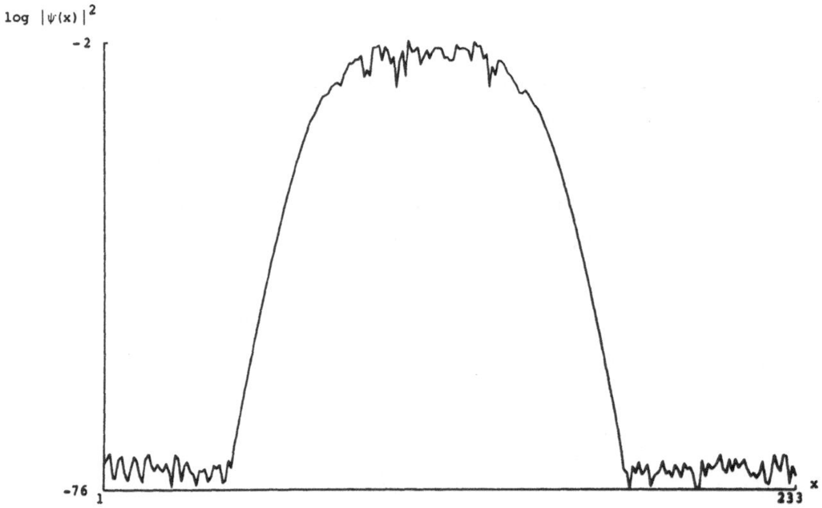

Figure 5b. Evolution of square of modulus of localised state wavefunction for Harper model for W = 1

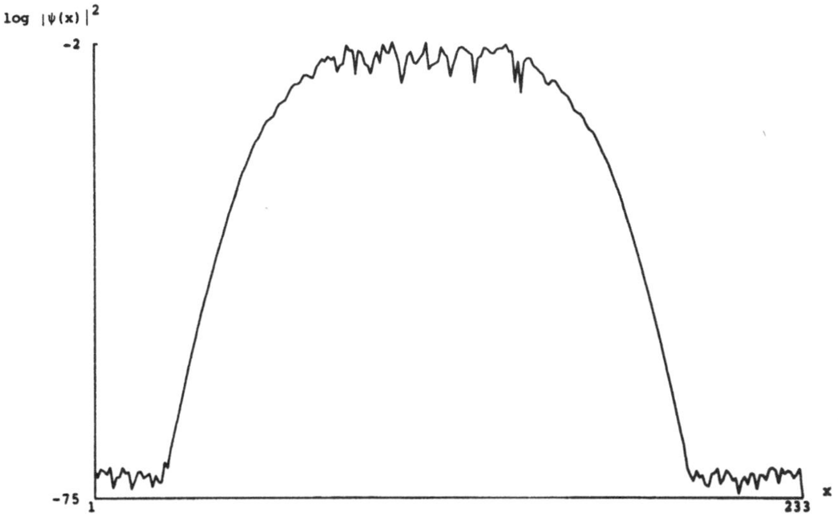

Figure 5c. Evolution of square of modulus of localised state wavefunction for Harper model for W = 1

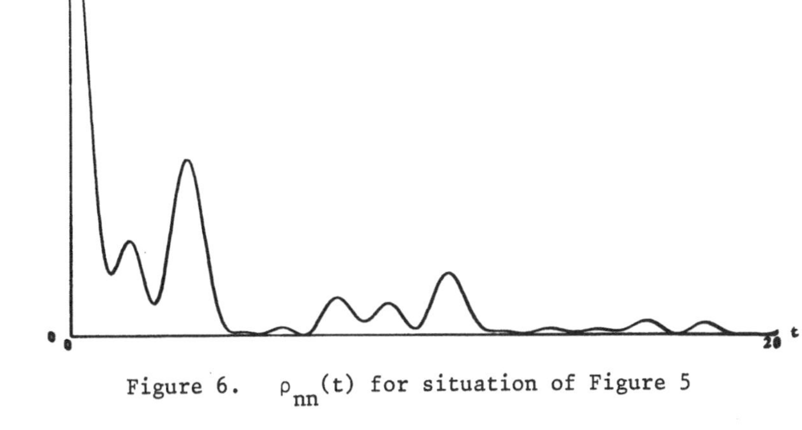

Figure 6. $\rho_{nn}(t)$ for situation of Figure 5

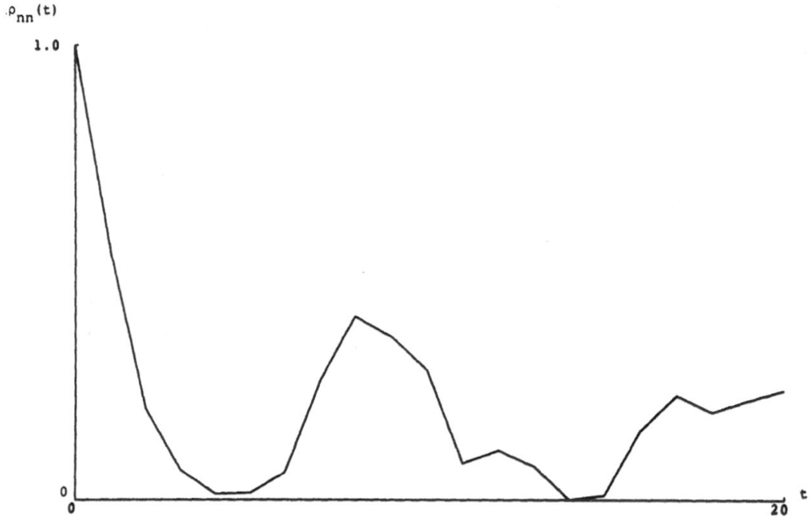

Figure 7. $\rho_{nn}(t)$ for kicked spin model with initial coherent state with θ = 2.34 and ϕ = - 0.8

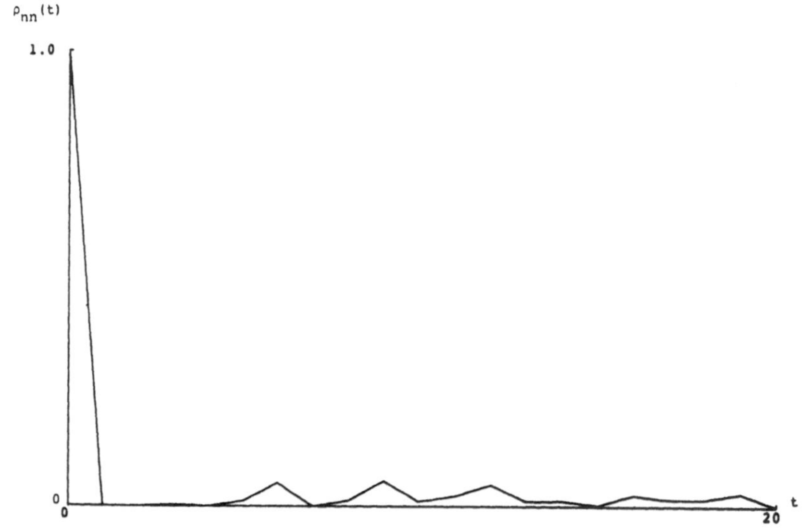

Figure 8. $\rho_{nn}(t)$ for kicked spin model with initial coherent state with θ = 2.8 and ϕ = - 0.8

the regular case is shown in Fig 9 and is characterised by large couplings to a small number of "lumpy" dressed states. In the chaotic region we have quite the opposite behaviour with a flat and uniform coupling to all extended states shown in Fig 10; only 10% of the total coupling is to lumpy states.

The level distribution g(E) of the "non-lumpy" dressed states is shown in Fig 11 and is essentially flat. The statistics of the lumpy state g(E) are too bad to make a good distribution but, as in the previous section, g(E) in this case is not very relevant to the motion. For completeness, again, we give the distribution of nearest neighbour levels in Fig 12.

The deduction to be drawn, finally, in the kicked spin case is that the fast decay in $\rho_{nn}(t)$ of a coherent state in the chaotic region is due to a flat spectrum in the energy level distribution with a uniform coupling as in the Brownian motion paradigm.

4.4 Coupled Spin Model

We made calculations for L = M = 10 with coupling constant 0.3 and first obtained similar results to Feingold and Peres for the decay of states in the chaotic region and for the quasi oscillatory behaviour of states in the regular region. The distribution of energy level differences showed, for the first time in this case, a slight maximum away from the origin, see Fig 13. The couplings in the chaotic region were rather uniform and the decay of the chaotic state can be ascribed to the form of g(E) with, perhaps, the peak giving rise to some oscillations in $\rho_{nn}(t)$.

Since there are no complications of quasi energies in this model we have looked also at the off-diagonal term $\phi_{nn}(t)$ in the master equation for $\rho_{nn}(t)$

$$\rho_{nn}(t) = \rho_{nn}(0) + \sum_m K_{nm}(t)[\rho_{nn}(0)-\rho_{mm}(0)] - \phi_{nn}(t)$$

which depends on initial off diagonal phases. The full form of this term is

$$\phi_{nn}(t) = \left[\sum_{n'\neq n} A_{nn'}(t) + cc\right] + \sum_{n'\neq m'} A_{nn'}(t)\, A^{*}_{nm'}(t)\, \rho_{n'm'}(0)$$

where

$$A_{nm}(t) = \sum_{M,m'} V_{nm'}\, C_M^{(m')} \left(C_M^{(m)}\right)^{*} \left(\frac{e^{\frac{i}{\hbar}(E_n-E_M)t} - 1}{E_n - E_M}\right)$$

and the same conventions are used for bare and dressed states as above.

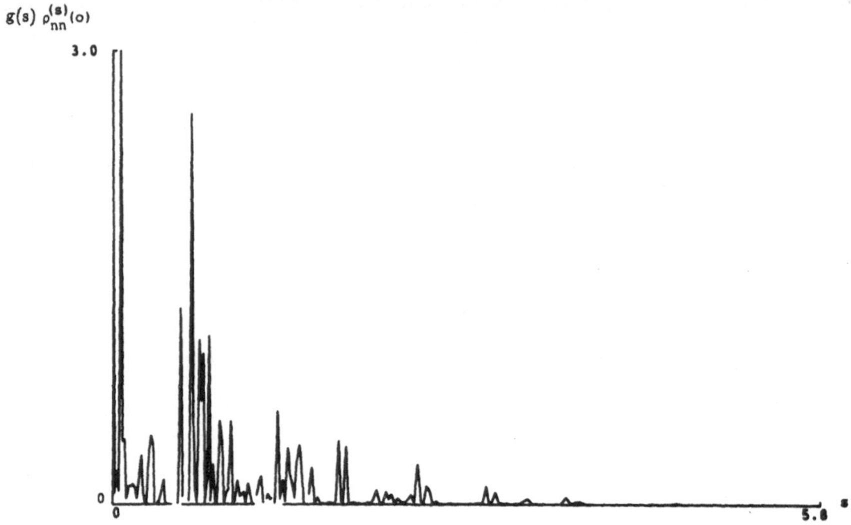

Figure 9. The coupling $\rho_{nn}^{(s)}(0)$ for the situation of Figure 7

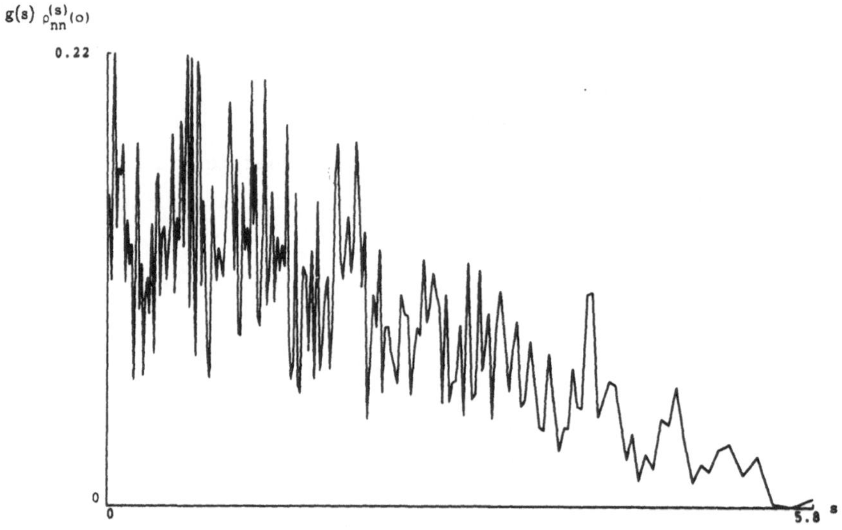

Figure 10. The coupling $\rho_{nn}^{(s)}(0)$ for the situation of Figure 8

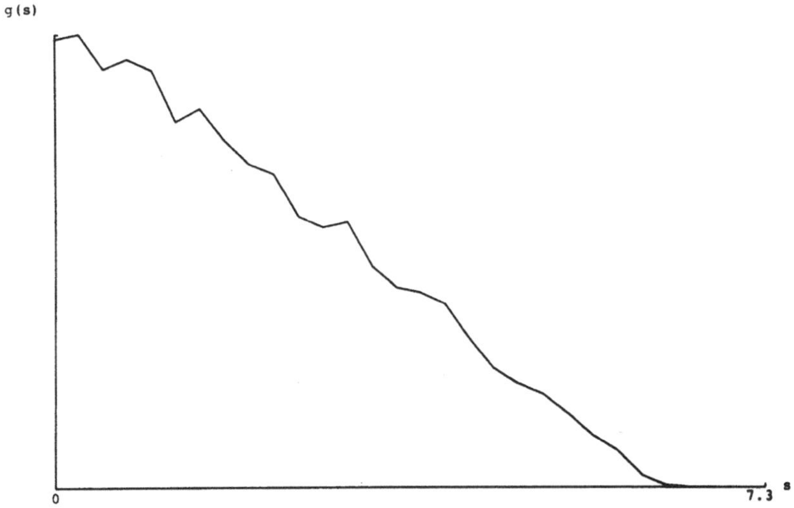

Figure 11. g(s) of 'non-lumpy' states for the kicked spin model

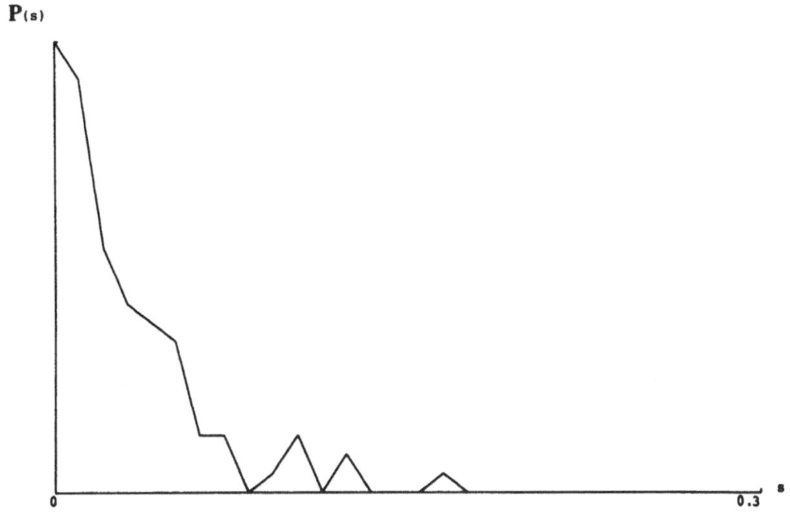

Figure 12. The distribution P(S) of nearest neighbour level spacing for the kicked spin model

It can be seen that the behaviour of this term depends on the differences in energy between the bare state $|n\rangle$ being considered and all the dressed energies E_M. For small times the term evolves linearly in t but in the large time limit the behaviour will be dominated by the smallest energy difference in the denominator.

It has been supposed that this off diagnonal component term will be unimportant for the time development of ρ_{nn} in the chaotic situation, leaving the diagonal terms to give rise to a Paulimaster equation. We have taken a simple initial state made up of a linear superposition of two bare states

$$|m\rangle = (|n\rangle + e^{i\phi} |n'\rangle)/\sqrt{2}$$

which just has one phase difference off the diagonal of the initial density matrix. Such a state was constructed for $|n\rangle, |n'\rangle$ both in the same regular region, for them in distant regular regions from each other and for them both in the same part of the chaotic region (see Figs 14, 15 and 16). As expected, in the first case $\phi_{nn}(t)$ grows from zero to significant values as t increases and cannot be neglected. In the second case, also as expected, $\phi_{nn}(t)$ has always very small values and plays a negligible role in the evolution of ρ_{nn}. The interesting case is the third and here, unfortunately, we have conflicting results. In some cases $\rho_{nn}(t)$ remains small but in others it does not, although in all cases a single diagonal pure state will decay in this region. Further work is required to understand these results.

Conclusions

We have considered a variety of models from those which are classically chaotic to chain models related to the quantum kicked rotator. Our aim was to see whether models which were classically chaotic could be differentiated in their dynamic behaviour from the other models. Although from Eqn 5 we have understood the general behaviour, the numerical results are not sufficiently detailed to draw very sharp distinctions in the rates of fall offs of the diagonal density matrix elements. We feel, however, that qualitatively our statements concerning g(S) and the extended nature of chaotic eigenstates has been vindicated by the numerical results. Moreover, we have numerically demonstrated that the initial hope that quantum chaos may lead naturally to a Pauli type master equation (without making a molecular chaos hypothesis) cannot be sustained, at least for the size of quantum numbers that we have considered.

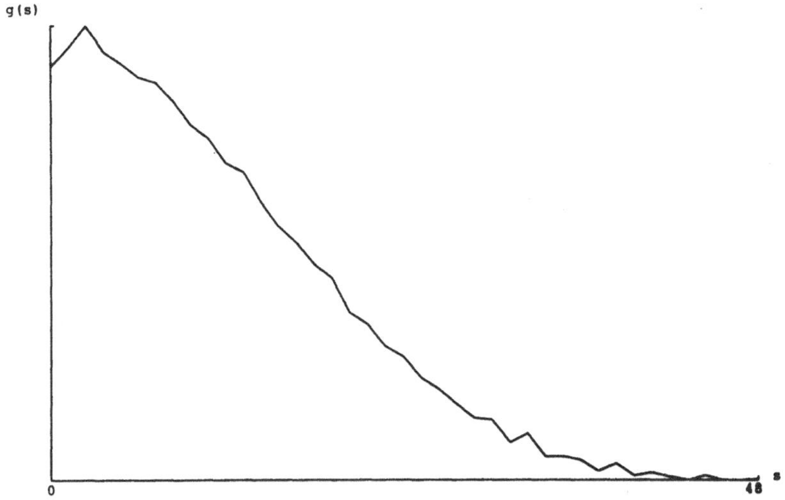

Figure 13. g(s) for coupled spin model

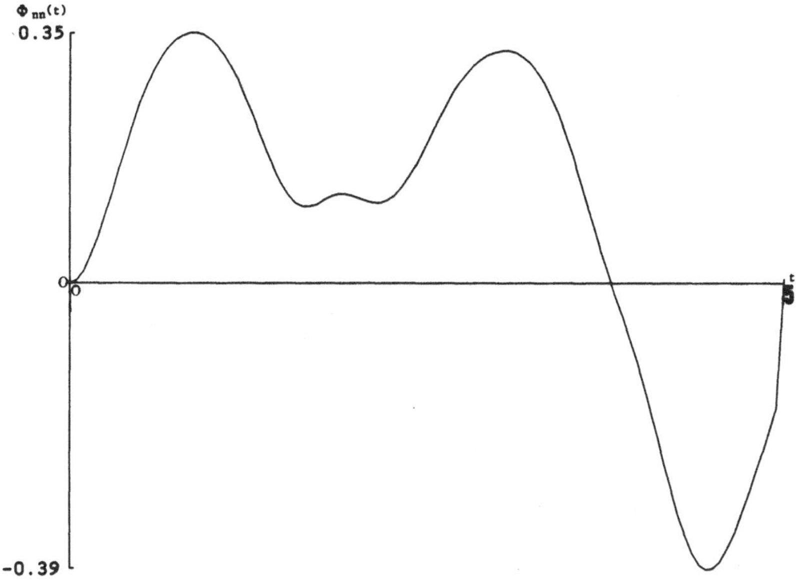

Figure 14. $\phi_{nn}(t)$ for coupled spin model with regular initial state located in one small region of the sphere

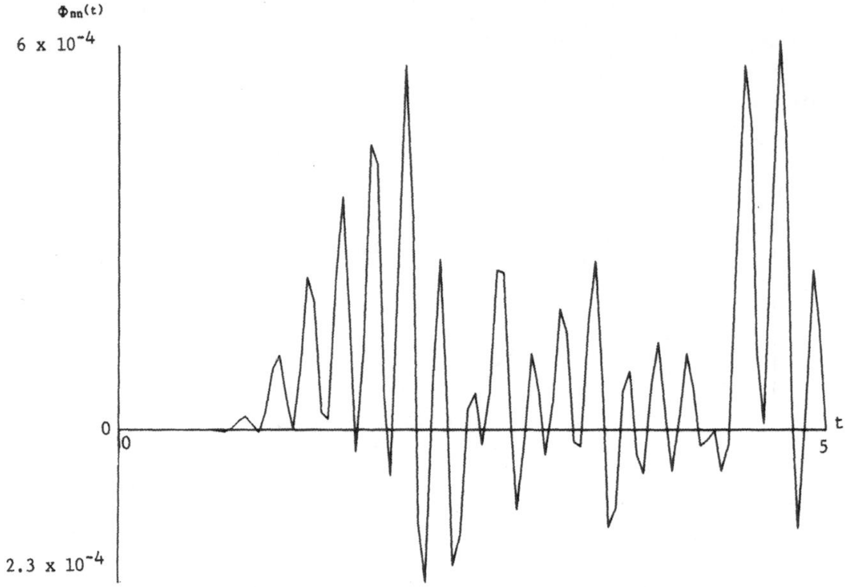

Figure 15. $\phi_{nn}(t)$ for coupled spin model with regular state not satisfying the criterion Figure 14

Figure 16. $\phi_{nn}(t)$ for coupled spin model with chaotic initial state

REFERENCES

1. P. Cvitanovic, Ed., Universality in Chaos (Adam Hilger, Bristol, 1984).
   A. J. Lichtenberg and M. A. Lieberman, Regular and Stochastic Motion (Springer-Verlag, New York, 1983).
   S. Sarkar, Ed., Nonlinear Phenomena and Chaos, (Adam Hilger, Bristol, 1986).

2. G. M. Zaslavsky, Phys. Repts. 80 157 (1981).

3. G. Casati, Ed., Chaotic Behaviour in Quantum Systems (Plenum, New York, 1985).

4. G. P. Berman and G. M. Zaslavsky, Physica A 97, 367 (1979).

5. E. R. Pike, Physica 31 461 (1965).
   R. Zwanzig, Physica 30 1109 (1964).
   I. Prigogine, Non-Equilibrium Statistical Mechanics (Interscience: New York 1962).
   R. J. Swenson, J. Math. Phys. 3 1017 (1962).
   L. Van Hove, Physica 23 441 (1957).
   G. V. Chester, Rep. Progr. Phys. (Physical Society, London) 26 411 (1963).

6. B. J. Berne and R. Pecora, Dynamic Light Scattering, (Wiley, New York, 1976).

7. A. Einstein, Ver. Deut. Phys. Ges. 19 82 (1917).
   I. C. Percival, J. Phys. B6 L229 (1973).
   M. V. Berry, J. Phys. A10 2083 (1977).
   E. J. Heller, J. Chem . Phys. 72 1337 (1980).

8. M. Feingold and A. Peres, Phys. Rev. A31 2472 (1985).

9. F. Haake, M. Kus, J. Mostowski and R. Scharf, in 'Coherence, Cooperation and Fluctuations', Eds. F. Haake, L. M. Narducci and D. F. Walls (Cambridge Univ. Press 1986).

10. R. E. Prange, D. R. Grempel and S. Fishman, Phys. Rev. B29 6500 (1984).

11. M. V. Berry, Physica 10D 369 (1984).

12. M. Ya Azbel, Sov. Phys. JETP 19 634 (1964).
I. M. Suslov, Sov. Phys. JETP 56 612 (1982).
M. Wilkinson, Proc. R. Soc. Lond. A 391 305 (1984).
A. D. Zdetsis, C. M. Soukoulis and E. N. Economou, Phys. Rev. B 33 4936 (1986).

13. S. Fishman, D. R. Grempel and R. E. Prange, Phys. Rev. B 29 4272 (1984).

QUANTUM THEORY OF CHAOTIC RABI OSCILLATIONS

R. Graham and M. Höhnerbach

Fachbereich Physik, Universität Essen GHS
D-4300 Essen, W. Germany

ABSTRACT

The Rabi oscillations of a single or a few two level atoms in a strong single-mode electromagnetic field are considered including the backreaction of the atoms on the field. In the rotating wave approximation the total energy of the system is the sum of two independently conserved quantities I_ψ and K. The problem has been exactly solved in this case by Jaynes and Cummings in 1963. Quantum mechanically, the two conservation laws in this case give rise to two different dynamical features, periodic 'revivals' of the Rabi oscillations, connected with the conservation and quantization of I_ψ and a splitting of the power spectrum of the mode into several distinct groups of discrete lines, connected with the conservation and quantization of K.

For sufficiently strong coupling and beyond the rotating wave approximation the separate conservation laws of I_ψ and K are destroyed and the oscillations become classically chaotic. Quantum mechanically, the revivals of the Rabi oscillations disappear and the power spectrum of the mode contains only one broad group of discrete lines. Both features may serve as indicators of ergodic dynamical behavior in the system. The present report is based on and extends earlier work by the authors on the same system.

1. INTRODUCTION

A two-level system described by spin-1/2 variables in interaction with a boson variable serves as a fundamental model in many branches of physics. E.g. in problems of quantum tunnelling between two local potential minima separated by a barrier, the positions $x=\pm 1$ of the two minima may be characterized by a spin-1/2 operator σ_x, the two tunnel-splitted lowest lying energy levels may be characterized by the Hamiltonian $H_0=\hbar\omega_0\sigma_z/2$, and the effect of a time-dependent external biassing field $E(t)$ may be incorporated into the Hamiltonian via

$$H = \frac{\hbar\omega_0}{2}\sigma_z + \hbar g E(t)\sigma_x \qquad (1.1)$$

Here σ_x , σ_z are Pauli matrices, ω_0 is the tunnelling frequency in the absence of the external field, g is a coupling constant and $E(t)$ is the external field in appropriate units. The model (1.1) occurs in this fashion e.g. in Holstein's model of a small polaron, where $E(t)$ describes a phonon field [1]. It occurs also as a basic model in the theory of spin resonance and in quantum optics. If $E(t)$ is monochromatic at frequency $\omega \approx \omega_0$ and the coupling g is sufficiently weak one may apply the rotating wave approximation and replace in eq. (1.1)

$$gE(t)\sigma_x \approx g\,E_0\,(\sigma_+ e^{-i\omega t} + \sigma_- e^{i\omega t}) \tag{1.2}$$

where $\sigma_x = \sigma_+ + \sigma_-$ and

$$\sigma_+ = \begin{pmatrix} 0 & 1 \\ 0 & 0 \end{pmatrix} \quad , \quad \sigma_- = \begin{pmatrix} 0 & 0 \\ 1 & 0 \end{pmatrix} \tag{1.3}$$

The Hamiltonian (1.1) in this approximation was first solved by Rabi [2] and leads to the familar Rabi oscillation of the expectation value of the population difference $\langle \psi(t) | \sigma_z | \psi(t) \rangle$ of the two levels with the Rabi frequency

$$\Omega = \sqrt{(\omega - \omega_0)^2 + 4g^2 E_0^2} \tag{1.4}$$

The corrections due to the neglected counterrotating terms in the Hamiltonian

$$g E_0\,(\sigma_+ e^{i\omega t} + \sigma_- e^{-i\omega t}) \tag{1.5}$$

have been first considered in an analytical approximation by Bloch and Siegert [3] and in a complete analytical and numerical analysis by Autler and Townes [4] . In this case the Hamiltonian (1.1) is still periodic and the Floquet theory applies [5] which implies that the expectation values $\langle \psi(t) | \sigma_{x,z} | \psi(t) \rangle$ must be quasi-periodic functions of time with frequencies $n\omega \pm \tilde{\Omega}$ where $\tilde{\Omega}$ must be determined numerically. The Hamiltonian (1.1) with quasi-periodic external driving field $E(t)$ has recently also been studied [6].

In order to study the back-reaction of the two level system on the field $E(t)$, the Hamiltonian of the latter must be added to (1.1). Assuming again a monochromatic field at frequency ω and quantizing it in terms of the boson raising and lowering operators b^+ and b with

$$[b, b^+] = 1 \tag{1.6}$$

we obtain

$$H = \frac{\hbar\omega_0}{2}\sigma_z + \hbar\omega b^+ b + \hbar g\,(b + b^+)\,\sigma_x \tag{1.7}$$

In the case where several (N) two-level systems (spins) interact with the same mode, eq. (1.7) is generalized to

$$H = \hbar\omega_0 S_z + \hbar\omega b^+ b + 2\frac{\hbar g}{\sqrt{N}}(b + b^+) S_x \tag{1.8}$$

where

$$S_z = \frac{1}{2}\sum_{\mu=1}^{N} \sigma_{z\mu} \quad , \quad S_x = \frac{1}{2}\sum_{\mu=1}^{N} \sigma_{x\mu} \tag{1.9}$$

and $\sigma_{x\mu}$, $\sigma_{z\mu}$ are the Pauli matrices for the individual two-level systems labelled by μ . The scaling factor $N^{-1/2}$ of g is introduced in order to render the classical equations of motion scale invariant with respect to N (cf. below). We note that

$$S^2 = S_x^2 + S_y^2 + S_z^2$$

is conserved and we shall consider the case where S^2 takes its maximal value

$$S^2 = \frac{N}{2}\left(\frac{N}{2}+1\right) \tag{1.10}$$

The Hamiltonian (1.8) furnishes a very interesting example for a quantum system exhibiting chaos in its classical counterpart and will be studied here from this point of view. Chaos in the classical version of the model was studied by a number of authors [7-10]. A classical version of the Hamiltonian (1.8) is obtained by using $S_z \rightarrow I_1$ as a canonical momentum and introducing the canonically conjugate azimuthal angle φ_1 via

$$\begin{aligned} S_x &\rightarrow \sqrt{S^2 - I_1^2}\,\cos\varphi_1 \\ S_y &\rightarrow \sqrt{S^2 - I_1^2}\,\sin\varphi_1 \end{aligned} \tag{1.11}$$

Similarly, we introduce $b^+b \rightarrow I_2$ as a classical action variable and φ_2, its canonically conjugate angle variable, via

$$\begin{aligned} b &\rightarrow \sqrt{I_2}\, e^{-i\varphi_2} \\ b^+ &\rightarrow \sqrt{I_2}\, e^{i\varphi_2} \end{aligned} \tag{1.12}$$

Thus, the classical Hamiltonian reads

$$\begin{aligned} H = \hbar\omega_0 I_1 + \hbar\omega I_2 &+ \frac{2\hbar g}{\sqrt{N}}\sqrt{I_2}\cdot\sqrt{S^2 - I_1^2}\,\cos(\varphi_1 - \varphi_2) \\ &+ \frac{2\hbar g}{\sqrt{N}}\sqrt{I_2}\cdot\sqrt{S^2 - I_1^2}\,\cos(\varphi_1 + \varphi_2) \end{aligned} \tag{1.13}$$

The dependence on N and $\hbar$ can be removed by the rescaling [7]

$$\begin{aligned} \frac{H}{\hbar N} &\rightarrow H \\ \frac{I_1}{N} &\rightarrow I_1 \\ \frac{I_2}{N} &\rightarrow I_2 \\ \frac{S^2}{N^2} &\rightarrow \frac{1}{4} \end{aligned} \tag{1.14}$$

which leaves the equations of motion invariant. The latter read

$$\dot{I}_1 = -\frac{\partial H}{\partial \varphi_1} \quad , \quad \dot{I}_2 = -\frac{\partial H}{\partial \varphi_2}$$
$$\dot{\varphi}_1 = \frac{\partial H}{\partial I_1} \quad , \quad \dot{\varphi}_2 = \frac{\partial H}{\partial I_2} \tag{1.15}$$

Thus, in the classical Hamiltonian (1.13), one may put $N=1$ without restriction of generality.

The classical limit leading to (1.13) is only reached in a unique way by taking $N \gg 1$ and $I_2 \gg 1$, and therefore does not apply to eq. (1.7). The non-uniqueness of the classical limit for the Hamiltonian (1.7) with $N=1$ has been discussed in ref. [13]. Despite of this non-uniqueness it is interesting to compare the results obtained for the quantum Hamiltonian (1.7) with the results for the classical Hamiltonian (1.13) in order to see the consequences of the presence of the extremely non-classical spin-1/2 variable in the Hamiltonian (1.7). Surprisingly, it turns out that prominent quantum effects in the dynamics described by (1.7) and their change under increasing coupling strength can be directly related to regular features of the classical dynamics (1.15) for weak coupling and their change to chaotic behavior for stronger coupling. In the following section the classical Hamiltonian (1.13) is studied in the case $I_2 \gg 1$, and the borderline in parameter space between regular behavior and large scale chaos for this case is estimated. Then, the results of a corresponding analysis for the quantum Hamiltonians (1.7) and (1.8) are presented with the aim to correlate the quantum mechanical results with regular or chaotic behavior in the classical system. In this discussion we extend our earlier results of refs. [11-15].

2. CLASSICAL ANALYSIS

The classical Hamiltonian (1.13) has been partially studied in refs. [16], [7]. However, [16] was only concerned with the integrable Hamiltonian obtained after the rotating wave approximation has been made, while in [7] the nonintegrable case has been studied, but only under the assumption that I_2 is small. Here we are interested in the case where $I_2 \gg I_1$. The Hamiltonian (1.13) may be written as

$$H = H_0 + H' \tag{2.1}$$

with

$$H_0 = \omega_0 I_1 + \omega I_2 + g\sqrt{I_2}\sqrt{1-4I_1^2}\cos(\varphi_1-\varphi_2)$$
$$H' = g\sqrt{I_2}\sqrt{1-4I_1^2}\cos(\varphi_1+\varphi_2) \tag{2.2}$$

Here we eliminated N applying (1.14).

In the rotating wave approximation H' is neglected. Then it is useful to introduce

$$\varphi = \varphi_1 - \varphi_2 \tag{2.3}$$

instead of φ_1, whereupon

$$\psi = \varphi_2 \tag{2.4}$$

becomes a cyclic variable. The new canonically conjugate action variables are

$$\begin{aligned} I_\varphi &= I_1 \\ I_\psi &= I_1 + I_2 \geq I_1 \end{aligned} \tag{2.5}$$

and we find

$$\begin{aligned} H_0 &= (\omega_0 - \omega) I_\varphi + \omega I_\psi + g\sqrt{I_\psi - I_\varphi}\sqrt{1-4I_\varphi^2}\cos\varphi \\ H' &= g\sqrt{I_\psi - I_\varphi}\sqrt{1-4I_\varphi^2}\cos(\varphi + 2\psi) \end{aligned} \tag{2.6}$$

For simplicity, in order to avoid cumbersome expressions involving elliptic functions, we shall limit our attention to the case $I_2 \gg |I_1|$ from which $I_\psi \gg |I_\varphi|$ follows. For the same reason of simplicity we restrict our attention to the case $\omega = \omega_0$. Then we have

$$\begin{aligned} H_0 &= \omega I_\psi + g\sqrt{I_\psi}\sqrt{1-4I_\varphi^2}\cos\varphi \\ H' &= g\sqrt{I_\psi}\sqrt{1-4I_\varphi^2}\cos(\varphi + 2\psi) \end{aligned} \tag{2.7}$$

Within the rotating wave approximation H_0, I_ψ and therefore also

$$K = g\sqrt{I_\psi}\sqrt{1-4I_\varphi^2}\cos\varphi \tag{2.8}$$

are conserved. The equations of motion are most easily solved by eliminating $\cos\varphi$ in favour of $\dot{I}_\varphi$ from the expression for H_0 using the equation of motion $\dot{I}_\varphi = -\partial H_0/\partial\varphi$. The resulting first order differential equation for I_φ is easily solved by quadrature. One finds

$$I_\varphi = \frac{1}{2}\sqrt{1 - \frac{4K^2}{\Omega^2}}\cos\Omega(t-t_0) \tag{2.9}$$

with the Rabi frequency

$$\Omega = 2g\sqrt{I_\psi} \tag{2.10}$$

From eq. (2.8) one obtains

$$\cos\varphi = \frac{2K}{\sqrt{\Omega^2\sin^2\Omega(t-t_0) + 4K^2\cos^2\Omega(t-t_0)}} \tag{2.11}$$

and

$$\tan\varphi = \sqrt{\frac{\Omega^2}{4K^2} - 1}\,\sin\Omega(t-t_0) \tag{2.12}$$

Finally we note that within the rotating wave approximation the phase changes uniformly in time according to

$$\dot{\psi} = \omega + K/(2I_\psi) \tag{2.13}$$

The conserved quantity K , according to eq. (2.8), satisfies

$$-\Omega/2 \leq K \leq \Omega/2 \tag{2.14}$$

The sign of K , according to eq. (2.11), separates the phase space into two different regions

$$\begin{aligned} K \geq 0 &: \quad -\frac{\pi}{2} \leq \varphi \leq \frac{\pi}{2} \\ K \leq 0 &: \quad \frac{\pi}{2} \leq \varphi \leq \frac{3\pi}{2} = -\frac{\pi}{2} \ (\mathrm{mod}\ 2\pi) \end{aligned} \tag{2.15}$$

which cannot communicate with each other within the rotating wave approximation. This makes explicit the non-ergodicity of the system in the rotating wave approximation. The value $\varphi = \frac{\pi}{2}, \frac{3\pi}{2}$ at the common borders of both parts of phase space is only reached asymptotically if $K \to 0$. In this case I_φ oscillates harmonically between $\pm$ 1/2 while $\cos\varphi = 0$, i.e. $\varphi = \frac{\pi}{2}$ or $\frac{3\pi}{2}$, except for small time intervals $\Delta t \propto 4|K|/\Omega^2$ around the times $t_n = t_0 + n\pi/\Omega$ where $\cos\varphi$ differs from zero and jumps to +1 or -1 at $t = t_n$ depending on the sign of K , and then jumps back to zero.

Going beyond the rotating wave approximation neither K nor I_φ are conserved. It is clear from the results of the rotating wave approximation that for sufficiently small $|K|$ already tiny changes of K can have drastic effects, because the signature of K at the times $t = t_n$ controls the signature of the jumps of $\cos\varphi$ to ± 1 at those times. Therefore, one has to expect, even if the rotating wave approximation is generally satisfied,that a stochastic layer exists in a small neighbourhood of $K = 0$ in phase space. The width of the stochastic layer is controlled by the size of $|K|$ for which a change of sign of K during a Rabi period is possible due to the influence of the counterrotating terms described by H' . To obtain at least an estimate of the width of this layer we consider the change of K in time by writing

$$\begin{aligned} \dot{K} = \frac{1}{2} g \frac{\sqrt{1-4I_\varphi^2}}{\sqrt{I_\psi}} \cos\varphi \, \dot{I}_\psi - 4g \frac{\sqrt{I_\psi} \cdot I_\varphi}{\sqrt{1-4I_\varphi^2}} \cos\varphi \, \dot{I}_\varphi \\ - g \sqrt{I_\psi} \sqrt{1-4I_\varphi^2} \sin\varphi \cdot \dot{\varphi} \end{aligned} \tag{2.16}$$

and use the equations of motion to express the time derivatives on the right hand side by partial derivatives of H . Of course only the derivatives of H' can contribute as K is conserved under H_0 . We obtain after some algebra

$$\dot{K} = -4g^2 I_\varphi I_\psi \sin 2\psi + g^2(1-4I_\varphi^2)\cos\varphi \sin(\varphi + 2\psi) \tag{2.17}$$

We note that the second term on the right hand side is smaller in order of magnitude than the first term by a factor of I_ψ^{-1} and therefore negligible. Using on the right hand side the results of the rotating wave approximation we obtain at least an estimate of the order of magnitude of the change of K with time. We find

$$\dot{K} \simeq -\Omega\sqrt{\frac{\Omega^2}{4}-K^2}\cos\Omega(t-t_0)\sin\left[2\omega(t-t_1)+\int_{t_1}^{t}\frac{K}{2I_\psi}d\tau\right] \tag{2.18}$$

By integration of eq. (2.18) neglecting $K/2I_\psi$ compared to 2ω, we obtain

$$K(t) \simeq \frac{\Omega}{2}\sin\left\{\frac{1}{2}\frac{\Omega}{\Omega+2\omega}\cos\left[(\Omega+2\omega)t+\int_0^t\frac{K}{2I_\psi}d\tau-\alpha\right]\right. \tag{2.19}$$

$$\left.-\frac{1}{2}\frac{\Omega}{\Omega-2\omega}\cos\left[(\Omega-2\omega)t-\int_0^t\frac{K}{2I_\psi}d\tau-\beta\right]+\gamma\right\}$$

where α,β,γ are constants of integration into which the constants t_0, t_1 have been absorbed. As the arguments of the cosine have to be taken mod 2π, the integral over $K/2I_\psi$ is not negligible there. It is clearly seen that the signature of K can change within a layer

$$|K| < \Delta K$$

of width

$$\Delta K = \frac{\Omega}{2}\sin\left[\frac{2\omega\Omega}{4\omega^2-\Omega^2}\right] \tag{2.20}$$

Thus, for $\Omega \ll \omega$ most K values in the domain (2.14) are outside this layer and chaos is confined to very small K values near zero. However, if Ω becomes comparable to ω chaos invades the whole phase space. According to eq. (2.20) this happens for

$$\mu \equiv \frac{2\omega\Omega}{4\omega^2-\Omega^2} \gtrsim \frac{\pi}{2} \tag{2.21}$$

The condition (2.21) should only be taken as an estimate of order of magnitudes in view of the simple approximations we have made in deriving it. In the chaotic domain of K values a two-valued symbolic dynamics [17] can be introduced very naturally by using as symbols the signatures of the jumps of $\cos\psi$ during the Rabi cycles, or, equivalently, the signatures of K at the times t_n where $|\cos\psi|$ takes its extrema. This means that we sample the signature of K roughly at time intervals $\Delta t = \pi/\Omega$. Neglecting the influence of $K/2I_\psi$ on the right hand side of (2.19) we obtain complicated but still quasi-periodic sequence of symbols in this way, due to a quasi-periodic beating of the sampling frequency $\Omega/2$ with the frequencies $2\omega\pm\Omega$. However, the quasi-periodicity of the sequence is destroyed if the influence of $K/2I_\psi$ on the right hand side of (2.19) is taken into account, because it causes a scatter of the phases of the cosines which makes the signatures of the latter and hence the signature of K unpredictable in time. Which variable shows most clearly the chaotic dynamics generated in this fashion? We note from eq. (2.9) that the Rabi oscillations of I_ψ are quite insensitive to this mechanism, at least as long as ΔK remains small. A similar conclusion is reached for the variable S_x. However, according to eq. (2.13), a signature of the chaotic dynamics of K should be visible in the power spectrum of the field which contains the frequency $\dot{\psi}$ given by (2.13). As K changes chaotically whithin the layer ΔK, the power spectrum of the field acquires a broadened line with center at ω and width $\sim \Delta K/2 \cdot I_\psi$.

The question remains whether any signatures of the classical chaos also show up in the dynamics of the fully quantized system. Quantum mechanically the two observables I_ψ and K are both quantized, namely, for a single two-level atom, $I_\psi = n - \frac{1}{2}$, $K = \pm g\sqrt{n}$, where n = 0, 1, ... integer. For N two level atoms, taking into account the rescaling of eq. (1.14), $I_\psi \simeq n/N$, and $K \simeq g\sqrt{n/N}, \dots, -g\sqrt{n/N}$ in N steps of equal size. In principle, therefore, two kinds of quantum effects are possible which are associated with the quantization either of I_ψ or of K. The quantization of I_ψ shows up directly in the Rabi oscillations of I_ψ (2.9), (2.10), where the Rabi frequency receives a fine structure of order $\Delta\Omega = g/\sqrt{I_\psi}$. This quantum effect manifests itself in beating phenomena on the time scale $T_R = 2\pi/\Delta\Omega$ which are well known under the name 'revivals' [18] and have recently been seen also experimentally [19]. The quantization of K, on the other hand, shows up directly in the frequency $\dot{\psi}$ given by (2.13). It is therefore seen in the power spectrum of the field which quantum mechanically has distinct maxima at the frequencies $\omega \pm g/\sqrt{n} \simeq \omega \pm \Delta\Omega$ in the case of a single two level atom. Clearly, the latter structure in the power spectrum can only exist as long as K is reasonably well conserved, which should be true as long as the quantum mechanically allowed values of K are outside the chaotic layer (2.20). Hence, we again obtain the condition (2.21) for mixing of positive and negative K-values which should become manifest by the disappearance of distinct maxima in the quantal power spectrum. Similarly, the quantum effect of the revivals can only exist as long as I_ψ is conserved with an accuracy better than $\pm$ 1. The fluctuations of I_ψ due to deviations from the rotating wave approximation can easily be estimated as in eqs. (2.16) - (2.19), with the conclusion that $|\delta I_\psi| \gtrsim 1$ and the revivals should disappear again if (2.21) is satisfied in order of magnitude. Hence, the revivals and the two-peak structure of the power spectrum are good indicators of regular non-ergodic behavior in the quantum system, and, vice versa, their disappearance serves as indicator for the transition from regular to irregular ergodic behavior in the quantum system.

3. QUANTUM ANALYSIS

We briefly recall the solution in the rotating wave approximation, first given by Jaynes and Cummings [16]. The Hamiltonian

$$H_0 = \frac{\hbar\omega}{2}(\sigma_z + 2b^+b) + \hbar g(b\sigma_+ + b^+\sigma_-) \tag{3.1}$$

is solved by the ansatz

$$|\psi(t)\rangle = \sum_{n=0}^{\infty} \Big(C_+(n,t)\, e^{-i(n+\frac{1}{2})\omega t} |n,\uparrow\rangle + C_-(n,t)\, e^{-i(n-\frac{1}{2})\omega t} |n,\downarrow\rangle \Big) \tag{3.2}$$

where $|n,\uparrow\rangle, |n,\downarrow\rangle$ are the simultaneous eigenstates of b^+b, σ_z with eigenvalues $n, 1$ and $n, -1$, respectively. The amplitudes $C_\pm(n,t)$ satisfy

$$i\hbar\dot{c}_+(n) = \hbar g\sqrt{n+1}\, c_-(n+1)$$
$$i\hbar\dot{c}_-(n+1) = \hbar g\sqrt{n}\, c_+(n) \tag{3.3}$$

from which they are easily determined. We now restrict attention to initial states with $\sigma_z = 1$. Assuming also an initially fixed quantum number $n=m$ we obtain

$$|\psi(t)\rangle_m = e^{-i(m+\frac{1}{2})\omega t} \cos\sqrt{m+1}\,gt\ |m,\uparrow\rangle$$
$$-i\,e^{i(m+\frac{1}{2})\omega t} \sin\sqrt{m+1}\,gt\ |m+1,\downarrow\rangle \tag{3.4}$$

For the Rabi flopping of $\langle\psi|\sigma_z|\psi\rangle$ we find

$$_m\langle\psi|\sigma_z|\psi\rangle_m = \cos 2g\sqrt{m+1}\,t \tag{3.5}$$

Choosing as initial state a coherent state $|\alpha,\uparrow\rangle$ eq. (3.5) is modified to

$$_\alpha\langle\psi|\sigma_z|\psi\rangle_\alpha = \sum_{m=0}^{\infty} \frac{|\alpha|^{2m} e^{-|\alpha|^2}}{m!} \cos 2g\sqrt{m+1}\,t \tag{3.6}$$

The evaluation of the sum leads to the well known collapse of the Rabi oscillations [20] on a characteristic time scale $\tau_c = g^{-1}$ given by the inverse width of the distribution of Rabi frequencies introduced by the initial state, and it also leads to the approximately periodic revivals with a frequency given by the average spacing of nearest neighboured Rabi frequencies [18] $\langle\omega_+\rangle \simeq g/|\alpha|$. The spacings of nearest neighboured Rabi frequencies $\omega_+(m) \simeq g/\sqrt{m+1}$ are again distributed in the initial state with a width $\Delta\omega_+ \simeq g/2|\alpha|^2$ which defines a decay time for the revivals $\tau_{rc} = 2|\alpha|^2/g = 2|\alpha|^2\tau_c$. On the other hand, due to the discreteness of the revival frequencies $\omega_+(m)$ one may again expect beating phenomena on time scales $\omega_{++} = \omega_+(|\alpha|^2+1) - \omega_+(|\alpha|^2)$ leading to revival of revivals and so on. All this structure is introduced into the Rabi oscillations by the existence of the conserved observable $I_\psi = b^+b + \sigma_z/2$ in the rotating wave approximation. As the observable σ_z we have considered commutes with this conserved observable, the different non-degenerate eigenstates of I_ψ contained in the initial state are never mixed in the expectation $\langle\psi|\sigma_z|\psi\rangle$. Hence, this expectation value is completely independent of the relative phases of the non-degenerate eigenstates of I_ψ contained in the initial state, which explains why revivals appear on time scales completely unrelated to and much shorter than Poincaré recurrence times.

We now go on to calculate the power spectrum

$$K(\Omega) = \int_{-\infty}^{+\infty} \frac{d\tau}{2\pi} e^{-i\Omega\tau} \lim_{T\to\infty} \frac{1}{2T}\int_{-T}^{+T} dt \langle\psi(0)|b^+(t+\tau)b(t)|\psi(0)\rangle \tag{3.7}$$

in the rotating wave approximation. In the definition (3.7) we employ the Heisenberg picture for simplicity. In the Schrödinger picture eq. (3.7) reads

$$K(\Omega)=\int_{-\infty}^{+\infty}\frac{d\tau}{2\pi}e^{-i\Omega\tau}\lim_{T\to\infty}\frac{1}{2T}\int_{-T}^{+T}dt\,\langle\psi(t+\tau)|b^{+}e^{-\frac{i}{\hbar}H_0\tau}b|\psi(t)\rangle \tag{3.8}$$

where the dependence on the initial state is less explicit. In order to evaluate (3.8) we compute $b|\psi(t)\rangle_m$ and $b|\psi(t+\tau)\rangle_m$ from eq. (3.4) and use $b|\psi(t)\rangle_m$ as a new initial state in eq. (3.2) to evaluate $\exp(-iH_0\tau/\hbar)b|\psi(t)\rangle_m$. We obtain by taking the matrix element

$$_{m'}\langle\psi(t+\tau)|b^{+}e^{-iH_0\tau/\hbar}b|\psi(t)\rangle_m=\delta_{mm'}e^{i\omega\tau}(F(t,\tau)+G(t,\tau)) \tag{3.9}$$

with

$$\begin{aligned}F(t,\tau)&=\cos g\sqrt{m+1}(t+\tau)\left[m\cos g\sqrt{m}\tau\cos g\sqrt{m+1}t\right.\\&\qquad\left.-\sqrt{m(m+1)}\sin g\sqrt{m}\tau\sin g\sqrt{m+1}t\right]\\G(t,\tau)&=\sin g\sqrt{m+1}(t+\tau)\left[\sqrt{m(m+1)}\sin g\sqrt{m}\tau\cos g\sqrt{m+1}t\right.\\&\qquad\left.+(m+1)\cos g\sqrt{m}\tau\sin g\sqrt{m+1}t\right]\end{aligned} \tag{3.10}$$

It remains to take the time average of $F(t,\tau)$ and $G(t,\tau)$ over t and then the Fourier transform over τ, which are both straight forward. We obtain

$$\begin{aligned}K(\Omega)&=\frac{1}{4}\left[(m+\tfrac{1}{2})-\sqrt{m(m+1)}\right]\left(\delta(\Omega-\omega-\sqrt{m+1}g-\sqrt{m}g)\right.\\&\qquad\left.+\delta(\Omega-\omega+\sqrt{m+1}g+\sqrt{m}g)\right)\\&+\frac{1}{4}\left[(m+\tfrac{1}{2})+\sqrt{m(m+1)}\right]\left(\delta(\Omega-\omega-\sqrt{m+1}g+\sqrt{m}g)\right.\\&\qquad\left.+\delta(\Omega-\omega+\sqrt{m+1}g-\sqrt{m}g)\right)\end{aligned} \tag{3.11}$$

i.e. a spectrum of 4 discrete lines, symmetrically placed around the central frequency ω. For $m\gg1$ the pair of lines furthest from ω are of intensity $1/(16m)$, while the pair of lines nearest to ω have intensity $(m+\frac{1}{2}-1/8m)/2$. Hence, the outer pair of lines is strongly suppressed and negligible. The frequency distance of the inner pair, for large m, is $\simeq g/\sqrt{m}$. If the initial state is not a number state $|m,\uparrow\rangle$ but again a coherent state $|\alpha,\uparrow\rangle$ the result (3.11) has to be weighted with the Poisson distribution

$$|\langle m|\alpha\rangle|^2=\frac{|\alpha|^{2m}}{m!}e^{-|\alpha|^2} \tag{3.12}$$

and summed over m. The inner pair of lines is thereby replaced by a pair of distributions of discrete lines around the two frequencies $\omega\pm g/2|\alpha|$, the width of the distributions being given by $g/4|\alpha|^2\approx\Delta\omega_r/2$ and the spacing between the adjacent discrete lines being given by $g/4|\alpha|^3\simeq\omega_{rr}/2$. The structure of the power spectrum therefore contains the same frequencies which determine the time scales of the

Rabi oscillations. This structure is a consequence of the conservation of the quantized observable $K = g(b^+\sigma_- + b\sigma_+)$ in the rotating wave approximation, as discussed at the end of the preceding section.

Now we turn to the solution of the quantum problem without making the rotating wave approximation [11-15]. The conservation laws for I_p and K then disappear. However, a quantum mechanical conservation law for the parity operator

$$P = \exp\left(i\left(b^+b + S_z + \frac{1}{2}\right)\right) \tag{3.13}$$

remains, which is sufficient, in the case of a single two-level system, to separate the spin variables from the boson variables. This is accomplished by the unitary transformation [21]

$$\begin{aligned} \tilde{b} &= (\sigma_+ + \sigma_-)\, b \\ \tilde{b}^+ &= (\sigma_+ + \sigma_-)\, b^+ \\ \tilde{\sigma}_z &= \sigma_z \cos \pi b^+b \end{aligned} \tag{3.14}$$

and we obtain the two boson Hamiltonians

$$H_\pm = \hbar\omega \tilde{b}^+\tilde{b} + \hbar g(\tilde{b} + \tilde{b}^+) \mp \frac{\omega_0}{2} : \exp(-2\tilde{b}^+\tilde{b}): \tag{3.15}$$

where H_+, H_- refer the subspaces of $P = 1$, $P = -1$, respectively, and : : denotes normal ordering of $\tilde{b}$ and $\tilde{b}^+$. The Hamiltonian (3.15) can be analyzed perturbatively in ω_0 and yields, in first order

$$E_{n\pm} = \hbar\omega n - \frac{\hbar g^2}{\omega} \mp \frac{\omega_0}{2} e^{-\frac{2g^2}{\omega^2}} (-1)^n L_n\left(\frac{4g^2}{\omega^2}\right) \tag{3.16}$$

where the $L_n(x)$ are the Laguerre polynomials. For $\omega \approx \omega_0$ the energy levels (3.16) furnish good approximations to the 'exact' numerically determined levels, except for $g^2 \ll \omega^2$, where the rotating wave approximation can be used [12]. The statistics of nearest neighboured level spacings resulting from (3.16) or the 'exact' numerical results is quite untypical for a quantum system displaying chaos, as has been discussed in refs. [22,15]. If several (say N) two-level systems are treated, the separation (3.15) is no longer possible, an analytical formula for the energy levels of comparable accuracy as eq. (3.16) is no longer available, and the level statistics approaches the typical form of the Wigner distribution of random matrix theory for increasing N [15].

Fig. 1 shows the lowest lying energy levels of positive parity eigenstates of the Hamiltonian (1.8) for $N = 9$ as a function of g^2/ω^2. In fig. 2 the nearest neighboured level spacing distribution for this spectrum for $g/\omega_0 = 0.2$ is shown and compared with the Wigner distribution of the Gaussian orthogonal ensemble of random matrix theory. A satisfactory agreement is found [15].

We now turn to the results for the dynamics. In figs. 3 and 4 we plot the occupation probability of the upper level of the two-level system as a function of time for an initial coherent state with $\alpha = 10$, and assuming that initially the upper level is occupied. These figures give the results of a numerical solution of the Schrödinger equation. In fig. 3 the parameter $\mu = 2\omega\Omega/(4\omega^2 - \Omega^2)$ is 0.1 and the revivals at times $t \approx 2\pi/\omega_+$ are clearly visible. In fig. 4 the parameter μ is 1.714 and larger than $\pi/2$. The revivals have disappeared, in agreement with the classical estimates presented before.

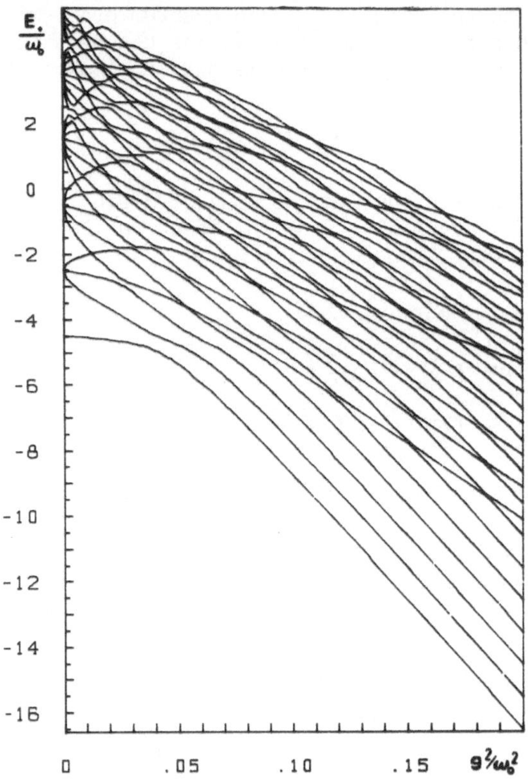

Fig. 1. Eigenvalues of the Hamiltonian (1.8) for N = 9, $\omega = \omega_0$ as function of g^2/ω^2 .

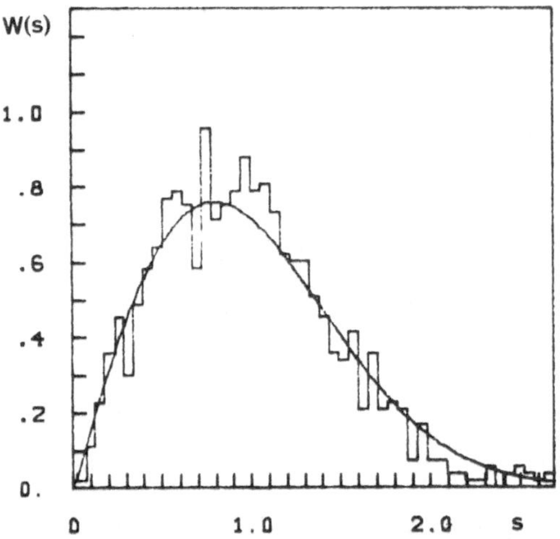

Fig. 2. Distribution $W(s)$ of neighbouring level spacings s of the spectrum of fig. 1 for g/ω = 0.2 and average level spacing normalized to 1. (Continuous line: Wigner distribution in the Gaussian orthogonal ensemble)

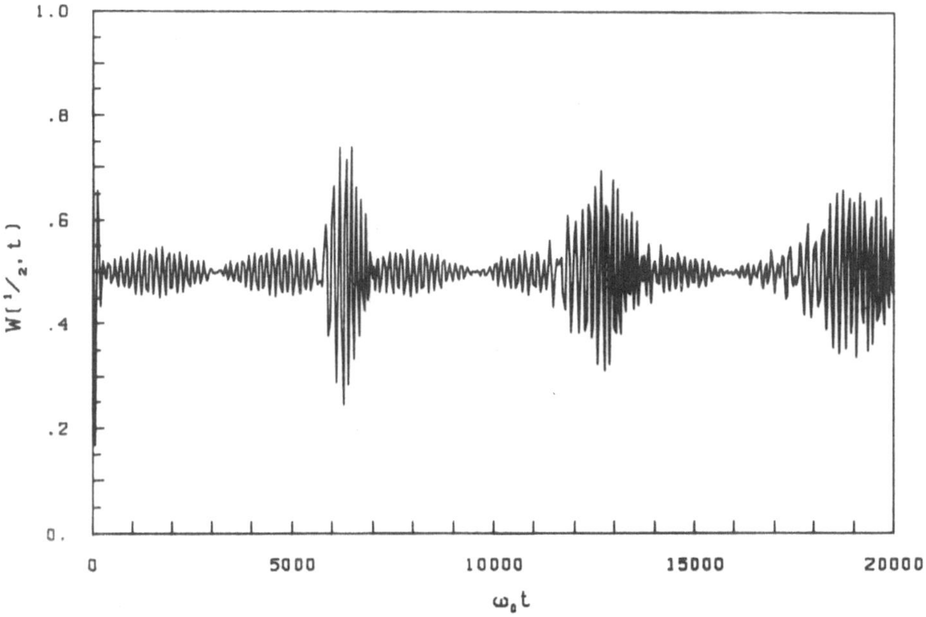

Fig. 3. Occupation probability $W(\frac{1}{2},t)$ of the upper atomic level as a function of time for $\alpha = 10, \omega=\omega_0, g/\omega = 0.01$.

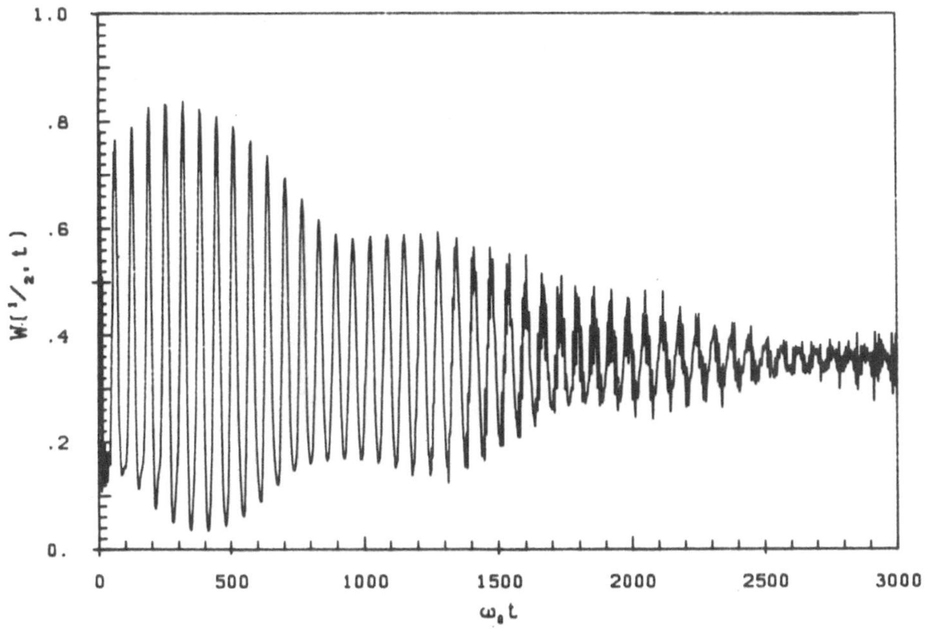

Fig. 4. As fig. 3 for $g/\omega = 0.075$

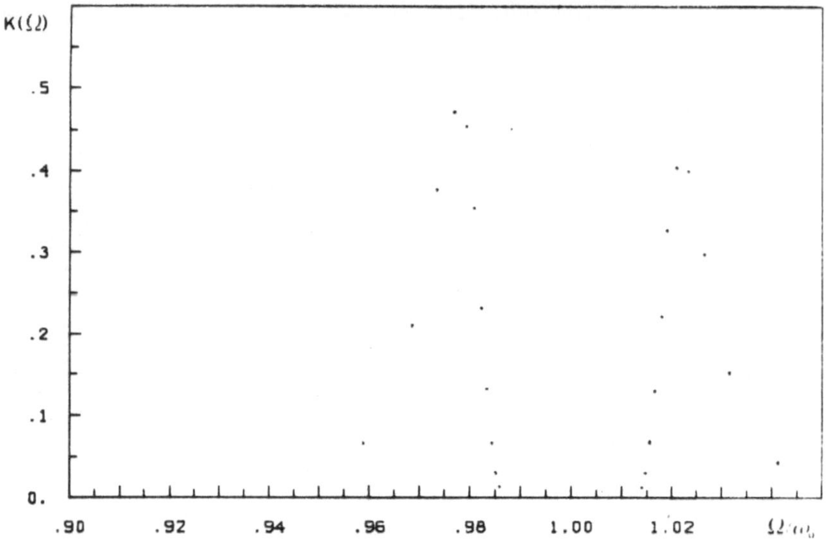

Fig. 5. Power spectrum (3.8) for N = 1, α = 2, g/ω = 0.03

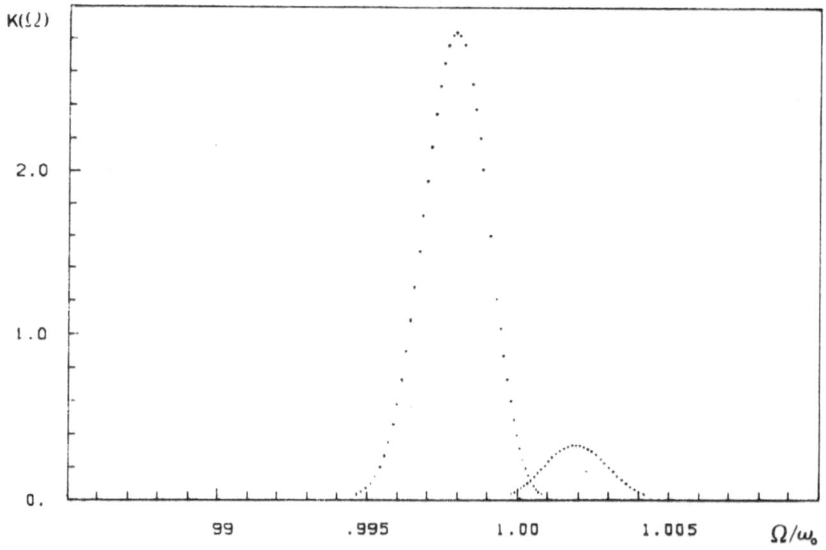

Fig. 6. Power spectrum (3.8) for N = 1, α = 8, g/ω = 0.1

Next we discuss numerical results for the power spectrum of the boson mode. In fig. 5 the power spectrum for the case μ = 0.21 is shown. The two groups of discrete lines symmetrically placed around $\Omega \approx \omega$, predicted within the rotating wave approximation is clearly visible. Fig. 6 shows the power spectrum for the case μ = 2.22. The group of lines corresponding to positive K-values has become much weaker and the power spectrum is now determined by only one group of discrete lines with its line center at a frequency smaller than ω . Thus, the classical expectation that only one group of lines should remain after the classical system becomes chaotic, is born out in fig. 6. The asymmetry with respect to the mode frequency ω depends on the phase of α in the initial state. Indeed, a dependence of the average value of K on the initial condition may also be inferred from the classical analysis (2.19).

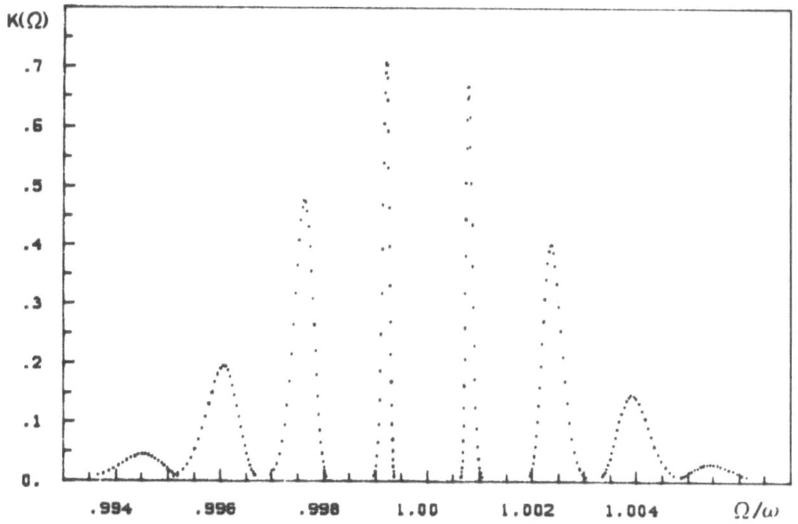

Fig. 7. Power spectrum (3.8) for N = 9, α = 6, g/ω = 0.03

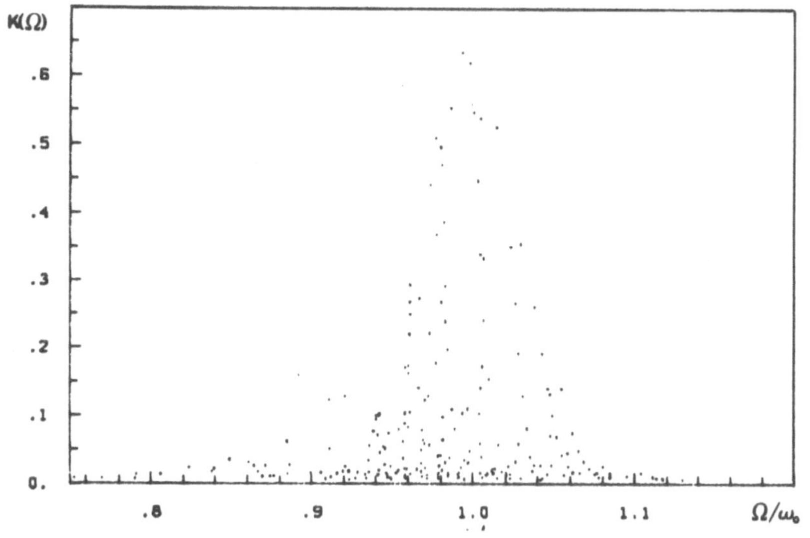

Fig. 8. Power spectrum (3.8) for N = 9, α = 4, g/ω = 0.3

For the case of $N = 9$ two level atoms we obtain the power spectrum of fig. 7 where $\mu = 0.06$. One can distinguish 8 groups of discrete lines, nearly symmetrical around ω. Two more groups in the wings have too little intensity and are therefore not discernible. Fig. 8 shows the corresponding case for $\mu = 0.48$. It is seen that the discrete groups of lines of fig. 7 have merged into a single broad spectral feature in fig. 8, symmetrical with respect to ω . That this should happen despite the comparatively small value of μ is not surprising, because the $(N+1)$ quantized values of K for each given value of the quantum number n are spaced by $\Delta K = \Omega/N$.

The critical value of μ arising from eq. (2.20) for this reduced value of ΔK is

$$\mu \approx \frac{2\Delta K}{\Omega} = \frac{2}{N} = 0.22$$

which is clearly below the value corresponding to fig. 8. Fig. 8, therefore, is in accordance with the expectations drawn from the classical analysis.

In summary: The conservation and quantization of I_ψ and K in the rotating wave approximation have strong dynamical consequences and lead to the appearance of revivals and characteristic structure in the power spectrum, respectively. The appearance of chaos in the full classical system is signalled, in the quantum system, by the disappearance of the revivals and the structure in the power spectrum. Both features may serve as dynamical indicators in the quantum system for the transition to ergodic dynamics unconstrained by exact or approximate conservation laws except for total energy and parity.

REFERENCES

1. Holstein, T., Ann. Phys. (NY) 8,343 (1959).
2. Rabi, I.I., Phys. Rev. 51,652 (1937).
3. Bloch, F., Siegert, A.J.F., Phys. Rev. 57,522 (1940).
4. Autler, S.H., Townes, C.H., Phys. Rev. 100,703 (1955).
5. Ya. B. Zeldovich, Zhurn. Eksper. Teor. Fiz. 51,1492 (1966) (Sov. Phys. JETP 24,1006 (1967).
6. Pomeau, Y., Dorizzi, B., Grammaticos, B., Phys. Rev. Lett. 26,681 (1986).
7. Beloborov, P.I., Zaslavski, G.M., Tartakovski, G.Kh., Zh. Eksp. Teor. Fiz. 71,1799 (1976).
8. Milonni, P.W., Ackerhalt, J.R., Galbraith, H.W., Phys. Rev. Lett. 50,966 (1983).
9. Eidson, J., Fox, R.F., Phys. Rev. A34,3288 (1986).
10. Munz, M., Z. Phys. B53,311 (1983); Kujawski, A., Munz, M., Z.Phys. B, to be published.
11. Graham, R., Höhnerbach, M., Phys. Lett. 101A,61 (1984).
12. Graham, R., Höhnerbach, M., Acta Physica Austr. 50,45 (1984).
13. Graham, R., Höhnerbach, M., Z. Phys. B57,233 (1984).
14. Graham, R., in "Optical Instabilities" ed. R.W. Boyd, M.G. Raymer, L.M. Narducci, Cambrdge University Press, Cambridge 1986.
15. Graham, R., Höhnerbach, M., Phys. Rev. Lett. 57,1378 (1986).
16. Jaynes, E.T., Cummings, F.W., Proc. IEEE 51,126 (1963).
17. Guckenheimer, J., Holmes, P., "Nonlinear Oscillations, Dynamical Systems, and Bifurcations of Vector Fields", Springer, Berlin 1983.
18. Eberly, J.H., Narozhny, N.B., Sanchez-Mondragon, J.J., Phys. Rev. Lett. 44,1323 (1980); Yoo, H.I., Sanchez-Mondragon, J.J., Eberly, J.H., J.Phys. A14,1383 (1981).
19. Meschede, D., Walther, H., Müller, G., Phys. Rev. Lett. 54,511 (1985).
20. Cummings, F.W., Phys. Rev. A140,1051 (1965).
21. Shore, H.B., Sander, L.B., Phys. Rev. B7,4537 (1973).
22. Kuś, M., Phys. Rev. Lett. 54,1343 (1985).

QUANTIZATION OF THE CHIRIKOV RESONANCE OVERLAPPING CRITERION (*)

Sandro Graffi

Dipartimento di Matematica
Università di Bologna
40127 Bologna, Italy

The Chirikov resonance overlapping criterion [1] is long known to represent the simplest, most effective criterion for the onset of stochastic behaviour in classical Hamiltonian systems. More recently it has been remarked [2] that it yields the best theoretical account so far obtained of a phenomenon [3–5], the sharp increase of the Hydrogen photoionization rate under the action of an oscillating, monochromatic electric field, which in principle should be best understood in quantum mechanics [6–9]. The issue has therefore attracted even more attention on the so-called "quantum chaos" [10], i.e. the problem of determining the manifestation in quantum mechanics, if any, of the chaotic behaviour of the corresponding classical Hamiltonian system. A prerequisite for any such manifestation is that the long time behaviour of the system should be sensibly affected. This happens if the quantum levels exhibit multiple avoided crossings, and therefore this phenomenon has been extensively studied in recent years, both theoretically [11] and numerically [12–13], in various systems including the standard one-dimensional model for hydrogen photoionization [14].

A major difficulty in this kind of investigation is the absence of any criterion predicting a threshold for the occurrence of avoided crossings at some values of the system parameters such as coupling constant, field strength, etc. The purpose of this talk is to describe in some detail an argument [21] pointing out that, at least in semiclassical quantum mechanics and for the particular case of the periodically driven rotators, the coupling constant threshold for avoided crossings among an arbitrarily large number of quantum levels (in this case, quasi-energy levels, or Floquet levels) is exactly the same threshold predicted by the classical resonance overlapping criterion.

The argument is based on writing the Schrodinger equation as a classical Hamilton-Jacobi equation plus corrections in powers of $\hbar$, and then integrating it by

(*) Work partially supported by NSF under Grant PHY–8502383.

the same algorithm of classical, canonical perturbation theory, in the non-degenerate case and in the degenerate case as well. In this way it will be possible to take over the classical results to all quantum states such that n is large and $\hbar$ small: all states for which $n\hbar = A$ is a KAM torus undergo avoided crossing as soon the perturbation is large enough to allow avoided crossing between levels for which $n\hbar = A$ lies on two consecutive classically resonant tori, exactly as in classical mechanics all KAM tori lying between two consecutive resonances are destroyed as soon the perturbation is large enough to allow overlapping between the librations around the two resonances [1,15].

Let us now turn to the description of the model, of the result and of the argument.

a) The Model. Consider in the action–angle variables (A, α), (B, β) the classical Hamiltonian:

$$K_\epsilon = A^2/2+\omega B+\epsilon V(\alpha, \beta) \tag{1}$$

where $\omega > 0$, and V is 2π–periodic with respect to both α and β and analytic with respect to α. K_ϵ is of course the quasi-energy (i.e., Floquet) representation in the extended phase space of the periodically driven rotator:

$$H_\epsilon(t) = A^2/2+\epsilon V(\alpha, \omega t) \tag{2}$$

The quantum counterpart of (1.1) is the Schrodinger operator acting on $L^2(\pi^2)$

$$T_\epsilon = -\frac{h^2}{2}\frac{\partial^2}{\partial\alpha^2}+\epsilon V(\alpha, \beta)-ih\omega\frac{\partial}{\partial t} \tag{3}$$

We assume that the maximal operator generated by (3) in $L^2(\pi^2)$ is self-adjoint and has pure point spectrum. The eigenvalues of T_0, denoted by $E_0(n, m, \hbar)$, are of course given by $E_0 = (n\hbar)^2/2+\omega m\hbar$, $(n, m) = 0, \pm1, \pm2, \ldots$
We note that in this case the Bohr-Sommerfeld quantization $A = n\hbar$, $B = m\hbar$ is exact, i.e. $E_0 = K_0(n\hbar, m\hbar)$.

b) Classical Resonances and Quantum Accidental Degeneracy.

Classically, a point $(A^*, B^*) = (A^*[p, q], B^*[p, q])$ in action space ("torus") is called a resonance ("resonant torus") if there are integers $(p, q) \neq (0, 0)$ such that:

$$p\frac{\partial K_0}{\partial A}\bigg|_{A=A^*}+q\frac{\partial K_0}{\partial B}\bigg|_{B=B^*} \equiv pA^*+\omega q = 0 \tag{4}$$

In quantum theory the unperturbed levels E_0 will be accidentally degenerate if there are integers (p, q) $\neq$ (0, 0) such that:

$$(n+p)^2\hbar^2/2+(m+q)\omega\hbar = (n\hbar)^2/2+\omega m\hbar \tag{5}$$

i.e., $np\hbar+\omega q+\hbar p^2/2 = 0$, which goes over to (4) at the classical limit $\hbar \to 0$, $n\hbar \to A^*$.

c) The Result. Let $A_1(p, q)$ be a classical resonance, A_2 the consecutive one, $\Delta A = |A_1-A_2|$, $V_{p,q}$ the resonant Fourier component of the potential, and $E(n, m, \hbar, \epsilon)$ the eigenvalues of T_ϵ, which reduce to $E_0(n, m, \hbar)$ as $\epsilon = 0$. Consider those eigenvalues $E^*(n, m, \hbar, \epsilon)$ of T_ϵ emanating from the unperturbed eigenvalues $E_0(n, m, \hbar)$ such that $n \gg 1$, $\hbar \ll 1$, $A_1 < n\hbar = A < A_2$. Then, if the classical resonance overlapping condition

$$\frac{2\sqrt{\epsilon V_{p,q}}}{\Delta A} > 1, \text{ i.e. } \sqrt{\epsilon} > \frac{\Delta A}{2\sqrt{V_{p,q}}} \tag{6}$$

is satisfied, all eigenvalues $E^*(n, m, \hbar, \epsilon)$ undergo avoided crossing up to order $\hbar$. Concerning this result, we remark: (i) Formula (6) is the condition of touching between the separatrices of the libration motions around the two consecutive resonances; see e.g. [1,15]. (ii) For the pulsed rotator $V(\alpha, \omega t) = V(\alpha)\cos \omega t$ we have $V_{p,q} = 0$ for $q \neq \pm1$. Hence only the resonances with $q = \pm1$ matter; for p large $\Delta A_1 = 2wq/p^2$, and (6) becomes:

$$\sqrt{\epsilon} > \frac{\omega}{p^2\sqrt{V_{p,q}}} \tag{7}$$

If the sequence $p^2(V_{p,q})^{1/2}$ is non decreasing in p, the condition (7) implies avoided crossing between eigenvalues emanating from any two consecutive classical resonant tori, exactly in the same way as in classical mechanics it implies global stochasticity (see e.g. [15]). (iii) The simplest pulsed rotator, i.e. $V_{p,\pm1} = 0$ for $p \neq \pm1$ is not covered by the forthcoming argument, and thus requires a higher order analysis, because the only two first order (i.e. those for which $V_{p,q} \neq 0$) resonances, $A_1 = \pm\omega$ have the same energy. (iv) The kicked rotator $V_{p,q} = 0$, $p,q \neq \pm1$, $V_{\pm1,\pm1} = 1/4$, yields $\Delta A_1 = \omega$ and therefore $\epsilon > 4\omega^2$, exactly as in the classical case. Strictly speaking, however, not even this case is covered because the spectrum of T_ϵ has a continuous component (16–17). We conjecture that, as in the Hydrogen photoionization case (18), the eigenvalues of T_0 turn for $\epsilon > 0$ into (quantum) metastable states, to which the present result might apply. Let us now turn to the argument, described in the following six steps.

Step 1. Schrodinger and Hamilton–Jacobi equations. The generating function $W(A, B, \alpha, \beta)$ of the canonical transformation integrating the Hamiltonian (1), i.e. transforming it into a function depending only on the actions (A, B), is of course given by the Hamilton-Jacobi equation:

$$\frac{1}{2}\left[\frac{\partial W}{\partial \alpha}(A,\ B,\ \alpha,\ \beta)\right]^2+\epsilon V(\alpha,\ \beta)+\omega\frac{\partial W}{\partial \beta}(A,\ B,\ \alpha,\ \beta)=E(A,\ B,\ \epsilon) \tag{8}$$

where both W and E have to be determined. For $\epsilon = 0$ we have $W_0 = A\alpha+B\beta$, which of course generates the identity transformation. Look now for solutions of the Schrodinger equation under the form $y = e^{W(A,B,\alpha,\beta,\epsilon)\ /i\hbar}$. Then we have:

$$\frac{1}{2}\left[\frac{\partial W}{\partial \alpha}(E,\ \alpha,\ \beta,\ \epsilon)\right]^2+\omega\frac{\partial W}{\partial \beta}(E,\ \alpha,\ \beta,\ \epsilon)+\frac{i\hbar}{2}\frac{\partial^2 W}{\partial \alpha^2}+\epsilon V(\alpha,\ \beta)=E(\epsilon) \tag{9}$$

which formally reduces to (8) for $\hbar = 0$. For $\epsilon = 0$ the spectrum of T_0 can be reobtained through (9) in the following way. Look for solutions in separated variables $E_0 = E_{0,1}+E_{0,2}$, $W_0 = W_{0,1}+W_{0,2}$. We immediately have $W_{0,2}=\omega^{-1}E_{0,2}\alpha$. Imposing the periodicity of the wave function y we get the quantization condition $E_{0,2} = m\hbar\omega$. Moreover $(dW_{0,1}/d\alpha)^2/2 - i\hbar d^2W_{0,1}/d\alpha^2 = E_{0,1}$, and the periodicity once more implies quantization, i.e. the only solution is $W_{0,1} = \pm(2E_{0,1})^{1/2}\alpha$, with $E_{0,1} = (n\hbar)^2/2$. Therefore $W_0 = n\hbar\alpha+m\hbar\beta$.

Step 2. Solution of the Schrodinger problem by classical perturbation theory. To solve eq. (9) by the algorithm of classical, canonical perturbation theory, we set $E(\epsilon) = E_0+E_1(\epsilon)+E_2(\epsilon)+\ldots$, $W(E,\epsilon) = W_0+W_1(E,\ \alpha,\ \beta)+\ldots$, and try to recursively solve for $(E_1,\ W_1)$, $(E_2,\ W_2)$ under the assumption $E_1 \ll E_0$, $W_1 \ll W_0$, $E_2 \ll E_1, \ldots$, to be justified a posteriori. As in the standard derivation of the Chirikov criterion we limit ourselves to first order corrections, i.e. to the determination of W_1 and E_1. Inserting (9) into (8) and expanding the 1. h. s. we get:

$$\begin{aligned}\frac{1}{2}\left(\frac{\partial W_0}{\partial \alpha}\right)^2+\frac{\partial W_0}{\partial \alpha}\frac{\partial W_1}{\partial \alpha}+\frac{1}{2}\left(\frac{\partial W_1}{\partial \alpha}\right)^2+i\frac{h}{2}\frac{\partial^2 W_1}{\partial \alpha^2}+\\ +\omega\left(\frac{\partial W_0}{\partial \beta}+\frac{\partial W_1}{\partial \beta}\right)+\epsilon V(\alpha,\ \beta)=E_0+E_1\end{aligned} \tag{10}$$

since $W_0 = A\alpha+B\beta = n\hbar\alpha+m\hbar\beta$ we have:

$$\frac{\partial W_0}{\partial \alpha}\frac{\partial W_1}{\partial \alpha}+\omega\frac{\partial W_1}{\partial \beta}+i\frac{h}{2}\frac{\partial^2 W_1}{\partial \alpha^2}+\frac{1}{2}\left(\frac{\partial W_1}{\partial \alpha}\right)^2+\epsilon V(\alpha,\ \beta)=E_1 \tag{11}$$

To solve (11) we look for the Fourier expansion of $W_1(E_1,\ \alpha,\ \beta)$:

$$W_1(E_1,\ \alpha,\ \beta)=\sum_{-\infty}^{+\infty} w_{r,s}^{(1)}(E_1)e^{ir\alpha+is\beta} \tag{12}$$

and try to determine the Fourier coefficients. We have:

$$\frac{\partial W_0}{\partial \alpha}\frac{\partial W_1}{\partial \alpha}+\omega\frac{\partial W_1}{\partial \beta}+i\frac{h}{2}\frac{\partial^2 W_1}{\partial \alpha^2}$$
$$=\sum_{-\infty}^{+\infty} i(Ar+\omega s+h\frac{r^2}{2})w^{(1)}_{r,s}\,e^{i(r\alpha+s\beta)} \tag{13}$$

We now distinguish two cases: (i) Non degenerate. i.e. classically non resonant. Since $Ar+\omega s \neq 0$ for all $(r, s) \neq (0, 0)$, for $\hbar \to 0$ the term $Ar+\omega s+\hbar r^2/2$ can vanish only for $|r| \to \infty$, and we can always assume $V_{r,s}$ decreasing so fast in $|r|$ to compensate this. From (13) we see that the second order term in (11) can be neglected, so that all Fourier coefficients of W_1, except the mean, are determined by the first order equation:

$$i(Ar+\omega s+hr^2)w^{(1)}_{r,s}+\epsilon V_{r,s}=0,\ (r,\ s) \neq (0,\ 0) \tag{14}$$

with vanishing quasi-energy correction at first order: $E_1 = V_{0,0} = 0$. Eq. (14) does not determine the mean of W_1. However this must be zero by periodicity, which as we have seen is equivalent to quantization. This procedure can be carried through to all orders in perturbation theory, provided $Ar+\omega s+\hbar s^2/2$ does not become too small; for a detailed discussion (in the framework of perturbations of the multidimensional harmonic oscillator), see [19]. The corresponding levels, i.e. all levels $E(n, m, \hbar, \epsilon)$ with $n\hbar = A$ and $Ar+\omega s+\hbar r^2/2$ not too small are, in this language and within the present approximation, the quantum counterpart of the KAM tori. (ii) Degenerate. classically resonant. Let (5) hold, so that $A^*p+\omega q = 0$, $(p, q) \neq (0, 0)$. We immediately see that the non-resonant components of W_1, i.e. those for which $r \neq jp$, $r \neq jq$, $j\in Z$, are determined as above, and yield zero energy correction at first order. However the resonant components, i.e. those with $r = jp$, $s = jq$, have to be determined by the quadratic part: in fact, formula (13) shows that the linear terms in the l.h.s. of (11) vanish. Setting now:

$$S_1(E_1,\ \alpha,\ \beta)=\sum_{-\infty}^{+\infty} w^{(1)}_{jp,jq}\,e^{i(jp\alpha+jq\beta)} \tag{15}$$

inserting in (11) we have:

$$\frac{1}{2}\left[\frac{\partial S_1}{\partial \alpha}(E_1,\ \alpha,\ \beta)\right]^2+i\frac{h}{2}\frac{\partial^2 S_1}{\partial \alpha^2}(E_1, \alpha,\ \beta)+\epsilon V'(\alpha,\ \beta)=E_1,$$
$$V'=\sum_{-\infty}^{+\infty} v_{jp,jq}\,e^{i(jp\alpha+jq\beta)} \tag{16}$$

If, once more as in the classical derivation, we neglect all terms with $j \neq \pm 1$ in V'

("primary resonances approximation") we get:

$$\frac{1}{2}\left[\frac{\partial S_1}{\partial\alpha}(E_1, \alpha, \beta)\right]^2 + i\frac{h}{2}\frac{\partial^2 S_1}{\partial\alpha^2}(E_1, \alpha, \beta) + 2\epsilon V_{p,q}\cos(p\alpha+q\beta) \tag{17}$$

Step 4. Quantum slow and fast variables. Let us now implement on our semiclassical approximation of the Schrodinger equation the linear canonical transformation decoupling slow and fast variables around resonance in classical mechanics (see e.g. [1,15] or Born [20]): Set:

$$\lambda = (p\alpha - q\beta)/2, \ \mu = (p\alpha+q\beta)/2; \quad \alpha = (\lambda+\mu)/p, \ \beta = (\lambda-\mu)/q \tag{18}$$

$$I = A/p - B/q, \ J = A/p + B/q; \quad A = p(I+J)/2, \ B = q(I-J)/2 \tag{19}$$

The unperturbed classical Hamiltonian takes then the form

$$K_0(I, J) = p^2(I+J)^2/8 + wq(I-J)/2 \tag{20}$$

and the resonance condition $pA^*+\omega q = 0$ takes the form $\partial K_0/\partial J|_{J=J^*} = 0$. Therefore J is the resonant action and μ the slow variable. Correspondingly, performing the change of variables (18) on the free equation, i.e. eq. (8) for $\epsilon = 0$, we get $W_0(I, J, \lambda, \mu) = I\lambda + J\mu$, $E_0 = K_0(I, J)$, where of course (I,J) are to be considered quantized through $A = n\hbar$, $B = m\hbar$ and the definition (19). The accidentally degenerate levels, i.e. those coming from the quantization of A^*, are expressed in these variables by $p^2(I^*+J^*)/2+\omega q = 0$. In this case we recover the well known fact (see e.g. Born [20]) that there is only one independent quantization condition instead of two: we have in fact $\partial W_0/\partial\mu|_{J=J^*} = J^* = -I^*+2\omega q/p^2$, so that at resonance the quantization of I implies the quantization of J. Therefore the independent quantization of J, the slow action near resonance, must take place at first order as we will now see.

Step 5. Quantization of the slow variable. By (16) and (18), $S_1(E_1, \alpha, \beta)$ transforms into:

$$S_1(E_1, \mu) = \sum_{-\infty}^{+\infty} W^{(1)}_{jp,jq}\, e^{2ij\mu} \tag{21}$$

which is a function independent of λ. Therefore eq. (17) becomes:

$$\frac{1}{8}\left[p\frac{\partial S_1}{\partial\mu}(E_1, \mu)\right]^2 + ih\frac{p^2}{4.}\frac{\partial^2}{\partial\mu^2}(E_1, \mu) + \epsilon V_{p,q}\cos 2\mu = E_1 \tag{22}$$

i.e. the above transformation decouples slow and fast variables (in a suitable neighbourhood of the accidentally degenerate quantum numbers) in semiclassical quantum mechanics as well. Performing the rescaling $\chi = 2\mu$ eq. (22) becomes:

$$\frac{1}{2}\left[p\frac{\partial S_1}{\partial \chi}(E_1, \chi)\right]^2 + ih\frac{\partial^2 S_1}{\partial \chi^2}(E_1, \chi) + \epsilon V_{p,q} \cos \chi = E_1 \tag{23}$$

Equation (23) can be of course reduced to the Schrodinger form. Setting $R(E_1, \chi) = \exp\left[i\, S_1(E_1, \chi)/\hbar\right]$ we get:

$$-\frac{(hp)^2}{2}\frac{d^2R}{d\chi^2}(E_1, \chi) + \epsilon V_{p,q} \cos \chi\, R(E_1, \chi) = E_1 R(E_1, \chi) \tag{24}$$

Since we are looking at large quantum numbers, we can take the quantized energy correction E_1 given by the standard WKB formula:

$$\pm\frac{2}{p}\int_{\chi_-(E_1)}^{\chi_+(E_1)} \sqrt{2\left(E_1 - \epsilon V_{p,q} \cos \chi\right)}\, d\chi = gh,\ g = 1, 2, \ldots . \tag{25}$$

$c_{\pm}(E_1)$ being the classical turning points, i.e. the endpoints of the libration motions of the simple pendulum (24) at energy E_1. Therefore, for $|A - A^*|$ suitably small, i.e. for all quantum numbers n such that $n\hbar$ is suitably close to the classically resonant torus A^*, and $\hbar$ suitably small, the energy levels are, to first order, $E = E_0 + E_1$, with E_1 given implicitly by (25). The spread in the quantum number g is obtained at the classical separatrix energy $E_1 = \epsilon V_{p,q}$:

$$\left|g_{max} h\right| = \frac{2}{p}\sqrt{\epsilon V_{p,q}}\int_0^{2\pi}\sqrt{2(1-\cos\chi)}\, d\chi = \frac{16}{p}\sqrt{\epsilon V_{p,q}} \tag{26}$$

and the corresponding maximum and minimum levels will be:

$$E_{\pm} = E_0(A^*, p, q) \pm \epsilon V_{p,q} \tag{27}$$

Step 6. Overlapping and Avoided Crossings. Consider two consecutive resonances A_1 and A_2, and, to simplify the exposition, suppose that their parameters are (p, q), (p+1, q), respectively. Let $p \gg 1$, so that $\Delta A \approx 4\omega q/p^2$, $V_{p,q} \approx V_{p+1}, q$. By (25) we have:

$$E_-(A_2, p+1, \epsilon) = E_0(A_2, p+1) - \epsilon V_{p,q}\ ;\ E_+(A_1, p, \epsilon) = E_0(A_1, p) + \epsilon V_{p,q} \tag{28}$$

Therefore these two levels will overlap as soon as

$$\epsilon V_{p,q} > \left| E_0(A_2,\ p+1) - E_0(A_1,\ p) \right| \tag{29}$$

This is nothing else than the classical condition (6). In fact, Steps 4 and 5 above are just a quantum version of the standard Chirikov quadratic approximation near resonance, which we now briefly recall. Perform the canonical transformation (18-19) on K_ϵ, average on the fast variable λ, set $(I,\ J) = (I_1,\ J_1+2G)$, expand in powers of G and keep only the primarily resonant Fourier components of the potential. After the rescaling $\chi = 2\mu$, which makes $\{G,\ \chi\}$ canonical variables, the result is:

$$K_\epsilon = K_0(p) + K_1(p,\ G,\ \chi,\ \epsilon) \tag{30}$$

where $K_0(p)$ is given by (20) computed at $(I,\ J) = (I_1,\ J_1)$ and:

$$K_1(p,\ G,\ \chi,\ \epsilon) = \frac{1}{2} p^2 G^2 + \epsilon V_{p,q} \cos \chi \tag{31}$$

Overlapping (in action) takes place as soon as the separatrices of K_1 (p) and K_1 (p+1) touch. This happens if $[G_{max}(p)+G_{max}(p+1)]/2\Delta J > 1$. Since the maximum elongation of G along the separatrix is $G_{max} = 2p^{-1}(\epsilon V_{p,q})^{1/2}$, by (19) we get the condition $(\epsilon V_{p,q})^{1/2}/\Delta A > 1$, which is (6). When $G(p) = G_{max}(p)$, we have of course $K_0(p)+K_1(p,\ G_{max},\ \epsilon) = K_0(p+1)-\epsilon V_{p,q}$, which is the minimum energy of $K_\epsilon(p+1)$, so that also the energies overlap. Now for the free problem the Bohr-Sommerfeld quantization is exact: hence $K_0(p+1) = E_0(A_2,\ p+1)$, $K_0(p) = E_0(A_1,\ p)$, so that if (6) holds, comparing with (29) we can conclude that not only the classical energies, but also the quantum levels $E_-(A_2,\ p+1,\ \epsilon)$ and $E_+(A_1,\ p,\ \epsilon)$ overlap. Therefore there will be overlapping also among all levels emanating from unperturbed ones for which $A_1 < n\hbar = A < A_2$. This is the quantum couterpart of the destruction of the KAM tori lying between two consecutive classical resonances. We remark that these overlappings yield avoided crossings rather than crossings. Apart from the correction of order $\hbar$, the real physical reason seems to be that if ϵ is larger than the classical threshold the maximum potential energy of the quantum pendulum near the pth resonance is lifted at the height of the minimum potential energy of the pendulum near the (p+1)th resonance, so that the states in the two wells can start tunneling. Hence there will be always a splitting of the order of magnitude of a barrier penetration, i.e. of order $\exp(-1/\hbar)$.

REFERENCES

1. B.V. Chirikov, Phys. Reports 52, 263 (1979).
2. R.V. Jensen, Phys. Rev. 30A, 386 (1984).
3. J.E. Bayfield and P.M. Koch, Phys. Rev. Letters 33, 711 (1974).

4. J.E. Bayfield, C.D. Gardner and P.M. Koch, Phys. Rev. Letters 39, 76 (1977).

5. J.N. Bardsley and B. Sundaram, Phys. Rev. 32A, 689 (1985).

6. J.N. Bardsley, B. Sundaram, L.A. Pinnaduwage and J.E. Bayfield, Phys. Rev. Letters 56, 1007 (1986).

7. G. Casati, B.V. Chirikov, D.L. Shepelyansky and I. Guarneri, Phys. Rev. Letters 56, 2437 and 57, 823 (1984).

8. R. Blumel and U. Smilansky, "On the Localisation of Floquet States in the RF Excitation of Fydberg Atoms" Preprint 1986.

9. G. Casati, B.V. Chirikov, D.L. Shepelyansky and I. Guarneri, To appear in Physics Reports.

10. See, e.g.: (a) Chaotic Behaviour in Quantum Mechanics, G. Casati, Editor (Plenum, New York 1984); (b) G.M. Zaslavsky; Physics Reports 80, 157 (1981); (c) W.P. Reinhardt, in The Mathematical Analysis of Physical Systems, R. Mickens, Editor (Van Nostrand, New York 1984); (d) Proceedings of the Workshop on Quantum Chaos, Trieste 1986, To appear in Physica Scripta.

11. See e.g. the papers of M.V. Berry and R.A. Marcus in [10a].

12. T. Uzer, D.W. Noid and R.A. Marcus, J. Chem. Phys. 79, 4412 (1983).

13. T. Uzer and R.A. Marcus, J. Chem. Phys. 81, 5013 (1984).

14. R. Blumel and U. Smilansky, "Subthreshold Ionization of Rydberg Atoms in a Radiation Field", Preprint 1986, Submitted to Phys. Rev. Letters.

15. A.J. Lichtenberg and M.A. Lieberman, Regular and Stochastic Motion, Springer Verlag (New York 1983).

16. F.M. Izrailev and D.L. Shepelyansky, Theor. Math. Phys. 43, 553 (1980).

17. G. Casati and I. Guarneri, Commun. Math. Phys. 95, 121 (1984).

18. S. Graffi, V. Grecchi and H.J. Silverstone, Ann. Inst. H. Poincare 42, 215 (1984).

19. S. Graffi and T. Paul, Commun. Math. Phys. 1986 (in press).

20. M. Born, Mechanics of the Atom (Bell, London 1960).

21. S. Graffi, T. Paul and H.J. Silverstone, "Classical Resonances Overlapping and Quantum Avoided Crossings. Preprint 1986.

4. J.E. Bayfield, L.D. Gardner and P.M. Koch, Phys. Rev. Letters 39, 76 (1977).

5. J.N. Bardsley and B. Sundaram, Phys. Rev. A 32, 689 (1985).

6. J.N. Bardsley, B. Sundaram, L.A. Pinnaduwage, and J.E. Bayfield, Phys. Rev. Letters 56, 1007 (1986).

[illegible]

ERGODIC ADIABATIC INVARIANTS

Edward Ott

University of Maryland
College Park, MD 20742, U.S.A.

We consider a conservative dynamical system characterized by a time-dependent Hamiltonian, $H(\underset{\sim}{p},\underset{\sim}{q},\varepsilon t)$, where $\underset{\sim}{p}$ and $\underset{\sim}{q}$ are N vectors, and the explicit time dependence of H is "slow". To emphasize this slowness, we have written the third argument of H as εt, where we shall formally take ε small. Alternatively, we can set $T = \varepsilon^{-1}$ and think of T as the time scale over which $H(\underset{\sim}{p},\underset{\sim}{q},\varepsilon t)$ goes through an order one change, $T^{-1} \sim H^{-1}\partial H/\partial t$. The statement that this time dependence is slow (or adiabatic) is equivalent to saying that T is much longer than any relevant characteristic time for the particle motion in the "frozen" Hamiltonian, $H(\underset{\sim}{p},\underset{\sim}{q},\varepsilon t_0)$, where t_0 is a constant.

For the case of periodic motion with one degree of freedom (N=1), $\left(p(t), q(t)\right) = \left(p(t+\tau_p), q(t+\tau_p)\right)$, this situation leads to the well-known adiabatic invariant

$$\mu = \oint p\,dq \ ,$$

where $\oint$ denotes integration over one period τ_p. The adiabatic invariant of periodic motion was discussed early on by Boltzmann and subsequently by Helmholz, Hertz, and Rayleigh, among others (cf. Jammer[1] for discussion and primary references). By the assertion that μ is an adiabatic invariant we mean that a particle orbit approximately conserves μ over a time interval large enough that H experiences an order one change provided that this change occurs slowly (in the sense already mentioned). For the case of one degree of freedom (N = 1) and periodic motion the conservation of μ has been shown to be very good[2-4] in that the error in the approximation is $O(\varepsilon^m)$, H is m times differentiable with respect to t. (In fact, in solvable examples[5] where H is analytic in t it is common for the error to be of the form $\exp(-k/\varepsilon)$, i.e., exponentially small.) The utility of the adiabatic invariant for periodic motion has long been recognized. For example, in plasma physics it forms the basis of the fundamental concept of mirror confinement of charged particles[6] and has also been extensively used in performing stability calculations.[7] In addition, in the early theory of quantum mechanics, Ehrenfest[8] argued that $\oint pdq$ was a proper quantity to quantize because it is an adiabatic invariant (e.g., for a harmonic oscillator, with frequency ω and energy E, $E/\omega = (n+1/2)\hbar$, quantum mechanically, while E/ω is an adiabatic invariant of the classical mechanics for slow variation of the oscillation frequency $\omega = \omega(\varepsilon t)$).

Here we shall consider another type of adiabatic invariant. We presume that the number of degrees of freedom is greater than 1, $N > 1$, and that motion in the frozen Hamiltonian is chaotic and ergodic on the constant energy surface, $H(\underline{p}(t), \underline{q}(t), \varepsilon t_0)$ = constant. Consequently, the motion has no additional isolating constant of the motion other than the frozen Hamiltonian itself. In this case, as shown subsequently, the volume enclosed within the surface of constant H is an adiabatic invariant.* (This presupposes, of course, that this volume is finite.) This case of an adiabatic invariant for $N \gg 1$ (statistical mechanics) was stated by Boltzmann.[9] The volume inside the constant H surface is

$$\mu(E,t) = \iint U[E-H(\underline{p},\underline{q},\varepsilon t)]d^N\underline{p}\, d^N\underline{q} \ , \qquad (1)$$

where U[...] denotes the unit step function and E is the energy. Thus, for example, given an initial condition and the corresponding energy $E = E_0$ at $t = 0$, calculation of $\mu(E,t)$ from (1) allows us to obtain an approximation to the energy E(t) at all subsequent times via $\mu(E,t) = \mu(E_0,0)$. We call $\mu(E,t)$ for $N > 1$ the <u>ergodic adiabatic invariant</u>. To see how the approximate invariance of the quantity given by (1) follows from Hamilton's equations, we note that if <u>any</u> surface is specified at $t = 0$ and each point on that surface is evolved in time using Hamilton's equations, then the new surface must enclose the same 2N dimensional phase space volume as the initial surface.[10] If a

*For $N = 1$, this reduces to the adiabatic invariant $\oint pdq$.

particle wanders ergodically over the $H(\underset{\sim}{p},\underset{\sim}{q},\varepsilon t_0) = E$ surface in a time short compared to T, then, as t increases, particles on an initial H = constant surface will all have qualitatively similar trajectories. In particular their subsequent energies will be approximately equal. Thus an initial H = constant surface evolves into another surface which is close to being an H = constant surface (cf. Fig. 1 and caption). Hence (1) is an adiabatic invariant.

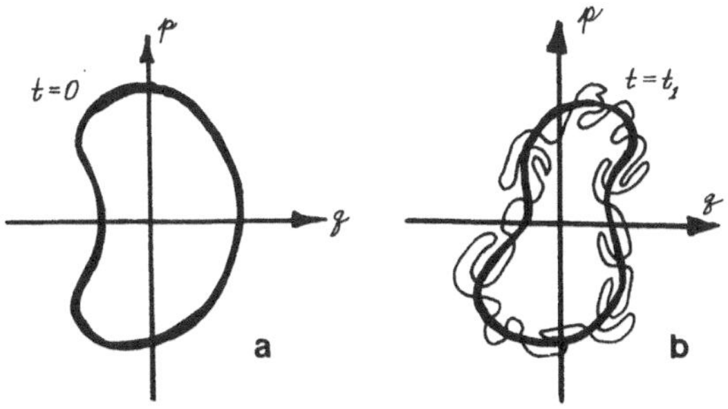

Fig. 1. (a) Initial $H = E_0$ surface at t = 0, evolves under the exact dynamics to the convoluted surface shown in (b) which is close to the surface $H = E(t_1)$, where $E(t_1)$ is obtained from the constancy of μ. The phase space volumes inside the initial smooth surface and inside the squiggly surface shown in (b) are exactly equal.

With the advent of computer solutions for particle motion, it has become more and more appreciated that low degree of freedom Hamiltonian systems can often behave chaotically in such a way that particle motion samples the surface of constant H, if not fully, at least nearly fully. Thus the ergodic adiabatic invariant is of interest not only for the $N \to \infty$ limit of statistical mechanics, but also for low N. This seems to have first been appreciated by Lovelace[11] (who used the $N = 2$ ergodic adiabatic invariant to analyze the compression of a plasma ring confined by large orbit gyrating ions) and by Wong et al.[12] (who used it in formulating a proposed magnetic plasma confinement concept).

At this point it may be instructive to discuss an example of the ergodic adiabatic invariant. Consider the situation shown in Fig. 2, where a point particle P moves in a two-dimensional square container with impenetrable walls of dimension L in the center of which is situated an impenetrable circular barrier of radius R. (The latter may also be thought of as a second large, and very massive, particle.) The dynamics is specified by the constancy of the particle velocity between encounters with the boundaries and by the law of specular reflection when encounters with the boundary occur. The motion of P in this "billiard" is known[13] to be chaotic and ergodic on the energy surface. Setting $E = mv^2/2$ and $A = L^2 - \pi R^2$ we see that the ergodic adiabatic invariant for this $N = 2$ example is $\mu = 2\pi mEA$. Thus, for example, if one of the dimensions, L or R, is varied slowly with time, then the variation of E would be determined by the constancy of μ,

$$E(t) = E(0)A(0)/A(t) \ . \tag{2}$$

This result also has an intimate connection to the adiabatic gas law, pV^γ = constant, where V is the volume, $\gamma = (N+2)/N$, N is the number of degrees of freedom, and p is the pressure $p = nk\tilde{T}$ (with n the particle density and $(N/2)k\tilde{T}$ the average energy of a gas particle). Now consider the particle in Fig. 2, and treat it as if it were a gas. Since $N = 2$, we have $\gamma = 2$ and $V = A$. Also, since we only have one particle, $n = 1/A$ and $k\tilde{T} = E$. Thus $pV^\gamma = EA$, so that constancy of pV^γ implies constancy of EA and hence μ. Thus the single chaotic particle behaves like a gas.

An essential question is <u>how good is the ergodic adiabatic invariant</u>; or, more specifically, what is the error incurred in the statement μ = constant? As we have already mentioned, in the case of $N = 1$ and periodic motion, the error is smaller than any power of ε for sufficiently smooth time variation of H. The case of the ergodic invariant with $N \geq 2$ has been considered theoretically by Ott[14] using a multiple time scale expansion. His main result is an estimate of the typical rms error incurred by the approximation. The error estimate result in Ref. 14 depends on two hypotheses:

(a) The particle orbit in the frozen Hamiltonian is ergodic on the energy surface; and

(b) A certain correlation function C(t) is integrable,

$$\int_0^\infty C(t)dt < \infty.$$

Thus, from (a), the derivation does not apply if islands are present, while, from (b), it does not apply if the relevant correlation function has a long time tail $C(t) \sim t^{-\xi}$ for $t \to \infty$ with $\xi \lesssim 1$. The

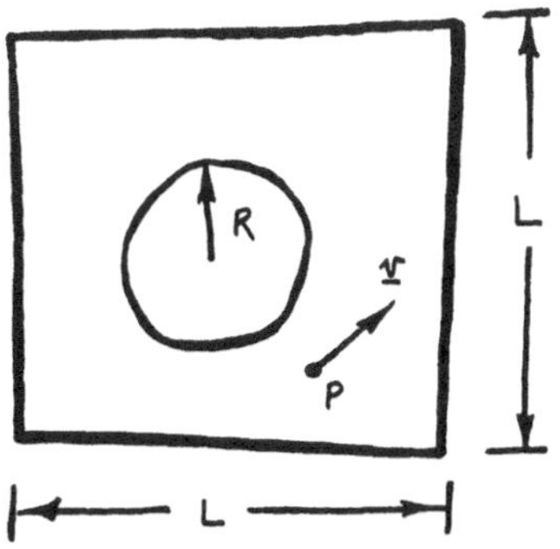

Fig. 2. Square billiard.

setting of Ref. 14 is that of Hamiltonians with smooth dependence on $\underset{\sim}{p}$ and $\underset{\sim}{q}$. However, it will be shown in a future publication (Ref. 17) that similar results apply for the case of chaotic billiard problems. Combining the theoretical results of Ref. 14 with those in Ref. 17, we have the following:

(1) If hypotheses (a) and (b) are satisfied, then the typical rms error in obtaining the value of H from the ergodic adiabatic invariant is of order $\varepsilon^{1/2}$, and this applies both for smooth $(\underset{\sim}{p},\underset{\sim}{q})$ variation (Ref. 14) and for billiards (Ref. 17).

(2) If hypothesis (a) is satisfied, but (b) is violated with $C(t) \sim 1/t$ for large t, then the typical rms error is of the order $(\varepsilon \ln \varepsilon^{-1})^{1/2}$ (Ref. 17).

Note, for example, that $C(t) \sim 1/t$ for the billiard example in Fig. 2. The existence of this type of long time tail for a correlation function in the situation of Fig. 2 has been shown by Zacherl, et al.[15,16] A modification of the billiard in Fig. 2 for which hypothesis (b) is satisfied is shown in Fig. 3.

Contrasting the theoretical results mentioned above (namely error scalings like $\varepsilon^{1/2}$ and $(\varepsilon \ell n \varepsilon^{-1})^{1/2}$) with the results for the case of periodic motion in N = 1 (where the error can be smaller than $O(\varepsilon^m)$ for all finite m), we see that the adiabatic invariant approximation is in general not as good for chaotic motion ($N \geq 2$) as compared to the case of periodic motion.

In Ref. 17, extensive numerical experiments on the error in the ergodic adiabatic invariant are reported for the cases illustrated in Figs. 2 and 3 (both of which satisfy hypothesis (a); cf. Ref. 13). As expected these experiments yield average errors scaling like $T^{-1/2}$ for the case of Fig. 3 and like $T^{-1/2}\log T$ for the case of Fig. 2. Also in Ref. 17 we treat a Hamiltonian $H = (2m)^{-1}(p_x^2 + p_y^2) + V(q_x,q_y,t)$ where V is a smooth function of $\underset{\sim}{q}$. In this case, invariant tori for the frozen Hamiltonian still exist amid chaotic trajectories, and the motion in the frozen Hamiltonian is thus not fully ergodic on the entire energy surface. Nevertheless, the numerical experiments indicate that the ergodic adiabatic invariant is still approximately conserved provided the island regions are small enough. In this case we find numerically that the error appears to scale as ε^α ($\alpha < 1$) for small ε. The numerical experiments to evaluate the average error in the adiabatic invariant are done as follows. At t = 0 a large number M of particles are given positions and momenta assigned randomly with a uniform distribution on the energy surface.

(In the case of billiards, Figs. 2 and 3, this corresponds to uniform distribution in the accessible area and uniform distribution in angular direction of the velocities; the velocity magnitudes are all equal, corresponding to a constant energy.) The Hamiltonian (or size and shape of the billiard) is then slowly varied and the particle orbits calculated. At some final time the particle energies, E_i, are calculated (here i is a particle label). For each E_i a corresponding μ_i is obtained

$$\mu_i = \iint U(E_i - H) d^N \underset{\sim}{p}\, d^N \underset{\sim}{q} ,$$

and compared to μ_0, the initial value of μ. The root mean squared error is then calculated from

$$\text{Error} = \{\frac{1}{M} \sum_{i=1}^{M} (\mu_i - \mu_0)^2\}^{1/2} .$$

We now discuss how the ergodic adiabatic invariant of classical mechanics carries over into a result in the field of quantum chaos. Say

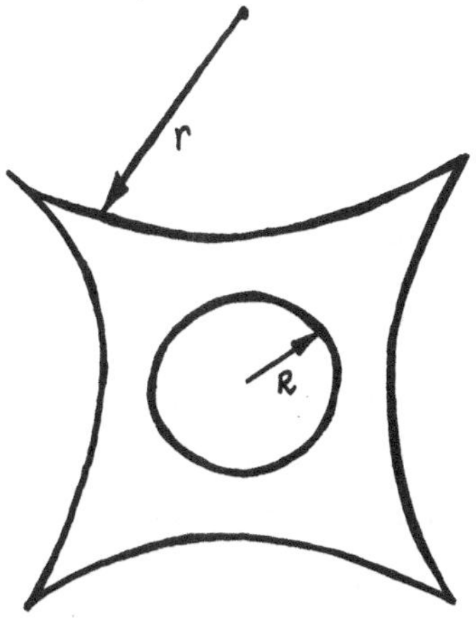

Fig. 3 Chaotic billiard without a long time t^{-1} tail. r is the radius of curvature of the outer walls.

we initialize a quantum system in a state with energy level E_0. We then evolve the system forward in time by slowly changing the Hamiltonian in time. Suppose energy level crossings are avoided for this system. Then, for very slow time variation, the system will remain in the state which is homotopic to its initial state (denote the energy of this state $E(t)$, $E(0) = E_0$). Furthermore, since energy level crossings are avoided, N(E(t)), the number of states with energies less than E(t), is constant in time. From the standard estimate,

$$N(E) \cong \frac{1}{h^N} \iint U[E - H(\underset{\sim}{p},\underset{\sim}{q},\varepsilon t)]d^N\underset{\sim}{p}\, d^N\underset{\sim}{q} ,$$

we can now state the following. The quantum version of the ergodic adiabatic invariant of classical mechanics is the avoidance of quantum energy level crossings for classically chaotic systems.

While we have here addressed the question of the error of ergodic adiabatic invariants only in the classical setting, a quantum treatment is also desirable. In particular, finite ε will induce transitions between states, particularly near the regions where the difference between the energy level E(t) and one of its neighboring states are local minima. Does the quantum calculated error agree with the classical result for $h \to 0$? Differences in this limit have occurred before (e.g., for the kicked rotor problem).

This work was supported by the Office of Naval Research.

The work reported here on the case $C(t) \sim 1/t$ as well as the numerical experiments was done in collaboration with R. Brown and C. Grebogi. Discussions with T. M. Antonsen, Jr. on the quantum correspondence with the classical invariant are gratefully acknowledged.

REFERENCES

1. M. Jammer, The Conceptual Development of Quantum Mechanics (McGraw-Hill, 1966), pp. 96-97.
2. A. Lenard, Ann. Phys. (N.Y.) 6, 261 (1959).
3. C. S. Gardner, Phys. Rev. 115, 791 (1959).
4. M. D. Kruskal, in Plasma Physics (International Atomic Energy Agency, 1965).
5. R. Kulsrud, Phys. Rev. 106 (1957).
6. T. G. Northrop, The Adiabatic Motion of Charged Particles (Interscience, 1963).
7. M. N. Rosenbluth and C. L. Longmire, Ann. Phys. 1, 120 (1957). For a recent example, see T. M. Antonsen, Jr. and Y. C. Lee [Phys. Fluids 25, 132 (1982)].

8. P. Ehrenfest, Phil. Mag. 33, 500 (1917); this paper is reprinted in Sources of Quantum Mechanics edited by B. L. van der Waerden (Dover, 1967).

9. For example, a statement within the context of statistical mechanics appears in the text by R. Kubo [Statistical Mechanics (North-Holland, 1965), p. 14].

10. H. Goldstein, Classical Mechanics (Addison-Wesley, 1959), p. 250.

11. R. V. Lovelace, Phys. Fluids 22, 542 (1979).

12. A. Y. Wong, Y. Nakamura, B. H. Quon, and J. M. Dawson, Phys. Rev. Lett. 35, 1156 (1975).

13. Ya. G. Sinai, Russ. Math. Surv. 25, 137 (1970).

14. E. Ott, Phys. Rev. Lett. 42, 1628 (1979).

15. A. Zacherl, T. Geisel, J. Nierwetberg, and G. Radons, Phys. Lett. 114A, 317 (1986).

16. For other work on long time tails in billiards see J. Machta [J. Stat. Phys. 32, 555 (1983)], and F. Vivaldi, G. Casati, and I. Guarneri [Phys. Rev. Lett. 51, 727 (1983)].

17. R. Brown, Ph.D. thesis, University of Maryland; R. Brown, C. Grebogi, and E. Ott (to be published).

SEQUENTIAL MEASUREMENTS IN QUANTUM MECHANICS

Willis E. Lamb, Jr.

University of Arizona
Tucson, AZ 85721, USA

None of the physicists who founded Quantum Mechanics in 1925-26, Heisenberg, Schrödinger, Dirac or Born, explained how measurements of physical quantities could be made. It is necessary to make additional assumptions to describe the measurement process in physical terms. Many schemes for doing this have been proposed in subsequent years.

In his 1927 paper on the uncertainty relationship, Heisenberg[1] mentioned the use of a "gamma ray microscope" to determine the position of a particle. He gave more details in his 1929 Chicago lectures[2]. However, neither he, or anyone else, ever gave a quantum mechanical theory of this device. Heisenberg did indicate how momentum of an electron in an atom might be measured: *Suddenly remove the Coulomb interaction between the electron and the nucleus, and make a time-of-flight measurement on the arrival of the now free electron at a suitably placed detector*. The momentum of the electron can then be calculated from the distance, time and mass. The method involved complete destruction of the system.

In his 1930 book, Dirac[3] took it for granted that measurements could be made, but was very vague about what was actually involved. He did assert that an observation of an observable always gave as a result one of the eigenvalues of that observable, and that two measurements of the same observable made in rapid succession would give equal results.

Von Neumann[4] introduced the reduction of wave function hypothesis in his 1932 book. The Dirac and the von Neumann formulations were essentially equivalent. In both of them, the measurement of an observable gave one of the eigenvalues of that observable. In some unexplained manner, the system was thereby brought into a state which was an eigenstate of the observable corresponding to the result of the measurement. Neither author discussed his measurements in physical terms. I regard this to be a fatal flaw. It is impossible to make a measurement by only talking about making a measurement. The experimental procedure has to be modelled by the theory. At some point in the argument, an addition to the laws of quantum mechanics has to be assumed in order to make possible a transition from the quantum to a classical description.

Pauli's 1933 Handbuch article[5] showed how an apparatus of the Stern-Gerlach type could be used to determine probability distributions for a wave function which was a linear combination of eigenstates of its Hamiltonian. The method was bound to destroy the system. The Stern-Gerlach technique could be used as a method to prepare a new state, but that is not needed here, and would completely miss the point of the exercise. We have already prepared the state of the system of interest, and only want to get statistical information about it.

In 1968-69, I considered quantum observations from an "operational" view point[6]. I made some assumptions about the experimental availability of a wide class of Hermitean operators which could be incorporated in the Hamiltonian for the the system under consideration. The discussion was mostly for one degree of freedom systems and was thoroughly nonrelativistic. Hence the velocity of light was taken infinite. There was no magnetic field, vector potential or spin. There were no questions about locality. I showed how any desired initial state of such a system could be prepared, and gave procedures for measuring "any" Hermitean observable. The Pauli (Stern-Gerlach) technique was adopted for the final stage of the measurement process.

The methods of the last two paragraphs are very destructive of the system. They permit one and only one measurement. As someone has said, "A cat can look at a Queen." Applied to the present situation, I would have to add, "but only once." Until recently, I did not attempt to discuss how a sequence of observations on a system could be made, and in fact, am still inclined to believe that there is no reason, except to make us feel good, why microscopic quantum theory should permit such a luxury.

There are several ways in which quantum mechanics might be applied to macroscopic problems. I mention only four. (1) A many body problem, such as that met in superconductivity theory, can be approximately manipulated to allow description by a "wave function", which is really not a wave function of a particle at all, but only the expectation value of a product of two quantum field operators. (2) Laser theory with bistability. (3) The kind of problem that celestial mechanicians would have if they had to treat the motion of a planet by quantum mechanics instead of Newtonian mechanics. (4) Someone interested in using a simple, large, and highly isolated, mechanical system to detect a very small force such as that due to gravity waves[7] is not primarily interested in checking out the predictions of quantum mechanics for a gravitational disturbance of known temporal dependence. Rather he is interested in using the behavior of the system to find out the time dependence of the pulse. I suspect that a problem of category (1) or (2) has lost some of its physics in the move from the many body to the few degree of freedom level. Hence work on quantum tunneling in Josephson junctions or laser physics might not be closely connected with the more fundamental problems of measurement in quantum mechanics. In (3), we have enormously important perturbations from interplanetary material, tidal effects and radiation pressure, etc. which are sufficient to overwhelm the quantum corrections. Only in problems of type (4) can one even *begin* to pretend that his system is sufficiently well isolated.

The present paper considers a number of ways in which quantum mechanics can be extended to allow a sequence of observations to be made on a carefully isolated, although rather large and otherwise macroscopic system. The object is not to find the best method of practical measurement, but only to examine methods which can be carefully and clearly explained, especially in respect to any proposed additions to the laws of quantum mechanics. The methods will be illustrated by considering a simple model for a gravity wave detector. This consists of a one dimensional harmonic oscillator with cartesian coordinate x, mass m, circular frequency ω, and momentum $p = m\,\frac{dx}{dt}$ which is acted on by a time dependent spatially uniform force field F(t). The Hamiltonian for this system is

$$H = H_0 - F(t)\,x\,,$$

where the unperturbed Hamiltonian $H_0 = T + V$ is the sum of two terms:

kinetic energy $T = \frac{1}{2}\frac{p^2}{m}$ and potential energy $V = \frac{1}{2} m\,\omega^2 x^2$.

The time dependent force F(t) is unknown, but hopefully it can be determined by a series of measurements of some observable quantity of the quantum mechanical system. We first take the unperturbed Hamiltonian H_0 as the quantity to be measured. Let the system's normalized wave function be $\psi(x)$ just before the time t_1 of the first measurement. In general, $\psi(x)$ is a linear combination of the eigenfunctions of H_0 of the form

$$\psi = \Sigma\, c_n\, u_n(x)\,.$$

I do not like the Dirac-von Neumann hypothesis. My reasons for this have been spelled out in a number of papers[8], and mainly involve the circumstance that measurement invariably enlarges the dynamical system under consideration, and there is no way to get back to a pure case wave function of the original system. However, to clarify ideas, I will begin the discussion by taking the Dirac-von Neumann assumption literally. According to this, the result of the measurement will be one of the eigenvalues of H_0, say

$$E_{n_1} = \left(n_1 + \frac{1}{2}\right)\hbar\,\omega\,,$$

where the quantum number n_1 can be any one of the integers 0, 1, 2, ... The wave function $\psi(x)$ is changed by the measurement into the n_1-th normalized eigenfunction of H_0, $u_{n_1}(x)$. This wave function then evolves in time according to the time dependent Schrödinger equation until the time t_2 of the second measurement of H_0. At this time the result of energy measurement is characterized by a quantum number which we denote by n_2. This process will be repeated many times. We have periods of causal time evolution of the wave function separated by instantaneous measurements of H_0. The results of each measurement are randomly selected from a probability distribution of n values given by $|c_n|^2$ where the c_n are the probability amplitudes of the current $\psi(x)$ in the H_0 representation.

With this measurement strategy, we will end up with a sequence of n values. Of course, if $F(t)$ is always zero, the n values will all be equal to each other. In the general case, each entry in the sequence will be chosen for us by a genuinely random process. It is easily possible to program a computer to make calculations of such a sequence of n values if we assume a knowledge of the time dependence $F(t)$ of the gravitational force. If we repeat the whole calculation we will get a different sequence of n values. If we wish, we can build up an ensemble of sequences. We are, of course, interested in an inverse problem: Given *one* sequence of n values, what can we learn about $F(t)$. (An ensemble of detector systems would be nice, but quickly runs into a cost barrier.) I don't claim to know the best solution of this problem, but I would try to use a least squares method based on working out n sequences for known force fields of the form $F(t; \alpha, \beta, \gamma, \ldots)$ where α, β, γ, etc. are parameters that characterize a plausible form for $F(t)$. Some kind of figure of merit function depending on the elements of the n sequence and the parameters would have to be maximized to determine the best values of α, β, γ, etc.

At present, this particular rather easy calculation has not been carried out. Instead I have considered measurement of the position coordinate x. The results of such calculations were reported at the 2nd International Symposium on Foundations of Quantum Mechanics[9] held in Tokyo early in September, 1986. The results of this treatment support the expectation that something could be learned about $F(t)$ from such measurements, if only one could make them. However, it became clear that the original Dirac-von Neumann hypothesis involves too strong a disturbance of the system by the measurement process. Instead, I will describe a better method which incidentally includes the problem discussed in Tokyo as a kind of special case.

For this purpose, I now turn to the use of a "meter" system. Of course, the Pauli's Stern-Gerlach method is an example of the use of a meter. A coordinate of the center of mass of the atomic system which is sent through the inhomogeneous electric or magnetic fields plays the role of a needle pointer for the meter. This method is clumsy, time-consuming, and it wipes out any past history of the system. Later authors have considered more general types of measuring instruments. Like many of them, I take the meter to have a cartesian coordinate X, and a (very large) mass M. Initially the meter system has been put in a state with a normalized wave function $\chi(X)$. The wave function of the combined x, X system is then a simple product $\psi(x)\,\chi(X)$. The two parts of the system are allowed to interact for a short, but not necessarily infinitesimal, time, and the combined system wave function becomes a non-product wave function $\Phi(x, X)$. One then turns off the interaction between the x and X parts of the system, and makes a measurement of X, thereby hoping to learn something worth knowing about the x system. Presumably, the nature of the interaction between x and X has been such that there is a meter-like quality hidden in the wave function $\Phi(x, X)$. There is a certain circularity in this plan, since if we knew how to measure something for the X system, we could use the method directly for the x system. To get around this difficulty, we can try to say that the X system has such a large mass that we can shift from a quantum treatment to a classical one when we find it convenient, and that then we can measure X by merely saying that we measure X. It is far better to make such an assumption for the X rather than for the x system!

Actually, in the last two pages of his 1932 book, von Neumann[4] did give an example of a specific model for the the kind of meter mentioned above. The interaction potential was taken to be pulse-like in time,

$$V = x P \delta(t) ,$$

and depended on a product of the coordinate x of the particle and the momentum P associated with the meter coordinate X. To get an idea what this interaction might do for us, consider the fully classical problem of the effect of the pulsed interaction potential, described by Hamilton's equations of motion,

$$\frac{\partial X}{\partial t} = x \delta(t) \qquad \frac{\partial P}{\partial t} = 0$$

$$\frac{\partial x}{\partial t} = 0 \qquad \frac{\partial p}{\partial t} = - P \delta(t)$$

After the pulse interaction, the dynamical variables x, p, X, and P have become

$$x = x_0,\ p = p_0 - P_0,\ X = X_0 + x_0, \text{ and } P = P_0 ,$$

instead of their initial values x_0, p_0, X_0 and P_0. We could learn the value of the initial coordinate x_0 by measuring the displacement $X - X_0$ of the meter pointer. Something like this same feature is found in the quantum treatment: By integrating the wave equation through the very short duration of the interaction, one finds that the wave function of the combined system has become[10]

$$\Phi(x, X) = \psi(x) \chi(X - x) ,$$

so that if $\chi(X)$ is a very sharply peaked function, a measurement of X will give a close approximation to the value of x.

There are a quite a few things about this model that spoil our pleasure. The delta function of time will have to be realized physically by use of a pulse of short but not zero time duration. That will complicate the problem, since the forces internal to the system will contribute to the time evolution during the pulse. We have to have a good supply of measuring systems in a prepared quantum state $\chi(X)$. The interaction potential can not be realized physically. It does remind us of a *part* of the Larmor interaction Hamiltonian for a charged particle in a magnetic field. I regard it as completely unacceptable. I have found a model with a more physically reasonable form of interaction, which will be described below. However, the von Neumann model is computationally simple, and easy and convenient to work with. It does serve to illustrate some important features of the measurement process. Hence, I will show you the working out of some numerical consequences of it in a situation which might be met in gravity wave detection.

In order to give an accurate measurement of x, the meter's initial wave function $\chi(X)$ has to be very sharply peaked. Then, as we will see, the x system after the measurement will have been very strongly disturbed.

A very serious difficulty is that after the interaction, we no longer have the original x system, but only the coupled x, X system. The X part of the system can be ignored dynamically, but not kinematically. We would have to say of the x system that it no longer has a wave function, but only a statistical distribution of wave functions which can be described by a density matrix, which, in the coordinate representation, is given by

$$\rho(x', x'') = \int dX\, \Phi(x'', X)^* \, \Phi(x', X) .$$

What can we do now? It seems pretty clear that to make measurements one will have to add some new postulate to the theory. Having rejected the Dirac-von Neumann reduction hypothesis as applied to the x system, we might still apply a variation of it to the x, X system. We assume that when the combined wave function was $\Phi(x, X)$ in the x, X configuration space, and we have measured X, obtaining a particular result X = b, the x system again gets a pure case wave function which is just equal to

$$\psi(x) = \Phi(x, (X)) ,$$

where (X) is no longer a dynamical variable, but merely a parameter with the value $X = b$ obtained in the measurement of X. The wave function $\psi(x)$ will have to be renormalized as for a single degree of freedom system. The meter system X has completely left the stage. Later on, its place will be taken by a new meter system X when the next measurement is to be made.

Clearly, if one thought that the above assumption was too invasive of the x system, it would be possible to introduce another meter system with coordinate Y for measurement of meter coordinate X. In that case, after measuring Y the x, X system would be assumed to have a wave function, say, $\Xi(x, X, (Y))$ and the decoupled x subsystem would be described by a density matrix. The system described by the density matrix would have to be followed by consideration of an ensemble of calculations with wave functions randomly selected from the distribution of wave functions of the mixture appropriate to the density matrix. If this complication was bearable, one might introduce a further system Z. Etc., etc. Fortunately, one can show that if the next meter is much heavier than the previous one, the new result is very close to the old one.

I first saw something like the above modified Dirac-von Neumann hypothesis in a paper by Arthurs and Kelly[11] of 1965, who applied it to "simultaneous" measurement of x and p.

For the present calculations, I have combined the von-Neumann meter model and our modified Dirac-von Neumann reduction hypothesis. For the system, we have one dimensional motion of a particle, either free, or moving in a force field described by a potential function of x with at most linear and quadratic terms. The terms in the potential can have any desired time dependence. Time will be divided up in a multilayer sandwich made of intervals of causal evolution under the (possibly time dependent) Hamiltonian, each separated by the short burst of a measurement processes of the von Neumann type. The time dependence of the wave function $\psi(x, t)$ during the causal periods can be dealt with easily, either analytically, or with a computer. It turns out that for our choice of Hamiltonian we can take advantage of the circumstance that some particular solutions of the time dependent wave equation, apart from a normalization factor, can be written in the form of a complex Gaussian exponential function of the form

$$\psi(x, t) = \exp(- \alpha (x - a)^2)$$

where α and a are complex functions of time. (The case of an infinite plane wave function can be approached in a limiting way, when α is real and small, and a is large and purely imaginary.)

Substitution of this ψ into the time dependent wave equation gives two coupled ordinary differential equations for the complex functions $\alpha(t)$ and $a(t)$. (It is possible to append a normalization factor $N(t)$ to $\psi(x, t)$, but we do not need to do this, because the normalization factor can be calculated easily in terms of $\alpha(t)$ and $a(t)$ whenever we want it.) The differential equations are

$$\frac{da}{dt} = \frac{i}{2 \alpha \hbar} (F - m \omega^2 a)$$

$$\frac{d\alpha}{dt} = \frac{i}{\hbar} \left[- \frac{2 \alpha^2 \hbar^2}{m} + \frac{1}{2} m \omega^2 \right] .$$

In some cases, these equations can be solved analytically, but even when F, (and possibly ω and m), depend on time, they are easily solved on a computer.

It should be pointed out that not all wave functions have the complex Gaussian exponential form assumed above. However, when that form is appropriate, the amount of calculation required to find the time evolution of a wave function is greatly reduced. Even if that short-cut were not available, the time dependent Schrödinger equation could be solved numerically with a computer.

The meter measurement process produces a discrete change in the wave function $\psi(x)$ as outlined above, and described in more detail in the following discussion. This change depends on the result of the measurement. The result of measurement is chosen at random using the appropriate probability distribution for getting that result. One feature of this model is that the system is described by a wave function at all times. However, each measurement involves a random process, so that if the calculations are repeated from the beginning, a different history of the time evolution will be obtained. One must decide whether a complete ensemble of histories has to be calculated, or whether one or a few histories will suffice for the purposes at hand.

It is also very convenient to use wave functions in Gaussian exponential form for the meter as well as for the system. Thus we will take the meter wave function to be

$$\exp(- \beta X^2)$$

where β is a real constant which measures the inverse square width of the meter's Gaussian. It will not be necessary to supply normalization factors in the following discussion. Such a Gaussian could be regarded as the ground state wave function for a simple harmonic oscillator problem for the X system. Before the meter is applied to the system, the spring mechanism exerting forces on X is thrown away. Only the mass M, which has then been given a Gaussian exponential wave function $\chi(X)$, is needed.

We now work out the effect of a measurement on the wave function of the x system. Let this wave function at the end of one of its causal intervals be

$$\psi(x) = \exp(- \alpha_1 (x - a_1)^2) .$$

The coupling of the x system with a meter system X gives a wave function

$$\Phi(x, X) = \exp(- \alpha_1 (x - a_1)^2) \exp(- \beta (x - X)^2)$$

for the combined system. According to the modified reduction hypothesis, when the result of the measurement of X gives the value $X = b$, we assume that the state of the x system becomes

$$\psi(x) = \exp(- \alpha_1 (x - a_1)^2) \exp(- \beta (x - b)^2) .$$

By completing squares, we find that this wave function is equivalent to

$$\psi(x) = \exp(- \alpha_2 (x - a_2)^2) ,$$

which has the same form as the wave function just before the measurement, except that the parameters α_1 and a_1 have changed to α_2 and a_2 given by the mappings

$$\alpha_2 = \alpha_1 + \beta \quad \text{and} \quad a_2 = \frac{\alpha_1 a_1 + \beta b}{\alpha_1 + \beta} .$$

The next stage in the calculation is to obtain from the wave function $\Phi(x, X)$ the probability density that a value $X = b$ is obtained in the measurement. This is given by

$$W(X) = \int dx \, \Phi(x, X)^* \Phi(x, X) ,$$

which is a Gaussian distribution in X with a peak at $X = X_p$ with

$$X_p = \frac{\mathrm{Re}(\alpha_1 a_1)}{\mathrm{Re}(\alpha_1)} .$$

The normalized probability distribution for the value of b is equal to

$$W(b) = \frac{1}{\sqrt{2 \pi}} \frac{1}{\sigma} \exp \left[- \frac{1}{2} \frac{(b - X_p)^2}{\sigma^2} \right] ,$$

where the standard deviation σ is given by

$$\sigma^2 = \frac{1}{4} \left[\frac{1}{\mathrm{Re}(\alpha_1)} + \frac{1}{\beta} \right] .$$

Each time a measurement is to be made, the computer uses its random number generator to select a value for $X = b$. The system wave function is changed according to the above rules, and the next stage of causal evolution occurs. At the end of the run, the whole history of the wave function has been recorded, together with the results of the sequence of measurements. For gravity wave detection, one would not know the force $F(t)$, and would have to solve the inverse problem of getting $F(t)$ from the sequences of the b values. As desecibed earlier, a reasonable prescription can be given for approximate solutions of such a problem.

Since we have no data to analyze yet, I will show results for calculations made on a massive free particle which has been acted on by a gravity wave force given by a pulse rectangular in time. The elapsed time is shown horizontally and the centroid of the wave packet is plotted vertically in each of the four figures. The following parameters are taken: $m = 10^6$ gm, $\omega = 0\ \text{sec}^{-1}$, with the initial $\alpha = 10^{32}\ \text{cm}^{-2}$. There is a pulse force of $F = 2.5 \times 10^{-12}$ dynes, lasting for 100 sec, applied at the mid point of the total elapsed time of 10,000 sec. Four cases are considered: Fig. 1 has $\beta = 10^{26}\ \text{cm}^{-2}$, Fig. 2 has $\beta = 10^{28}\ \text{cm}^{-2}$, Fig. 3 has $\beta = 10^{30}\ \text{cm}^{-2}$, and Fig. 4 has $\beta = 10^{32}$. In each case, five histories are shown. One of the five shows the simple history of the centroid of the wave packet obtained when no disturbing measurements are made. In the other four runs, a von Neumann measurement of x is made once every 100 seconds. The four runs differ from each other because of the use of random numbers in the computational algorithm. Incidentally, we could have plotted the measurement values b instead of the wave packet centroid without changing the figures very much.

As I have pointed out, the von Neumann pulsed meter interaction is physically unsatisfactory. This difficulty can be surmounted as follows: We had taken the initial wave function for the meter to be a ground state simple harmonic oscillator wave function. Now we endow the meter with the same kind of dynamics given to the system. We had two coordinates, x and X. We will now replace them by x_1 and x_2, respectively. The quantities like m, x, p, ω, F, α, a, etc. will be given subscripts $I = 1, 2$. The Hamiltonian H for the combined system is taken to be $H = H_1 + H_2 + V(x_1, x_2)$. The interaction between the two systems, $V(x_1, x_2)$, will be specified below. The uncoupled Hamiltonian for each of the two systems is essentially what we had before for the x system:

$$H_I = \frac{p_I^2}{2m_I} + \frac{1}{2} m_I \omega_I^2 x_I^2 - F_I(t) x_I$$

for $I = 1, 2$. The coupling V between system and meter is taken to be

$$V = -\kappa x_1 x_2 ,$$

with a coupling constant κ which could depend on time. The V expression could just as well be written as $\frac{\kappa}{2}(x_1 - x_2)^2$. That would merely make some compensating changes in the coefficients of the x_I^2 terms.

The wave function is taken to be a two particle complex Gaussian

$$\psi(x_1, x_2, t) = \exp(-f(x_1, x_2))$$

where

$$f(x_1, x_2) = \sum_{I=1}^{2} \alpha_I (x_I - a_I)^2 - 2\gamma x_1 x_2 .$$

The time dependent wave equation for $\psi(x_1, x_2, t)$ implies a system of five ordinary differential equations for the time dependence of the five complex parameters: two $\alpha_I(t)$, two $a_I(t)$ and $\gamma(t)$. These equations are written below. The index I can be 1 or 2. If $I = 1$, the index J is 2, and if $I = 2$, $J = 1$.

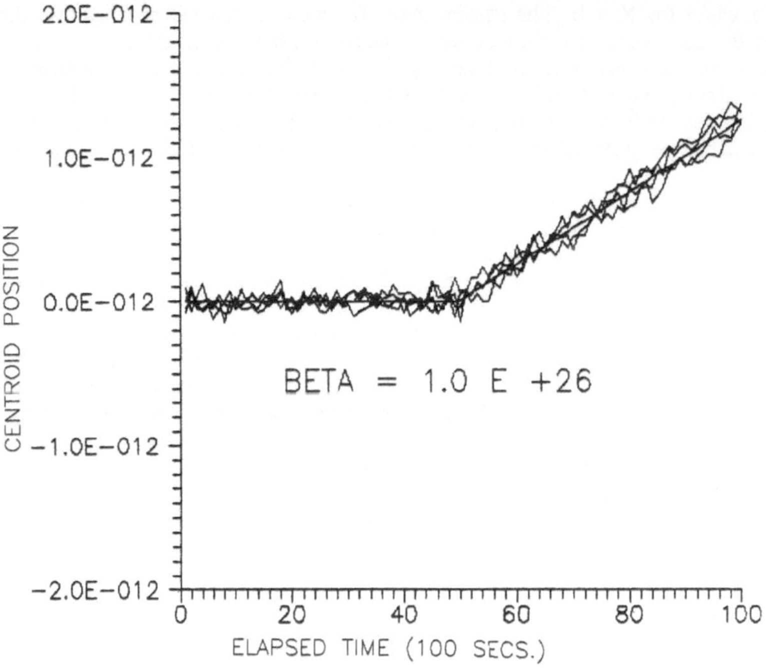

Figure 1.

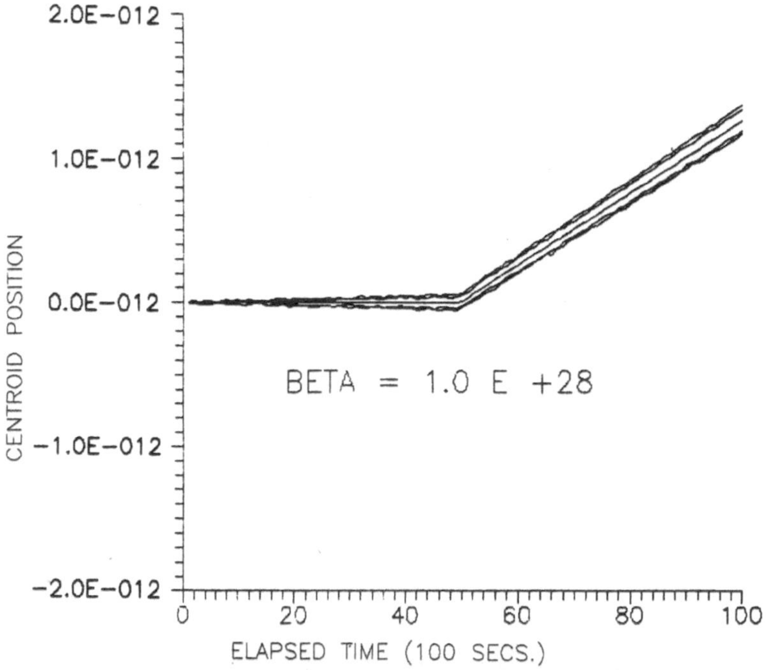

Figure 2.

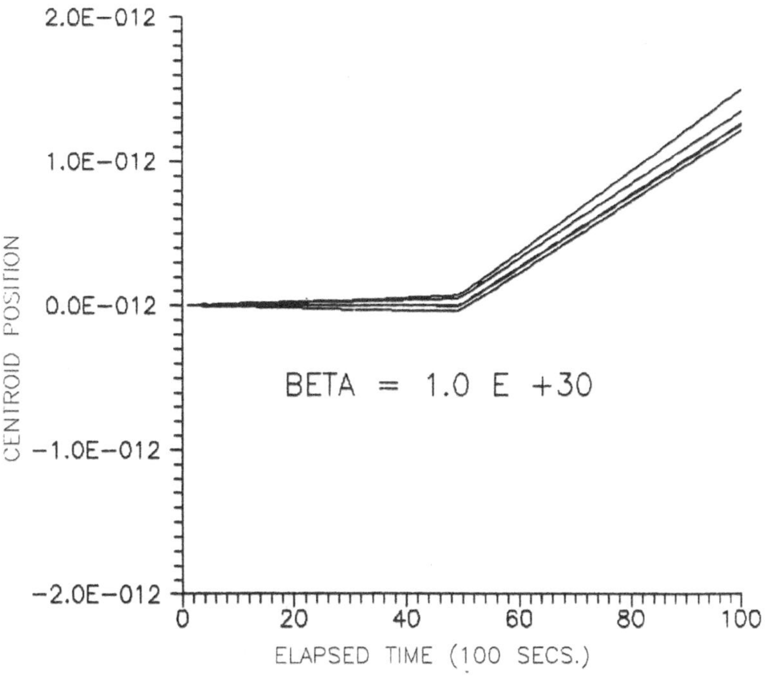

Figure 3.

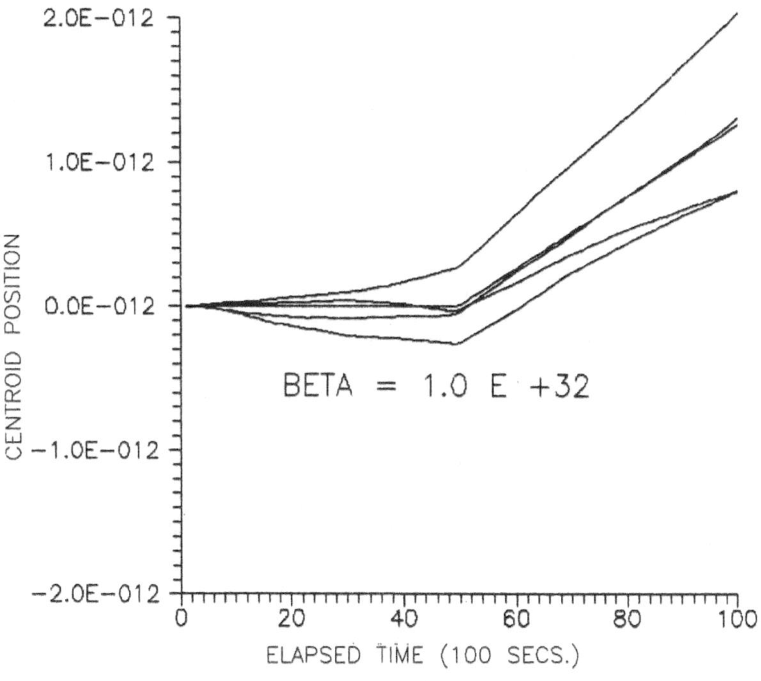

Figure 4.

$$\frac{d\alpha_I}{dt} = \frac{i}{2\hbar}\left[m_I\,\omega_I^2 - 2\,\hbar^2 \left[\frac{\alpha_I^2}{m_I} + \frac{\gamma^2}{m_J} \right]\right]$$

$$\frac{da_I}{dt} = \frac{i}{2\,\hbar\,\alpha_I}\left[F_I - m_I\,\omega_I^2\,a_I + \hbar^2\,\frac{4\gamma}{m_J}(\,\alpha_J + \gamma\,) \right]$$

$$\frac{d\gamma}{dt} = -\frac{i}{2}\frac{\kappa}{\hbar} + 2\,i\,\hbar\left[\frac{\alpha_1}{m_1} + \frac{\alpha_2}{m_2} \right]\gamma\,.$$

These five equations determine what happens to the combination of the x_1 system and the x_2 meter during their time of interaction. The force F_1 describes the gravity wave force on the system. We might, or might not, want to have a similar force F_2 on the meter during the relatively short time it was in place. The model can allow for a number of different types of meters during the sequence of measurements, and various choices of the meter variable to be read can be made. In no case do we need there to be any *direct* observation of the system variables. It might be mentioned in passing that it is easy to extend the above treatment to allow for more than two coupled oscillators.

A number of people have worried about the instantaneous measurements made by the von Neumann meter. We can easily make provision in the new model for any desired timing strategy by slight changes in the computer program. With a little thought, we can see that the exponential factor in the wave function which involves γ can give a correlation between x and X of the same form met with the von Neumann meter. Consider the differential equation for $\gamma(\,t\,)$. At the beginning of the system-meter interaction, $\gamma = 0$. If the solution is expanded in a Taylor's series in t, the first order term is $\gamma_1(\,t\,) = -\frac{i}{2}\frac{\kappa}{\hbar}\,t$ which is purely imaginary. The second order term $\gamma_2(\,t\,)$ can be more generally complex, depending on the values of the α's. If γ were purely imaginary, the probability distribution W(b) would be a property of the meter wave function parameters α_2 and a_2 only, and independent of the parameters α_1 and a_1 which contain information about the system. As a result, although a measurement would have a disturbing effect on the wave function of the system, nothing about the system would be learned in the measurement process.

On the other hand, when γ is complex, the meter and system have the same *kind* of correlation involving a Gaussian exponential of x - X similar to that met with the von Neumann meter. For this reason, the plots that were shown for the von Neumann meter are qualitatively pretty much like those that would be obtained for the harmonic oscillator meter model.

What conclusion can be drawn about the possibility of making sequential measurements? I still think it is likely that quantum mechanics simply cannot describe such things. At least, the simple harmonic meter model seems to be vastly better than the von Neumann model, and it ought to be a lot easier to realize experimentally. Of course, we are still making the modified Dirac-von Neumann hypothesis as a new assumption of quantum mechanics. Very likely, the question of its validity can only be answered by experiments in a laboratory.

The whole discussion has been non-relativistic. We don't know what a future theory generalizing quantum mechanics will look like, and certainly there will be many surprises. Classical physics looked very good in 1886, and certainly there are many things that look very good about quantum mechanics in 1986.

REFERENCES

1. W. Heisenberg, "On the Intuitive Content of the Quantum Mechanical Kinematics and Mechanics", Zeit. Phys. *43*, 172-198 (1927).
2. W. Heisenberg, *The Physical Principles of the Quantum Theory*, (Chicago University Press, 1930).
3. P. A. M. Dirac, *The Principles of Quantum Mechanics*, 1st ed., (Oxford, 1930).
4. J. von Neumann, *Mathematical Foundations of the Quantum Mechanics*, (Springer, 1932); also reprinted in translation, (Princeton, 1955).
5. W. Pauli, "The General Principles of Wave Mechanics", *Handbuch der Physik*, vol. 24/1, pp. 83-272, ed. H. Geiger and K. Scheel, (Springer, 1933).
6. W. E. Lamb, Jr., "An Operational Interpretation of Non-Relativistic Quantum Mechanics", Physics Today *22*, April, 23-28 (1969).
7. It may avoid some confusion if it is pointed out that *quantum* aspects of gravity can be completely neglected in the present discussion.
8. W. E. Lamb, Jr., "Von Neumann's Reduction of the Wave Function", in *Proceedings of Fourth International Conference on the Unity of the Sciences*, pp. 297-303 (International Cultural Foundation, Tarrytown, N. Y., 1975);

   "Remarks on the Interpretation of Quantum Mechanics", in Ta-Yu Wu Festschrift: *Science of Matter*, pp. 1-8, ed. S. Fujita (Gordon and Breach, 1979);

   "A Letter to Kip Thorne", in *CCNY Physics Symposium in Celebration of Melvin Lax's Sixtieth Birthday*, pp. 40-43, ed. H. Falk (CCNY, New York, 1983);

   "Schrödinger's Cat", in Dirac Festschrift volume, eds. B. N. Kursunoglu and E. P. Wigner (Cambridge, to be published).
9. W. E. Lamb, Jr., "Theory of Quantum Mechanical Measurement", in *Proceedings of the 2nd International Symposium on the Foundations of Quantum Mechanics*, September, 1986. (To be published by the Physical Society of Japan, 1987); See also, "Quantum Theory of Measurement", in a conference *New Techniques and Ideas in Quantum Measurement Theory*, January 21-24, 1986, New York, N. Y. (To be published by the New York Academy of Sciences, 1987).
10. For the derivation of this equation, see the last two pages of Ref. 4.
11. E. Arthurs and J. Kelly, "On the Simultaneous Measurement of a Pair of Conjugate Observables", Bell System Technical J. *44*, 725-729 (1965).

MEASUREMENTS DISTRIBUTED IN TIME

Carlton M. Caves

Theoretical Astrophysics 130-33
California Institute of Technology
Pasadena, California 91125

I. INTRODUCTION

Consider a measurement that provides information about the behavior of a quantum system during a nonzero time interval. The result of such a measurement is a single number, but this single number conveys some sort of averaged information about the behavior of the system. I have referred to such measurements as "measurements distributed in time";[1,2] they have also been called "multiple-time measurements" or "measurements of multiple-time observables."[3–6] Time-distributed measurements should be contrasted with instantaneous measurements, which provide information about a system at a single time.

Why consider time-distributed measurements? Because real measurements are never instantaneous; they are always distributed in time. To make an instantaneous measurement requires coupling the system to a measuring apparatus with infinitely strong coupling. Moreover, aside from the practical impossibility of making instantaneous measurements, one is often interested not in instantaneous information, but rather in a particular behavior of some quantity as a function of time. A good example, which often occurs in practice, is that the desired information is contained in a temporal Fourier component of some quantity.

The job of quantum mechanics is to provide probabilities for the results of measurements. For an ideal instantaneous measurement, the desired probabilities can be generated directly from the state vector of the system. Not so for a time-distributed measurement. The system state vector at a particular time contains all information about the system at that time, but it contains no multiple-time information. There is no way to generate the statistics of a time-distributed measurement from the system state vector alone.

On the other hand, one knows perfectly well how to proceed. Extend the quantum-mechanical description to include a measuring apparatus, which interacts with the system, and choose the interaction so that some observable of the measuring apparatus collects and stores the desired multiple-time information about the system. One can then "read off" the multiple-time information about the system by measuring the appropriate observable of the measuring apparatus. A time-distributed measurement on the system becomes an instantaneous measurement on the measuring apparatus. The statistics of the measurement come directly from the joint state vector of the system and the measuring apparatus.

In this paper I sketch a model for time-distributed measurements. The model involves coupling the system to a very simple measuring apparatus, which I call a "meter." Though not very realistic, the model has the advantage that one can identify precisely what multiple-time quantity is being measured.

Despite its idealization, the model is useful because I believe it corresponds to the "best possible" measurement of a multiple-time observable—"best possible" meaning that the measurements described by the model are the least intrusive, that they disturb the system least. It is interesting that the description of these best possible time-distributed measurements can be reformulated in a language that involves the system alone and avoids explicit reference to the meter. This reformulation does not, of course, use the language of a system state vector involving in time. Instead, it is cast in terms of a "sum over histories" of the system. A sum over histories is ideally suited to describing time-distributed measurements because the histories—more precisely, the probability amplitudes for the histories—contain the desired multiple-time information about the system.

Sections II and III develop the measurement model and the sum-over-histories formulation; they are based on two recent papers,[1,2] which the reader may consult for further details. Similar ideas have been explored by Aharonov, Albert, and D'Amato.[3–6] Section IV applies the sum-over-histories formulation to linear measurements on a harmonic oscillator.

II. MEASUREMENT MODEL

Consider a quantum-mechanical system and an observable A of the system. Associated with A is a Hermitian operator $\hat{A}$, which has eigenvalues labeled by A and eigenvectors $|A\rangle$. Suppose one wishes to measure the multiple-time quantity

$$y \equiv \int_0^T dt\, Y_t(A(t)) , \tag{2.1}$$

where Y_t is an arbitrary real-valued function which depends on t and which vanishes identically for $t<0$ or $t>T$. A measurement of y clearly yields information about the behavior of A during the time interval $[0,T]$; it is a time-distributed measurement on the system.

It is trivial to devise a coupling to a measuring apparatus which realizes a measurement of y. Let the measuring apparatus be a one-dimensional, quantum-mechanical system with canonical coordinate $\overline{y}$ and canonical momentum $\overline{p}$. The associated operators, $\hat{\overline{y}}$ and $\hat{\overline{p}}$, obey the canonical commutation relation $[\hat{\overline{y}},\hat{\overline{p}}]=i\hbar$. I call this measuring apparatus a "meter," and I assume that the meter has no dynamics of its own—i.e., its self-Hamiltonian is zero. Suppose further that the interaction of the meter with the system is described by an interaction Hamiltonian $\hat{\overline{p}}Y_t(\hat{A})$. Then the total Hamiltonian for the system and the meter is

$$\hat{H}_{\rm tot}=\hat{H}+\hat{\overline{p}}Y_t(\hat{A}) , \tag{2.2}$$

where $\hat{H}$ is the Hamiltonian of the system.

Why does this total Hamiltonian realize a measurement of y? The reason is most evident in the Heisenberg picture, where the equation of motion for the meter coordinate is

$$\frac{d\hat{\overline{y}}}{dt}=-\frac{i}{\hbar}[\hat{\overline{y}},\hat{H}_{\rm tot}]=Y_t(\hat{A}) . \tag{2.3}$$

For times $t \geqq T$, the meter coordinate becomes

$$\hat{\bar{y}}(t \geqq T)=\hat{\bar{y}}(0)+\int_0^T dt\, Y_t(\hat{A}(t)) . \tag{2.4}$$

Thus the meter coordinate is displaced by the quantity one wants to measure. The meter coordinate collects (due to the interaction with the system) and stores (due to its conservation in the absence of the interaction) the desired multiple-time information about the observable A. One can "read off" this multiple-time information by measuring the meter coordinate at any time $t \geqq T$.

Notice that the accuracy of this time-distributed measurement is degraded by the initial uncertainty in the meter coordinate. In addition, the system is disturbed by the interaction with the meter—a disturbance characterized by the uncertainty in the meter momentum. Unless A happens to be a "quantum nondemolition observable,"[7] this disturbance affects the Heisenberg operator $\hat{A}(t)$—and, hence, disturbs the time-distributed measurement. Thus, in general, this model of a time-distributed measurement involves a trade-off between an "additive noise" due to the uncertainty in the meter coordinate and a "back-action disturbance" due to the uncertainty in the meter momentum.

The statistics of the time-distributed measurement are, of course, displayed most easily in terms of a joint wave function for the system and the meter. At $t=0$ let $|\psi\rangle$ be the state of the system, and prepare the meter in a state $|\Upsilon\rangle$. Thus the joint initial state vector is $|\Psi\rangle=|\Upsilon\rangle\otimes|\psi\rangle$. The joint wave function for the system and the meter at $t=T$ is given by

$$\Phi(\bar{y};A)\equiv\langle\bar{y},A\,|\hat{U}_{\text{tot}}(T,0)|\Psi\rangle, \tag{2.5a}$$

where $\hat{U}_{\text{tot}}(T,0)$ is the evolution operator generated by the total Hamiltonian (2.2). From $\Phi(\bar{y};A)$ one derives the probability distribution of the meter coordinate,

$$P(\bar{y})=\sum_A |\Phi(\bar{y};A)|^2 , \tag{2.5b}$$

which contains the complete quantum statistics of the time-distributed measurement.

It should be pointed out that the probability distribution (2.5b) can equally well be derived from the joint wave function $\Phi(\bar{y};A,t)$ at any time $t \geqq T$:

$$P(\bar{y})=\sum_A |\Phi(\bar{y};A,t)|^2 .$$

This is a consequence of the fact that the interaction turns off at $t=T$, which implies that

$$\Phi(\bar{y};A,t\geqq T)=\sum_{A'}\langle A\,|\hat{U}(t,T)|A'\rangle\Phi(\bar{y};A') ,$$

where $\hat{U}(t,T)$ is the evolution operator generated by the system Hamiltonian $\hat{H}$.

It should also be noted that the model can be generalized quite easily to describe many time-distributed measurements performed during the interval $[0,T]$. For each measurement one introduces a meter, which is coupled to the system in such a way that the meter coordinate is displaced by the quantity one wants to measure. Needless to say, each of these measurements in general disturbs the others.

One of the diagnostics for classical chaos is to examine an appropriate power spectrum. Thus, of possible interest for quantum chaos is a special case of the above model, the measurement of a Fourier component of

A. One possibility is a measurement of the Fourier cosine component

$$y_n \equiv \int_0^T dt\, A(t) \cos\omega_n t\ , \quad \omega_n \equiv \pi n/T \tag{2.6a}$$

$[Y_t(A(t)) = A(t)\cos\omega_n t]$, which can be modeled by choosing a total Hamiltonian

$$\hat{H}_{\text{tot}} = \hat{H} + \hat{\bar{p}}\hat{A}\cos\omega_n t\ . \tag{2.6b}$$

The interaction Hamiltonian $\hat{\bar{p}}\hat{A}\cos\omega_n t$ clearly probes the behavior of the system at frequency ω_n. Another possibility is the measurement of the "digital" Fourier cosine component

$$y_n \equiv \sum_{j=0}^{N} A(j\tau)\cos(\pi nj/N) \tag{2.7a}$$

$[\tau \equiv T/N,\ Y_t(A(t)) = \sum_{j=0}^{N} A(t)\cos\omega_n t\,\delta(t - j\tau)]$. The corresponding total Hamiltonian,

$$\hat{H}_{\text{tot}} = \hat{H} + \hat{\bar{p}}\hat{A}\sum_{j=0}^{N}\cos(\pi nj/N)\delta(t - j\tau)\ , \tag{2.7b}$$

looks a lot like the "kicked" systems discussed at this Workshop, except that the kick strength is proportional to an operator $\hat{\bar{p}}$ and is modulated sinusoidally by the factor $\cos(\pi nj/N)$. As noted above, the model can be generalized to describe the measurement of several Fourier components.

III. SUM-OVER-HISTORIES FORMULATION

The model of Sec. II can be rewritten in terms of a "sum over histories." To illustrate how this is done, I specialize to the case where y is a sum of contributions from discrete times $t_1, \ldots, t_N$:

$$y = \sum_{j=1}^{N} Y_j(A(t_j)) \tag{3.1}$$

$[Y_t(A(t)) = \sum_{j=1}^{N} Y_j(A(t))\delta(t - t_j)$; cf. Eq. (2.1)]. Here Y_j is an arbitrary real-valued function, and the times t_j are temporally ordered within the interval $[0,T]$–i.e., $0 \leq t_1 < t_2 < \cdots < t_N \equiv T$. The case where y is an integral over contributions from a continuous time interval might be treated as a limit of the discrete-time case. When A is the position of some system, such a limit can be taken, and the sum over histories becomes a Feynman path integral (see Sec. IV).

For the discrete-time case the appropriate total Hamiltonian is

$$\hat{H}_{\text{tot}} = \hat{H} + \hat{\bar{p}}\sum_{j=1}^{N} Y_j(\hat{A})\delta(t - t_j)\ , \tag{3.2}$$

with corresponding total evolution operator

$$\hat{U}_{\text{tot}}(T,0) = \prod_{j=1}^{N} \exp[-(i/\hbar)\hat{\bar{p}}Y_j(\hat{A})]\hat{U}(t_j, t_{j-1}) \tag{3.3}$$

$(t_0 \equiv 0,\ t_N \equiv T)$. In Eq. (3.3), $\hat{U}(t,t')$ is the evolution operator for the system Hamiltonian $\hat{H}$, and the product is time-ordered–i.e., increasing values of j on the left.

Now let $\psi(A) \equiv \langle A|\psi\rangle$ be the initial $(t=0)$ system wave function, and let $\Upsilon(\bar{y}) \equiv \langle \bar{y}|\Upsilon\rangle$ be the initial coordinate-space wave function for the meter. The corresponding momentum-space wave function for the meter is

$$\Upsilon(k) \equiv \langle k|\Upsilon\rangle = \int \frac{d\bar{y}}{(2\pi)^{1/2}} e^{-ik\bar{y}}\Upsilon(\bar{y}) \tag{3.4}$$

Here $|k\rangle$ denotes a δ-function normalized eigenstate of $\hat{p}/\hbar$ ($\hat{p}\,|k\rangle=\hbar\, k\,|k\rangle$).

The objective now is to write the joint wave function (2.5a),

$$\Phi(\bar{y};A)=\langle\bar{y},A\,|\hat{U}_{\text{tot}}(T,0)\,|\Psi\rangle, \tag{3.5}$$

as a sum over histories. I refer to $\Phi(\bar{y};A)$ as the fundamental amplitude; it is the joint probability amplitude that the measurement of y yields the result $\bar{y}$ and that the observable A has value A at time T.

Begin by transforming to the momentum basis for the meter:

$$\Phi(\bar{y};A)=\int\frac{dk}{(2\pi)^{1/2}}e^{ik\bar{y}}\langle k,A\,|\hat{U}_{\text{tot}}(T,0)\,|\Psi\rangle. \tag{3.6}$$

Then notice that

$$\begin{aligned}\langle k,A\,|\hat{U}_{\text{tot}}(T,0)\,|\Psi\rangle&=\Upsilon(k)\left\langle A\left|\left[\prod_{j=1}^{N}e^{-ikY_j(\hat{A})}\hat{U}(t_j,t_{j-1})\right]\right|\psi\right\rangle\\&=\Upsilon(k)\sum_{A_0,A_1,\ldots,A_{N-1}}\exp\left[-ik\sum_{j=1}^{N}Y_j(A_j)\right]\\&\quad\times\left\{\prod_{j=1}^{N}\langle A_j\,|\hat{U}(t_j,t_{j-1})\,|A_{j-1}\rangle\right\}\psi(A_0),\end{aligned} \tag{3.7}$$

where $A_N\equiv A$. The latter equality in Eq. (3.7) follows from inserting the unit operator, written as $\hat{1}=\sum_{A_j}|A_j\rangle\langle A_j|$, between the jth and $(j+1)$th terms in the product $\prod_{j=1}^{N}e^{-ikY_j(\hat{A})}\hat{U}(t_j,t_{j-1})$, for $j=0,\ldots,N-1$.

The quantity in curly brackets in Eq. (3.7), which I denote by

$$\mathcal{A}(A_1,\ldots,A_N\,|A_0)\equiv\prod_{j=1}^{N}\langle A_j\,|\hat{U}(t_j,t_{j-1})\,|A_{j-1}\rangle, \tag{3.8}$$

is the key to writing a sum over histories. It is a product of propagators, the jth of which, $\langle A_j\,|\hat{U}(t_j,t_{j-1})\,|A_{j-1}\rangle$, is the probability amplitude for the observable A to have value A_j at time t_j, given that it had value A_{j-1} at time t_{j-1}. According to the usual quantum-mechanical rules for combining conditional probability amplitudes, $\mathcal{A}(A_1,\ldots,A_N\,|A_0)$ is the amplitude for the sequence of values $A_1,\ldots,A_N$ at times $t_1,\ldots,t_N$, given initial value A_0 at time $t_0=0$. The sequence $A_0,A_1,\ldots,A_N$ can be regarded as a "history" for the system; $\mathcal{A}(A_1,\ldots,A_N\,|A_0)$ is the probability amplitude for the history, given the initial value A_0.

One now has all the ingredients to write the fundamental amplitude $\Phi(\bar{y};A)$ as a sum over histories:

$$\Phi(\bar{y};A)=\sum_{A_0,A_1,\ldots,A_{N-1}}\Upsilon\Big(\bar{y}-\sum_{j=1}^{N}Y_j(A_j)\Big)\mathcal{A}(A_1,\ldots,A_N\,|A_0)\psi(A_0) \tag{3.9}$$

($A_N\equiv A$). The quantity $\Upsilon(\bar{y}-\sum_{j=1}^{N}Y_j(A_j))$ in Eq. (3.9) is the meter's initial coordinate-space wave function, displaced by a distance $\sum_{j=1}^{N}Y_j(A_j)$—i.e., displaced by the value of y corresponding to the history $A_0,A_1,\ldots,A_N$ [cf. Eq. (3.1)]. Within the model one interprets $\Upsilon(\bar{y}-\sum_{j=1}^{N}Y_j(A_j))$ as the probability amplitude that the meter coordinate has value $\overline{y}$, given displacement by a distance $\sum_{j=1}^{N} Y_j(A_j)$. One can reinterpret $\Upsilon(\overline{y} - \sum_{j=1}^{N} Y_j(A_j))$ as the amplitude that the measurement of y yields the result $\overline{y}$, given that y has precisely the value $\sum_{j=1}^{N} Y_j(A_j)$–i.e., given the history $A_0, A_1, \ldots, A_N$. I call $\Upsilon(\overline{y} - \sum_{j=1}^{N} Y_j(A_j))$ the resolution amplitude, because it describes the irresolution of the measurement of y; in the language of the model, this irresolution is a consequence of the initial uncertainty in the meter coordinate.

The sum-over-histories expression (3.9) for the fundamental amplitude has a simple and appealing interpretation. Select a history $A_0, A_1, \ldots, A_N$, and begin with the amplitude $\mathcal{A}(A_1, \ldots, A_N | A_0)$, the amplitude for the history, conditioned on the initial value A_0. Multiply by the initial wave function $\psi(A_0)$, the amplitude for the initial value A_0; thereby obtain the unconditioned amplitude $\mathcal{A}(A_1, \ldots, A_N | A_0)\psi(A_0)$ for the history $A_0, A_1, \ldots, A_N$. Further, multiply by the resolution amplitude $\Upsilon(\overline{y} - \sum_{j=1}^{N} Y_j(A_j))$, the amplitude to obtain $\overline{y}$ as the result of the measurement of y, given the history $A_0, A_1, \ldots, A_N$; thereby find the joint amplitude

$$\Upsilon(\overline{y} - \sum_{j=1}^{N} Y_j(A_j))\mathcal{A}(A_1, \ldots, A_N | A_0)\psi(A_0)$$

for the result $\overline{y}$ and for the history $A_0, A_1, \ldots, A_N$. Finally, compute the fundamental amplitude by summing over unobservable quantities–i.e., by summing over all histories such that $A_N = A$.

Not only does the resolution amplitude describe the "additive noise"–the irresolution which, in the model, is a consequence of the initial uncertainty in the meter coordinate–it also describes the "back-action disturbance." In the model the back-action disturbance arises from the coupling of the system to the meter. In the sum-over-histories formulation (3.9), the back-action disturbance appears in a different, but equivalent guise. It is manifested in the restriction of the sum over histories produced by the resolution amplitude–a restriction which forces the system away from unitary evolution.

IV. LINEAR MEASUREMENTS ON A HARMONIC OSCILLATOR

The formalism sketched in Secs. II and III is somewhat abstract. A good way to flesh out the formalism–to see it in action–is to apply it to an example where it can be worked out analytically. Such an example arises when the system is a harmonic oscillator and the time-distributed measurements provide information which is linear in the position of the oscillator. Additional motivation for considering this example is that it allows one to address an important practical question: how does quantum mechanics limit one's ability to detect a classical force which acts on an oscillator? This question is important for efforts to detect gravitational waves and for other high-precision measurements.[7]

Consider then a harmonic oscillator, with mass μ, frequency ω_0, and position $x(t)$. When a force $F(t)$ acts on the oscillator, its Lagrangian is given by

$$L(x, \dot{x}; t) = \tfrac{1}{2}\mu(\dot{x}^2 - \omega_0^2 x^2) + xF(t) . \tag{4.1}$$

Consider further a sequence of Q time-distributed measurements, each of which provides information which is linear in $x(t)$. Denote the Q measured quantities by $y(t_q)$, for $q = 1, \ldots, Q$; the times $t_1, \ldots, t_Q$ which label the measured quantities are assumed to be temporally ordered–i.e.,

$t_1 < t_2 < \cdots < t_Q$. Choose the measured quantities ("multiple-time observables") to have the form

$$y(t_q) \equiv \int_{t_0}^{t_q} dt\, g(t_q - t) x(t)\,, \quad q = 1, \ldots, Q\,, \tag{4.2}$$

where $g(t_q - t)$ is a time-stationary filter function, and t_0 is an initial time which I eventually allow to go to $-\infty$.

The filter function $g(t-t') \equiv g(u)$ is assumed to be real $[g(u) = g^*(u)]$ and causal $[g(u) = 0$, if $u < 0]$. It is useful in the following to work in the frequency domain, so I introduce the Fourier transform of the filter function,

$$g(\omega) \equiv \int_0^\infty du\, g(u) e^{i\omega u}\,. \tag{4.3}$$

Moreover, it is convenient to normalize the filter function so that the maximum value of $|g(\omega)|$ is 1. Then $y(t_q)$ has the same units as $x(t)$, and a natural measure of the filter's bandwidth is

$$\Delta f \equiv \int_0^\infty \frac{d\omega}{2\pi} |g(\omega)|^2 = \tfrac{1}{2} \int_0^\infty du\, g^2(u)\,. \tag{4.4}$$

The result of the qth measurement is a number, which I label by $\bar{y}_q$. As before, introduce a resolution amplitude $\Upsilon(\bar{y}_q - y_q)$, which is the amplitude to obtain the result $\bar{y}_q$ in the qth measurement, given that the measured quantity $y(t_q)$ has precisely the value y_q. Notice that I choose the resolution amplitude to have the same functional form for all the measurements. In the example discussed below the resolution amplitude is chosen to be a real Gaussian,

$$\Upsilon(\bar{y}_q - y_q) \equiv (2\pi\sigma^2)^{-1/4} \exp[-(\bar{y}_q - y_q)^2/4\sigma^2]\,. \tag{4.5}$$

The variance σ^2 is the resolution of the measurements; it characterizes the "additive noise."

One further ingredient is required—the initial wave function $\psi(x)$ for the oscillator at time t_0.

The fundamental amplitude for the Q time-distributed measurements, denoted by $\Phi(\bar{y}_1, \ldots, \bar{y}_Q; x)$, is the joint probability amplitude that the measurements yield the sequence of results $\bar{y}_1, \ldots, \bar{y}_Q$ and that the oscillator is at position x at time t_Q. This fundamental amplitude can be derived from a Feynman path integral (sum over paths) in the following way:[1,2]

$$\Phi(\bar{y}_1, \ldots, \bar{y}_Q; x) = \int_{t_0}^{(x,t_Q)} \mathcal{D}x(t) \left[\prod_{q=1}^{Q} \Upsilon(\bar{y}_q - y(t_q)) \right] e^{(i/\hbar) S[x(t)]} \psi(x(t_0))\,. \tag{4.6a}$$

Here the integral denotes a sum over all paths $x(t)$ on the interval $[t_0, t_Q]$ such that $x(t_Q) = x$ [initial positions $x(t_0)$ summed over]; $y(t_q)$ is calculated from each path $x(t)$ using Eq. (4.2); and

$$S[x(t)] \equiv \int_{t_0}^{t_Q} dt\, L(x, \dot{x}; t)$$

is the action for a forced harmonic oscillator.

The joint probability distribution to obtain the sequence of results $\bar{y}_1, \ldots, \bar{y}_Q$ follows immediately from the fundamental amplitude:

$$P(\bar{y}_1, \ldots, \bar{y}_Q) = \int dx\, |\Phi(\bar{y}_1, \ldots, \bar{y}_Q; x)|^2\,. \tag{4.6b}$$

Interpretation of Eqs. (4.6) arises naturally from the usual quantum-mechanical rules for combining conditional probability amplitudes. Select a path $x(t)$, and begin with the quantity $e^{(i/\hbar)S[x(t)]}$, the familiar quantum-mechanical amplitude for the path, conditioned on the initial value $x(t_0)$. Multiply by the initial wave function $\psi(x(t_0))$, the amplitude for the path's initial value $x(t_0)$; thereby obtain the unconditioned amplitude $e^{(i/\hbar)S[x(t)]}\psi(x(t_0))$ for the path $x(t)$. For each $q = 1, \ldots, Q$, multiply by a resolution amplitude $\Upsilon(\bar{y}_q - y(t_q))$, the amplitude to obtain the result $\bar{y}_q$ in the qth measurement, given the path's value for $y(t_q)$—i.e., the amplitude to obtain $\bar{y}_q$, given the path $x(t)$; thereby find the joint amplitude

$$\left[\prod_{q=1}^{Q} \Upsilon(\bar{y}_q - y(t_q))\right] e^{(i/\hbar)S[x(t)]}\psi(x(t_0))$$

for the sequence of results $\bar{y}_1, \ldots, \bar{y}_Q$ and for the path $x(t)$. Now compute the fundamental amplitude $\Phi(\bar{y}_1, \ldots, \bar{y}_Q; x)$ by summing over all paths such that $x(t_Q) = x$. Why not sum over final values $x(t_Q)$ as well? Because the oscillator's final position is potentially observable by an independent measurement. Hence, first take the absolute square of Φ to obtain a probability distribution, and then integrate over final values x to get the joint probability distribution (4.6b).

The path-integral formulation (4.6) is equivalent to a simple measurement model.[1,2] For each measurement introduce a meter, the qth of which has canonical coordinate $\bar{y}_q$ and canonical momentum $\bar{p}_q$. The initial ($t = t_0$) state of the oscillator and the meters is

$$|\Psi\rangle = |\Upsilon_1\rangle \otimes \cdots \otimes |\Upsilon_Q\rangle \otimes |\psi\rangle , \tag{4.7}$$

where $|\psi\rangle$ is the initial oscillator state $[\psi(x) = \langle x|\psi\rangle]$, and $|\Upsilon_q\rangle$ is the initial state for the qth meter. The initial coordinate-space wave function for each meter is chosen to be the resolution amplitude—i.e., $\langle \bar{y}_q | \Upsilon_q\rangle = \Upsilon(\bar{y}_q)$.

Now assume that the total Hamiltonian for the oscillator and the Q meters is

$$\hat{H}_{\text{tot}} = \hat{H} + \sum_{q=1}^{Q} \hat{\bar{p}}_q \hat{x} g(t_q - t) , \tag{4.8}$$

where $\hat{H}$ is the Hamiltonian for a forced harmonic oscillator. Using an approach similar to that in Sec. III, one can easily show that the fundamental amplitude (4.6a) is the joint wave function for the oscillator and the Q meters, evolved to time t_Q using the total Hamiltonian (4.8):[1,2]

$$\Phi(\bar{y}_1, \ldots, \bar{y}_Q; x) = \langle \bar{y}_1, \ldots, \bar{y}_Q, x | \hat{U}_{\text{tot}}(t_Q, t_0) | \Psi\rangle . \tag{4.9}$$

Here $\hat{U}_{\text{tot}}(t_Q, t_0)$ is the evolution operator generated by $\hat{H}_{\text{tot}}$.

The path-integral formulation of Eqs. (4.6) treats the oscillator as an isolated system; there is no systematic dissipation. It is not difficult to include dissipation by coupling the oscillator to a heat reservoir. The heat reservoir can be a large collection of harmonic oscillators, and the coupling can be linear in the position of the primary oscillator and in the position variables of the heat-reservoir oscillators.[8] I assume that the heat reservoir and its coupling to the primary oscillator are chosen so that the oscillator undergoes standard exponential damping with amplitude damping constant β. Yurke[9] has analyzed a nice realization of such a damped harmonic oscillator; the primary oscillator—a mass on a spring—is attached to a semi-infinite string, whose transverse vibrational modes are the heat-reservoir oscillators.

To incorporate dissipation into the path-integral formulation (4.6), one simply writes down a path integral for the composite system consisting of the primary oscillator and the heat reservoir; one includes the time-distributed measurements via a product of resolution amplitudes, just as in Eq. (4.6a); and one "traces out" the degrees of freedom of the heat reservoir.

Assume now that at initial time $t_0 = -\infty$ the oscillator and the heat reservoir are in thermal equilibrium at temperature T, and further assume that the resolution amplitude has the (real) Gaussian form (4.5). Then one can do the path integral, or alternatively one can integrate the equations of motion of the equivalent measurement model. Not surprisingly, one finds that the joint probability distribution $P(\bar{y}_1, \ldots, \bar{y}_Q)$ is a multivariate Gaussian, specified by a set of mean values and a covariance matrix.

The mean value for the result of the qth measurement,

$$\langle \bar{y}_q \rangle = \int_{-\infty}^{t_q} dt\, g(t_q - t) \int_{-\infty}^{t} dt'\, R_x(t-t') F(t') , \tag{4.10}$$

is simply the classical value of the measured quantity $y(t_q)$; it contains the information about the force $F(t)$. In Eq. (4.10)

$$R_x(t-t') \equiv R_x(u) \equiv \frac{1}{\mu\Omega} H(u) e^{-\beta u} \sin\Omega u , \quad \Omega \equiv (\omega_0^2 - \beta^2)^{1/2} , \tag{4.11}$$

is the oscillator's temporal response function [$H(u)$ is the Heaviside step function]; its Fourier transform,

$$R_x(\omega) \equiv \int_{-\infty}^{\infty} du\, R_x(u) e^{i\omega u} = \frac{1}{\mu(\omega_0^2 - \omega^2 - 2i\beta\omega)} \equiv \Delta_x(\omega) + i\Gamma_x(\omega) , \tag{4.12}$$

can be conveniently decomposed into a real part $\Delta_x(\omega)$, the dispersive part, and an imaginary part $\Gamma_x(\omega)$, the dissipative part.

What one wants to know is how quantum and thermal noise limit one's ability to detect the force. The noise is quantified by the covariance matrix

$$\langle \Delta\bar{y}_q \Delta\bar{y}_r \rangle = C_{qr} + \sigma^2 \delta_{qr} + \frac{1}{4\sigma^2} \sum_{s=1}^{Q} \mathcal{A}_{qs} \mathcal{A}_{rs} , \tag{4.13}$$

where $\Delta\bar{y}_q \equiv \bar{y}_q - \langle \bar{y}_q \rangle$. The symmetric matrix C_{qr} is defined by

$$\begin{aligned} C_{qr} &\equiv \int_{-\infty}^{t_q} dt\, g(t_q - t) \int_{-\infty}^{t_r} dt'\, g(t_r - t') C_x(t-t') \\ &= \int_0^{\infty} \frac{d\omega}{2\pi} |g(\omega)|^2 S_x(\omega) \cos\omega(t_q - t_r) , \end{aligned} \tag{4.14}$$

where

$$C_x(t-t') \equiv \tfrac{1}{2} \int_{-\infty}^{\infty} \frac{d\omega}{2\pi} S_x(\omega) e^{-i\omega(t-t')} = \int_0^{\infty} \frac{d\omega}{2\pi} S_x(\omega) \cos\omega(t-t') , \tag{4.15a}$$

$$S_x(\omega) \equiv 2\hbar\, \Gamma_x(\omega) \coth(\hbar\omega/2kT) , \tag{4.15b}$$

and the matrix $\mathcal{A}_{qr}$ is defined by

$$\begin{aligned} \mathcal{A}_{qr} &\equiv \int_{-\infty}^{t_q} dt\, g(t_q - t) \int_{-\infty}^{t_r} dt'\, g(t_r - t') [-\hbar\, R_x(t-t')] \\ &= \int_{-\infty}^{\infty} \frac{d\omega}{2\pi} |g(\omega)|^2 [-\hbar\, R_x(\omega)] e^{-i\omega(t_q - t_r)} \\ &= \int_0^{\infty} \frac{d\omega}{2\pi} |g(\omega)|^2 [-2\hbar\, \Delta_x(\omega) \cos\omega(t_q - t_r) - 2\hbar\, \Gamma_x(\omega) \sin\omega(t_q - t_r)] . \end{aligned} \tag{4.16}$$

The covariance matrix (4.13) has also been derived by Schmid.[10]

Each term in the covariance matrix has a simple interpretation. The first term, C_{qr}, is the contribution one would expect from fluctuation-dissipation theory; it arises from the standard two-point correlation function $C_x(t-t')$ and associated spectral density $S_x(\omega)$ for an oscillator in contact with a heat reservoir at temperature T. The second and third terms are additional "quantum noise" due to the measurements. The second term, $\sigma^2\delta_{qr}$, is the "additive noise"; it comes from the resolution σ^2 of the measurements. The third term,

$$\frac{1}{4\sigma^2}\sum_{s=1}^{Q}\mathcal{A}_{qs}\mathcal{A}_{rs}\ ,$$

is the "back-action disturbance"; it becomes progressively more important as the resolution improves. The matrix element $\mathcal{A}_{qr}$ characterizes how the rth measurement affects the qth measurement.

Equations (4.10) and (4.13) can be used to analyze how thermal and quantum noise set limits on detection of the the force $F(t)$. To gain insight into the irreducible quantum limits, it is instructive to specialize to a case where the matrix C_{qr} vanishes. For that purpose, consider an undamped oscillator—i.e., $\beta=0$—for which

$$\Delta_x(\omega)=\frac{1}{\mu}\mathrm{P}\left[\frac{1}{\omega_0^2-\omega^2}\right], \tag{4.17}$$

$$\Gamma_x(\omega)=\frac{\pi}{2\mu\omega_0}[\delta(\omega-\omega_0)-\delta(\omega+\omega_0)] \tag{4.18}$$

(P denotes a principal value at the poles $\omega=\pm\omega_0$). Thus, for an undamped oscillator, the spectral density

$$S_x(\omega)=2\pi\delta(\omega-\omega_0)\frac{\hbar}{\mu\omega_0}\left[\frac{1}{2}+\frac{1}{e^{\hbar\omega_0/kT}-1}\right] \tag{4.19}$$

[Eq. (4.15b)] becomes a δ function at the resonant frequency ω_0, and consequently the matrix C_{qr} vanishes if one chooses a filter that has a zero at ω_0—i.e., $g(\omega_0)=0$.

In real situations, where the oscillator is damped, the matrix C_{qr} introduces additional noise, because $S_x(\omega)$ is not concentrated precisely at the resonant frequency. For any particular measurement scheme one must formulate a precise criterion for neglecting C_{qr}, but, crudely speaking, the noise added by C_{qr} is small when one is making wide bandwidth measurements on a low-temperature, low-loss oscillator—i.e., $\beta\coth(\hbar\omega_0/2kT)\ll\Delta f$. Here I am interested in the best performance allowed by quantum mechanics, so I specialize henceforth to the case $C_{qr}=0$.

With these same assumptions [$\beta=0$ and $g(\omega_0)=0$], $\mathcal{A}_{qr}$ becomes a symmetric matrix, given by

$$\mathcal{A}_{qr}=\mathcal{A}_{rq}=\int_0^\infty\frac{d\omega}{2\pi}\,|g(\omega)|^2[-2\hbar\,\Delta_x(\omega)]\cos\omega(t_q-t_r)\ . \tag{4.20}$$

One further specialization is useful in exploring the quantum limits. Suppose the measured quantities $y(t_q)$ are given by Eq. (4.2) with $t_0=-\infty$ and that the times t_q are given by $t_q=q\tau$, where q now runs from $-\infty$ to $+\infty$. This corresponds to the experimental situation where one measures a time-distributed quantity repeatedly at regularly spaced times, the intervals between measurements being τ. In this situation, $\mathcal{A}_{qr}$ depends only on the distance from the diagonal—i.e.,

$$\mathcal{A}_{qr}=\int_0^\infty\frac{d\omega}{2\pi}\,|g(\omega)|^2[-2\hbar\,\Delta_x(\omega)]\cos(q-r)\omega\tau\equiv\mathcal{A}_{q-r}=\mathcal{A}_{r-q}\ , \tag{4.21}$$

and the covariance matrix (4.13) becomes

$$\langle\Delta\bar{y}_q\Delta\bar{y}_r\rangle=\sigma^2\delta_{qr}+\frac{1}{4\sigma^2}\sum_{s=-\infty}^{\infty}\mathcal{A}_{q-s}\mathcal{A}_{r-s}=\sigma^2\delta_{qr}+\frac{1}{4\sigma^2}\sum_{s=-\infty}^{\infty}\mathcal{A}_{q-r+s}\mathcal{A}_s \tag{4.22}$$

Insight into the quantum limits comes from considering situations where one of the elements $\mathcal{A}_p$ dominates. Suppose, for example, that one knows that the force $F(t)$ has Fourier components almost entirely on one side of the resonant frequency. Then one uses a filter function such that $|g(\omega)|^2$ has support mainly on the same side of ω_0, with bandwidth Δf

matched to the bandwidth of the force, and one chooses $\tau \simeq (\Delta f)^{-1}$. Under these circumstances, $\mathcal{A}_0$ is the dominant element of $\mathcal{A}_{qr}$.

To analyze this situation, I assume formally that $|\mathcal{A}_0| \gg |\mathcal{A}_p|$, for $p \neq 0$. Keeping only the dominant terms in the covariance matrix (4.22), one finds that

$$\langle(\Delta\bar{y}_q)^2\rangle = \sigma^2 + \frac{(\mathcal{A}_0/2)^2}{\sigma^2} \to |\mathcal{A}_0| = \left| \int_0^\infty \frac{d\omega}{2\pi} |g(\omega)|^2[-2\hbar\, \Delta_x(\omega)] \right| , \tag{4.23a}$$

$$\begin{aligned} \langle \Delta\bar{y}_q \Delta\bar{y}_r \rangle &= \frac{1}{2\sigma^2} \mathcal{A}_0 \mathcal{A}_{q-r} \\ &\to \frac{\mathcal{A}_0}{|\mathcal{A}_0|} \mathcal{A}_{q-r} = \frac{\mathcal{A}_0}{|\mathcal{A}_0|} \int_0^\infty \frac{d\omega}{2\pi} |g(\omega)|^2[-2\hbar\, \Delta_x(\omega)] \cos\omega(t_q - t_r) , \quad q \neq r . \end{aligned} \tag{4.23b}$$

The terms following the arrows correspond to assuming that σ^2 takes on its optimum value: $\sigma^2 \to \sigma^2_{\text{opt}} = |\mathcal{A}_0|/2$.

The surprising conclusion contained in Eqs. (4.23) is that when $\sigma^2 = \sigma^2_{\text{opt}}$, one can calculate the covariance matrix by treating

$$\mathcal{S}_x(\omega) \equiv \mp 2\hbar\, \Delta_x(\omega) = \pm \frac{2\hbar}{\mu} \mathrm{P}\left[\frac{1}{\omega^2 - \omega_0^2}\right] \tag{4.24}$$

just as though it were a classical spectral density for the noise in $x(t)$. In Eq. (4.24) the upper (lower) sign holds if $\mathcal{A}_0$ is positive (negative). Of course, $\mathcal{S}_x(\omega)$ is not a true spectral density because it changes sign at the resonant frequency; I call it a *pseudo-spectral-density* for the quantum noise in the position of a harmonic oscillator. Notice that for a free mass ($\omega_0 = 0$),

$$\mathcal{S}_x(\omega) = \frac{2\hbar}{\mu} \mathrm{P}\left[\frac{1}{\omega^2}\right] \tag{4.25}$$

becomes a true spectral density, corresponding to diffusion with diffusion constant $\hbar/\mu$.

Shift gears a bit now, and consider the situation where $\mathcal{A}_1 = \mathcal{A}_{-1}$ dominates. This situation can arise when $|g(\omega)|^2$ is nearly symmetric about ω_0 and one chooses $\tau \simeq (\Delta f)^{-1}$. To analyze this situation, I assume formally that $|\mathcal{A}_1| = |\mathcal{A}_{-1}| \gg |\mathcal{A}_p|$, for $p \neq \pm 1$. Again keeping only the dominant terms in the covariance matrix (4.22), one finds that

$$\langle(\Delta\bar{y}_q)^2\rangle = \sigma^2 + \frac{\mathcal{A}_1^2/2}{\sigma^2} \to 2^{1/2}|\mathcal{A}_1| = \left| \int_0^\infty \frac{d\omega}{2\pi} |g(\omega)|^2[-2^{3/2}\hbar\, \Delta_x(\omega) \cos\omega\tau] \right| , \tag{4.26a}$$

$$\begin{aligned} \langle \Delta\bar{y}_q \Delta\bar{y}_r \rangle &= \frac{1}{2\sigma^2} \mathcal{A}_1 (\mathcal{A}_{q-r+1} + \mathcal{A}_{q-r-1}) \\ &\to 2^{-1/2} \frac{\mathcal{A}_1}{|\mathcal{A}_1|} (\mathcal{A}_{q-r+1} + \mathcal{A}_{q-r-1}) \\ &= \frac{\mathcal{A}_1}{|\mathcal{A}_1|} \int_0^\infty \frac{d\omega}{2\pi} |g(\omega)|^2[-2^{3/2}\hbar\, \Delta_x(\omega) \cos\omega\tau] \cos\omega(t_q - t_r) , \quad q \neq r , \end{aligned} \tag{4.26b}$$

where the terms following the arrows correspond to assuming that $\sigma^2 \to \sigma^2_{\text{opt}} = |\mathcal{A}_1|/2^{1/2}$. In this situation one finds that the appropriate quantum-mechanical pseudo-spectral- density is

$$\mathcal{S}_x(\omega) \equiv \mp 2^{3/2}\hbar\, \Delta_x(\omega) \cos\omega\tau = \pm \frac{2^{3/2}\hbar}{\mu} \mathrm{P}\left[\frac{1}{\omega^2 - \omega_0^2}\right] \cos\omega\tau , \tag{4.27}$$

where the upper (lower) sign holds if $\mathcal{A}_1$ is positive (negative).

The pattern should now be clear. Suppose $\mathcal{A}_p=\mathcal{A}_{-p}$ ($p\neq 0$) dominates—i.e., $|\mathcal{A}_p|=|\mathcal{A}_{-p}|\gg|\mathcal{A}_{p'}|$, for $p'\neq p$. Then one finds that

$$\langle(\Delta\bar{y}_q)^2\rangle=\sigma^2+\frac{\mathcal{A}_p^2/2}{\sigma^2}\rightarrow 2^{1/2}|\mathcal{A}_p|=\left|\int_0^\infty\frac{d\omega}{2\pi}|g(\omega)|^2[-2^{3/2}\hbar\,\Delta_x(\omega)\cos p\,\omega\tau]\right|, \tag{4.28a}$$

$$\begin{aligned}\langle\Delta\bar{y}_q\Delta\bar{y}_r\rangle&=\frac{1}{2\sigma^2}\mathcal{A}_p(\mathcal{A}_{q-r+p}+\mathcal{A}_{q-r-p})\\&\rightarrow 2^{-1/2}\frac{\mathcal{A}_p}{|\mathcal{A}_p|}(\mathcal{A}_{q-r+p}+\mathcal{A}_{q-r-p})\\&=\frac{\mathcal{A}_p}{|\mathcal{A}_p|}\int_0^\infty\frac{d\omega}{2\pi}|g(\omega)|^2[-2^{3/2}\hbar\,\Delta_x(\omega)\cos p\,\omega\tau]\cos\omega(t_q-t_r)\,,\quad q\neq r\,,\end{aligned} \tag{4.28b}$$

where the arrows correspond to $\sigma^2\rightarrow\sigma^2_{\text{opt}}=|\mathcal{A}_p|/2^{1/2}$. The appropriate pseudo-spectral-density is

$$\mathcal{S}_x(\omega)\equiv\mp 2^{3/2}\hbar\,\Delta_x(\omega)\cos p\,\omega\tau=\pm\frac{2^{3/2}\hbar}{\mu}\text{P}\left(\frac{1}{\omega^2-\omega_0^2}\right)\cos p\,\omega\tau\,, \tag{4.29}$$

where the upper (lower) sign holds if $\mathcal{A}_p$ is positive (negative).

This discussion in terms of quantum-mechanical pseudo-spectral-densities is meant to serve two purposes. First, the pseudo-spectral-densities provide a heuristic way of thinking about quantum noise in a language familiar from the theory of classical noise. Second, they can be, in specific cases, a useful tool for calculating quantum limits on detecting a force.

None of this means that the pseudo-spectral-densities provide a general way of analyzing measurements on a harmonic oscillator. Many assumptions were made in getting to Eqs. (4.21) and (4.22), which served as the starting point for introducing the pseudo-spectral-densities. Even having gotten to Eqs. (4.21) and (4.22), it is not necessarily true that one of the elements $\mathcal{A}_p$ dominates, and if several of the elements $\mathcal{A}_p$ have comparable values, there seems to be no way to introduce a suitable pseudo-spectral-density.

Just as important, one can readily find situations where all the matrix elements $\mathcal{A}_{qr}$ vanish. Suppose, for example, that the filter function is given by

$$g(u)=\tfrac{1}{2}[\delta(u)-\delta(u-\tau_0)]\,,\qquad g(\omega)=-ie^{i\omega\tau_0/2}\sin(\omega\tau_0/2)\,, \tag{4.30}$$

where $\tau_0\equiv 2\pi/\omega_0$ is the oscillator's period. Then the matrix $\mathcal{A}_{qr}$ [Eq. (4.20); $\beta=0$] becomes

$$\mathcal{A}_{qr}=\begin{cases}-(\hbar/4\mu\omega_0)\sin\omega_0|t_q-t_r|\,, & |t_q-t_r|\leqq\tau_0\\ 0\,, & |t_q-t_r|\geqq\tau_0\end{cases}. \tag{4.31}$$

Thus all the matrix elements $\mathcal{A}_{qr}$ vanish if one chooses $t_q=q\,\tau_0/2$, or if one chooses $t_q=q\,\tau$ where $\tau\geqq\tau_0$. This measurement scheme is an example of a stroboscopic quantum nondemolition measurement;[7] there is no quantum noise to hinder detection of a force.

ACKNOWLEDGMENTS

This work was supported in part by a Precision Measurement Grant from the US National Bureau of Standards [Grant No. NB83-NADA-4038].

REFERENCES

[1] C. M. Caves, "Quantum mechanics of measurements distributed in time. A path-integral formulation," Phys. Rev. D **33**, 1643 (1986).

[2] C. M. Caves, "Quantum mechanics of measurements distributed in time. II. Connections among formulations," Caltech Preprint GRP-073 (August 1986), submitted to Phys. Rev. D.

[3] Y. Aharonov and D. Z. Albert, "Is the usual notion of time evolution adequate for quantum-mechanical systems? I," Phys. Rev. D **29**, 223 (1984).

[4] Y. Aharonov, D. Z. Albert, and S. S. D'Amato, "Multiple-time properties of quantum-mechanical systems," Phys. Rev. D **32**, 1975 (1985).

[5] S. S. D'Amato, *A Multiple-Time Description of Quantum Mechanical Systems*, Ph.D. Thesis, University of South Carolina, 1984.

[6] S. S. D'Amato, "Multiple-time measurements on quantum mechanical systems," in *Fundamental Questions in Quantum Mechanics*, edited by L. M. Roth and A. Inomata (Gordon and Breach, New York, 1985), p. 225.

[7] C. M. Caves, K. S. Thorne, R. W. P. Drever, V. D. Sandberg, and M. Zimmermann, "On the measurement of a weak classical force coupled to a quantum-mechanical oscillator. I. Issues of principle," Rev. Mod. Phys. **52**, 341 (1980).

[8] A. O. Caldeira and A. J. Leggett, "Path-integral approach to quantum Brownian motion," Physica **121A**, 587 (1983).

[9] B. Yurke, "Quantizing the damped harmonic oscillator," Am. J. Phys., in press.

[10] A. Schmid, "Repeated measurements on dissipative linear quantum systems," Ann. Phys. (N.Y.), in press.

QUANTUM MECHANICS IN STRONG TIME DEPENDENT EXTERNAL FIELDS

Y. Pomeau
Laboratoire de Physique de l'ENS
24 rue Lhomond 75231
Paris, France

Introduction

In textbooks as well as in most theoretical papers the analytic study of quantum systems submitted to time dependent external perturbations is often limited to a perturbative approach. This may be insufficient for understanding the behaviour of systems with infinitely many quantum levels in the unperturbed state when an external perturbation is turned on: even when the amplitude of this perturbation is very small, small denominators appear, leading to phenomena that are as yet not fully understood. So it seems interesting to try to study the other limit, that is very large external time dependent perturbations. This limit has its own mathematical interest and could be relevant for some experimental situations too.

Stated in a very general form, the problem under consideration is the following one: the Hamiltonian operator is the sum of an unperturbed and time independent part $\mathbf{H}_0$ and of a time dependent part in the form $g\lambda(t)\mathbf{H}_1$ were $\mathbf{H}_1$ does not depend on time, $\lambda(.)$ is a given real function of time and g is a "coupling" constant assumed to be large. In dimensionless notations the Schrodinger equation for this system reads:

$$i\psi_t = (\mathbf{H}_0 + g\lambda(t)\mathbf{H}_1)\,\psi,$$

Taking formally $g\int^t \lambda(t')\,dt'$ as new time variable it would seem that this problem is exactly equivalent to a formal perturbation with a small term $\mathbf{H}_0/(g\lambda(t))$. This could perhaps be exploited to get in some way or another a duality "a la Aubry - Andre" with a convenient choice of λ, $\mathbf{H}_0$ and $\mathbf{H}_1$. However this is not the way we shall follow: this leads to difficulties either because the change of time variable is singular when $\lambda(.)$ vanishes or because $\mathbf{H}_1$ has a continuous spectrum extending down to zero. Indeed usual perturbation theory could be amended in order to deal with those complications, but in both cases to be considered later on it is more simple to extract information directly by keeping the "perturbation" as large.

The first example I will study is the behaviour of a two level system, the two levels being coupled by a strong time dependent external field. In the large perturbation limit, one finds a WKB like solution depending on two amplitudes that are adiabatic invariants. This WKB limit has problems whenever the c-number function of time $\lambda(.)$ in front of the large perturbation crosses zero as time goes on. When this happens, a local "turning point" analysis shows that the adiabatic invariants exchange at this time some amount of amplitude, of order $g^{-1/2}$, g being the large parameter. In the case of a random $\lambda(.)$, this leads to a diffusion process on the Hermitian 2d sphere where the amplitudes are bound to stay because of unitarity.

The next problem I will examine is the behaviour of an atom (= one electron in a central force field) submitted to a large uniform external field depending sinusoidally on time. In the absence of interaction between the electron and the central ion, the external field can be forgotten thanks to an elementary canonical transformation. With the interaction turned on one can get rid of the large time dependent external field thanks to a Kapitza averaging: the vicinity of the center of force is visited during very short periods in the back and forth motion in the external field. Thus this force is felt as rapidly varying and one may apply the Kapitza averaging. But this has to be done with care: the effective interaction resulting from this averaging has not only a constant part but also Fourier components at multiple frequencies of the external field with an amplitude of the same order as the one of the constant part, and one has to show that this time dependent part of the force field can be neglected. This cannot be done if the external field is a noise, because then the equivalent force field has components at frequencies close to zero, although if the external field is periodic, one may replace it and the central force field by an equivalent central force field. I will consider in this framework first a central force field that is short ranged in 1d, then the more realistic case of an atom in 3d with a central Coulombic field.

The Two Level System

It is a matter of elementary transformations to get the second order differential equation:

$$\phi_{tt}+\phi-2ib(t)\phi_t=0 \qquad [1]$$

where the subscript t denotes time derivatives, from the equations of motion of a two level system:

$$i\psi_{1,t}=a_1\psi_1+b(t)\psi_2\,,$$

$$i\psi_{2,t}=a_2\psi_2+b(t)\psi_1\,,$$

where $a_{1,2}$ are two real constants although b(.) is the large time dependent coupling between the two states indexed as 1 and 2, so that $\psi_{1,2}$ are the complex amplitudes of those two states. As pointed out by M. Berry the above written second order equation can be transformed into an equation formally identical to the familiar 1d Schrodinger equation. Moreover it appears [1] in the Landau-Zener theory for dynamical level crossings. If b is large and so can be seen as almost constant over short time intervals of order 1/b, it is natural to seek a solution of [1] in a WKB like form. Putting $\phi \simeq e^{i\sigma t}$

into [1] and taking b as constant one finds:

$$\sigma = b \pm (b^2+1)^{1/2},$$

so that the WKB solution of [1] reads:

$$\phi = A' + B' \exp[-2i \int^{t} dt' \, b(t')]$$

where A and B are- for the moment -arbitrary complex numbers. Pursuing the expansion at next order, one finds:

$$A' = A(1 + \int^{t} dt' \, (2ib(t'))^{-1} + \ldots.)$$

and

$$B' = B/2b(t)(1+ \ldots)$$

where now A and B are constant on the time scale of the variations of b(.). Moreover one may verify that $|A|^2+|B|^2$ is the total probability. From the above formulae it is clear that this WKB solution loses its validity near turning points when b approaches 0, a situation that I will analyse now, in the generic case where b expands near this zero as:

$$b(t) \simeq \beta t + \beta' t^2 + \ldots,$$

β as well as β' being large, and the origin of time being such that $b(0) = 0$. The method of study of such a turning point situation is to formulate a local inner problem by using stretched variables and then by asymptotic matching to get the transformation relating the values of the adiabatic invariants on the two sides of the singular region. The stretching here is done by taking as time variable $t|\beta|^{1/2}$ and the solution of [1] is obtained [2] near $t = 0$ in the form of Fresnel integrals. From the asymptotic study of those integrals the transformation of the pair of amplitudes A and B at the crossing of b(.) through 0 may be represented as follows: let A and B be the two components of a spinor Σ in the spin 1/2 representation and let Σ_+ (resp. Σ_-) be the values of this spinor after (resp. before) the zero crossing of b(.). At the dominant order in the large β limit those two spinors are related through the linear transform:

$$\Sigma_+ = (1+|\pi/2\beta|)^{1/2}(\sigma_x + \mathrm{sgn}\beta \, \sigma_y) \, \Sigma_- ,$$

where $\sigma_{x,y}$ are the usual Pauli matrices and sgn is the sign function. As β is large this corresponds to small rotations. If b(.) is a periodic function of time, this constant rotation is simply related to the Floquet eigenvalues of the original problem. If on the contrary b(.) is a random function of time one can shown that this rotation leads to a random diffusion on the complex Hermitian sphere where A and B must stay due to the constraint of unitarity. This random diffusion is described mathematically by a Chapman-Kolmogoroff equation wherein the coefficients have [2] simple expression in terms of statistical properties of the random b(.) function. The "intermediate" case of a quasiperiodic b(.) cannot be answered so simply. This leads [3] to a stochastic behaviour of the quantum amplitudes at least in a certain sense. Heuristically this might be related to the weak randomness found for instance in the discrepancy for the irrational rotations on a circle.

A Particle in a Constant Potential and a Strong External Field

Below I consider the following class of problems: a particle is submitted to a constant potential v(x) vanishing at infinity and to a large time dependent external field. The time dependent Schrodinger equation for this case reads in dimensionless notations:

$$i\psi_t = -1/2\ \psi_{xx} + v(x)\ \psi + xb(t)\psi \qquad [2]$$

This has been written with one space dimension only, and at the end I will give some results for the 3d case with an attracting Coulombic potential. The last term on the r.h.s. of [2] represents the effect of the large external field. This one is supposed to be homogeneous on any space scales that will appear later on, whence the simple x-dependence of this term.

The general strategy for this problem (of a large amplitude of b in [2]) is to get rid formally of this external field by a canonical transformation. This one is elementary and introduces the classical displacement $\Lambda(t)$ solution of:

$$\Lambda_{tt} = b(t),$$

and the equation for the wavefunction in the transformed coordinates reads:

$$i\psi_t = -\psi_{xx}/2 + v(x-\Lambda(t))\psi \qquad [3]$$

This is exactly equivalent to the first equation. If the amplitude of Λ is much larger than the range of v(x), then the time dependent potential $v(x-\Lambda(t))$ at a given x is made of short bursts separated by long time intervals during which it vanishes almost completely. For such a rapidly varying potential one may apply the general method of Kapitza [4,5] that amounts to replace this potential by an effective steady force field. However in the present case the validity of this approximation is not totally obvious and so I shall discuss this point more specifically.

This discussion is partly self consistent because one has to know some scaling properties of the bound states in the effective potential. Let us assume a sinusoidal dependence of $\Lambda(t)$ in the form $\Lambda(t) = \lambda \sin(2\pi t/T_0)$ so that the effective potential takes the form

$$V_\lambda(x) = 2/\pi \int_{-1}^{1} du\ v(x-\lambda u)\ (1-u^2)^{-1/2}.$$

If λ is much larger than the range of v(.) one has approximately $V_\lambda(x) = w\ (\lambda^2-x^2)^{-1/2}$ for $|x| < \lambda$ and zero otherwise, w being half the mean value of v(.), i.e. $1/2 \int_{\mathbb{R}} dy\ v(y)$, supposed to be convergent and negative. The effective potential V_λ has two inverse square root singularities at $x = \pm\lambda$. Nearby these singularities it behaves as $w(\lambda z)^{-1/2}$, and a scaling of the Schrodinger equation for the bound states shows that they have an energy of order $\lambda^{-1/3}$ and a range over the space variable x of order $\lambda^{+2/3}$ nearby $x = \pm\lambda$. This is valid for the more deeply bounded states, there are also states with energies of order $1/\lambda$ and a momentum of order $\lambda^{-1/2}$, which can be analysed by a WKB method. This leads [2] to slightly unusual Bohr-Sommerfeld quantization since the inverse square root singularities in the potential are also classical turning points.

Now we can discuss the applicability of the Kapitza averaging. As said before one has to show that the time dependent part of $v(x-\Lambda(t))$ has a small effect upon the bound states in the effective potential. Let us remark first that in a classical picture the change of momentum due to the impulse received during one period of the external field is of order $F\ \delta t$ where F is of the order of the attractive force in the potential v, and δt is the duration of the crossing of the force field, that is of order $1/\lambda$. This is thus negligible compared to the mean momentum in bound states which is of order $\lambda^{-1/3}$. Let us consider now the question of stability of these bound states from a quantal point of view. As a function of time the potential $v(x-\Lambda(t))$ is periodic and can be expanded in Fourier series of time. The first term in this expansion is indeed the constant effective potential. As the force field is a set of short impulses many Fourier components of this force are almost independent of their order at least for not too large frequencies (i.e. for frequencies less than λ, which is the inverse of the time needed to cross the potential v during the large oscillations), which implies that they have the same magnitude and x-dependence as the effective constant potential. Suppose thus that this time dependent part of the potential acts as a perturbation in the effective Kapitza problem introduced before. Thus the effect of this time dependence may be treated in the usual framework of perturbation theory, and each Fourier component of the potential is responsible for a perturbation if the particle is supposed to be in a bound state of the effective potential. To be such in the true sense of the word some dimensionless quantity measuring this perturbation has to be small. As already said the Fourier spectrum of the potential is almost flat for small frequencies, so that the perturbation is measured at first order by the matrix element $\Gamma = \int dx\, e^{ikx}\, V_\lambda(x)\, \psi_0(x)$, where k is the wave number of the particle in the coupled state. This has to be smaller than the energy of the ground state which is of order $E_0 = \int dx\, V_\lambda(x)\, |\psi_0(x)|^2$. By the energy conservation the wave number k is of the order of the momentum of a particle with energy of the order of the frequency of the external field, which is independent of λ. Thus in the estimation of Γ, e^{ikx} has to be considered as a rapidly varying function so that the integral for Γ is dominated by the vicinity of the inverse square root singularity of $V_\lambda(x)$ and is of order $k^{-1/2}$. Accordingly the dimensionless ratio Γ/E_0 is of order $(ka)^{-1/2}$, where a is a typical width of the ground state wavefunction, that is $\lambda^{1/3}$. This shows that this ratio is small and of order $\lambda^{-1/6}$ in the large λ limit. Indeed there are other Fourier components and one might wonder if the interaction with all of them will increase the effective Γ, compared to the case where a single Fourier component is acting. This does not seem to be the case because this effective width would remain of the same order of magnitude if one include all "low" frequency components of the varying part of the potential.

Another, perhaps more convincing manner of proving the validity of the method that we have used is to examine the next order contribution to the effective potential and to show that this yields effectively a small correction to $V_\lambda(x)$. Thus let us assume that the Fourier expansion of $v(x-\Lambda(t))$ is $\sum v_n(x) \sin(n\omega t+\phi_n)$ where ω is $2\pi/T_0$, T_0 being again the period of the external field. As shown by Kapitza, this yields an effective potential equal to $\sum v_{n,x}^2/(n\omega)^2$, n = 0 being excluded from the sum. As the series $\sum 1/n^2$ converges and as the Fourier components of $v(x-\Lambda(t))$ are n-independent for n smaller than λ, the first correction to the effective potential is of order $v'(x) = V_{\lambda,x}^2$ where $V_{\lambda,x}$ (x) is the x derivative of the dominant contribution to the effective potential considered before. This potential $v'(x)$ is of order λ^{-2} for x arbitrary between $-\lambda$ and $+\lambda$, but not close to these two values. Accordingly it is negligible there compared to $V_\lambda(x)$ which is of order $1/\lambda$. Near $x = \pm\lambda$, $v'(x)$ diverges as $1/(\lambda z^3)$ where z is the distance of x to the boundary value. However this divergence is cut at short distance by the range of the potential v(x), so that even when $v'(x)$ is the biggest it is of order $1/\lambda$, still negligible compared to $V_\lambda(x)$ that is of order $\lambda^{-1/2}$ in the same range.

All this can be extended to 3d potentials v(r). In the case of an attracting Coulomb field one gets [2] the result that for a linearly polarised external uniform field of amplitude λ, they are still bound states in an effective average potential and by some coincidence the binding energies are still of order $\lambda^{-2/3}$, as in the 1d case. If the external strong potential is circularly polarised and still with a large amplitude of order λ the energies of the more deeply bound states in the equivalent force field are of order $-\ln\lambda/\lambda$.

Conclusion

We have shown that the problem of large time dependent perturbations constitutes in some "good" cases a tractable limit. In particular, in the case of the two level system one may get a rather detailed picture of what is quantum chaos when the external perturbation is a random function of time. There is certainly a need to extend this sort of result to much more difficult - and yet not totally understood - problems such as the famous kicked rotator.

References

1. C. Zener, Proc. Roy. Soc. Vol. A137, 1932, p.696

2. Y. Pomeau, Annales de l'IHP, Vol. 45, 1986, p.29

3. Y. Pomeau, B. Dorizzi and B. Grammaticos, Phys. Rev. Lett., Vol 56, 1986, p.681

4. P. Kapitza, Zh. Exp. Teor. Fiz., Vol. 38, 1951, p.588

5. R.J. Cook, D.G. Shankland and A.L. Wells, Phys. Rev. A, Vol. 31, 1985, p.564

A MODEL FOR A QUANTUM-MECHANICAL MEASUREMENT

N.G. van Kampen

Institute for Theoretical Physics of the University
Utrecht, The Netherlands

ABSTRACT

Section 1 points out that quantum mechanics works, and analyzes how it works. Speculations about the real meaning of the wave function and about occurrence of probability are redundant. Section 2 describes and criticizes the orthodox description of the measuring process due to von Neumann. Section 3 presents a rather realistic model for measuring the position of an electron. The Schrödinger equation for the combined system of electron and measuring apparatus results automatically in a collapse of the electron wave function. Section 4 shows why density matrices are needed and draws conclusions about the entropy.

1. COMMENTS ON QUANTUM MECHANICS

The formalism of quantum mechanics deals with mathematical quantities, such as the wave function, that are less directly linked to observed physical objects than is the case in classical theory. This fact has given rise to numerous discussions and theories about the interpretation of quantum mechanics [1], in which one can often detect an atavistic urge to attribute more direct physical meaning to these mathematical quantities than is warranted by their role in the formalism. These discussions continue unabated, while quantum mechanics is successfully applied every day. Apparently the question of how to relate the mathematical formalism to observed phenomena is not a problem for the working man, but only for the philosophers. Let us therefore analyze how quantum mechanics is actually used.

Theorem 1. Quantum mechanics works. It explains the phenomena for which it was designed, such as black-body radiation, photo-electric effect, and spectra. It also enables one to deduce numerous other phenomena, such as specific heat and superconductivity. These phenomena are macroscopic, objectively and permanently fixed on a photographic plate or as a curve printed in the Physical Review. Thus we have

Theorem 2. Quantum mechanics deals with macroscopic phenomena, which are unperturbed by observation. Speculations about the influence of the observer on the system do not refer to actual observations as they are done, but to an imagined underlying microscopic system with a microscopic observer.

A typical quantum phenomenon is the diffraction pattern produced by a beam of electrons passing through a suitable crystal. Again the phenomenon consists of a photographic picture, macroscopic, indelibly registered and reproducible. To compute this effect one computes the wave function ψ of a single electron and identifies $|\psi|^2$, times the number of electrons in the beam, with the blackening on the plate. The square itself is interpreted as the probability distribution of the single electron, but it cannot be observed. The only way to observe a probability density is to repeat the experiment so often that it becomes an actual density, which is precisely what the beam of electrons does. Hence

Theorem 3. Probabilities are not observed but merely serve as an intermediate step to facilitate the calculation. Speculations involving lack of knowledge or absence of information attach too much reality to the probabilities and have nothing to do with the photographs of Davisson and Germer. The wave function ψ has merely a mathematical role in the process of computing the picture on the plate. Any meaning attached to it beyond this is unauthorized.

Theorem 4. Whoever reads into ψ more physical meaning than that of a computational aid is responsible for the consequences. In this respect ψ is analogous to the vector potential in electromagnetism and the coordinates x^μ in general relativity, but somehow ψ seems to be more conducive to metaphysical speculations.

Much of what has been said in this section resembles the "Copenhagen interpretation of quantum mechanics", as formulated by Bohr 59 years ago here in Como. However, I emphasize that my purpose was not to construct or propagate some "interpretation" of quantum mechanics, but merely to analyze what is actually done in practice when quantum mechanics is applied to physical phenomena.

2. THE THEORY OF MEASUREMENT

Let a system be at a given time in a state described by a wave function (or vector in Hilbert space) ψ. I want to measure an observable given by a self-adjoint operator A having a complete set of eigenfunctions χ_n with eigenvalues λ_n, which for convenience I suppose discrete and non-degenerate:

$$A\chi_n = \lambda_n\chi_n \ .$$

According to von Neumann [2] the possible outcomes are λ_n, each being realized with the probability $|(\chi_n|\psi)|^2$. If a certain λ_m has been observed the wave function just after the measurement is no longer ψ but χ_m. This is the reduction or collapse of the wave function.

This collapse is supposed to be the effect of the physical act of applying a measuring apparatus. If the apparatus is applied but the result is not read, the system is left in one of the states χ_n, each with probability $|(\chi_n|\psi)|^2$. This is simply a statistical probability, caused by my lack of knowledge, just as when a die is cast and I fail to look at it. The resulting situation cannot be expressed in terms of a single wave function - no more than the hidden die can be described by a single set of classical coordinates. Rather one needs a probability distribution, represented by an ensemble, which may be described by a density matrix, namely

$$\rho = \sum_n |\chi_n)|(\chi_n|\psi)|^2(\chi_n| \ . \tag{1}$$

On the other hand, when no measurement is performed, the wave function evolves in time according to the Schrödinger equation, which generates a unitary transformation group of the Hilbert space with parameter t. One would expect that the collapse is a consequence of the Schrödinger equation of the system and measuring apparatus together. Von Neumann, however, emphasizes that the collapse and the Schrödinger evolution are basically different processes ("grundsätzlich verschieden").

Nonetheless he also shows that there exists a unitary transformation U in the Hilbert space spanned by the combined wave functions $|\psi\omega\rangle$ of object system and apparatus such that

$$U|\psi\omega\rangle = \sum_n \langle\chi_n|\psi\rangle|\chi_n\omega_n\rangle \ . \tag{2}$$

The ω_n constitute a complete orthogonal set of states of the apparatus, whose labels n correspond to the various outcomes of the measurement ("positions of the pointer"). If the density matrix of the pure state (2) is averaged over the apparatus states one recovers (1). His existence proof is purely formal: it merely shows that U violates no mathematical rules, but the relation with any physical measuring apparatus is ignored.

The fundamental flaw of the argument, however, is that the measuring apparatus is supposed to perform its task by ending up in one state ω_n of a preassigned complete orthogonal set of wave functions. This leads to the question: How does the observer read the result of his measurement? That can only be done by a second measuring apparatus, and so on. Von Neumann and others [3] have concluded that this chain can only end in the consciousness of the observer, where presumably quantum mechanics ceases to be applicable. Such flights of fancy are clearly beside the point, since quantum mechanics is a physical theory designed to explain physical phenomena such as spectra and specific heats (Theorem 2). In fact, quantum mechanics also applies to parts of the Universe where no observers live.

An actual measurement registers the result in a macroscopic way, so that it is fixed objectively, that is, it is not perturbed by any further observation. A measuring apparatus, such as a pointer on an ammeter, is a macroscopic object. Macroscopic systems have an enormously dense energy spectrum; what is called a "state" in macrophysics is a subspace of Hilbert space spanned by an enormous number of microscopic eigenstates of the system [4]. When the pointer indicates a value n it does <u>not</u> mean that it is in an eigenstate ω_n [5], but that it is somewhere in a very large subspace [4,6]. When I subsequently want to look at the pointer it has to scatter photons, but the only effect on the pointer is that it moves around a bit in this subspace; the probability that it is moved to another mark on the scale is entirely negligible - as every experimenter knows. In fact, that is precisely the reason why macroscopic objects obey the laws of classical physics [7].

How is it possible for a microscopic system to affect the macroscopic state of the measuring apparatus? The answer is that the initial macrostate of the apparatus must be <u>metastable</u>, in such a way that the microscopic object merely needs to trigger the transition into a stable macrostate. Think of the AgBr crystals of the photographic plate, the Geiger counter, and the supersaturated vapor of a cloud chamber.

Theorem 5. <u>A measuring apparatus is a macroscopic object prepared in a metastable state</u>. Notice that a metastable state decaying into a stable state makes sense only for macroscopic systems subject to statistical mechanics and behaving in an irreversible manner.

Before you think of counterexamples let me insert the following remark. If a microscopic system is highly excited into an energy region where the levels are dense, it is virtually macroscopic and behaves classically. This is precisely the contents of Bohr's correspondence principle. Such a highly excited system can be observed and measured by classical means, without the need of a prepared metastable apparatus. An example is the detection of particles by Cerenkov radiation [8], the apparatus being the electromagnetic field. Bohm's analysis of the Stern-Gerlach experiment [6] also makes use of this fact. Of course such a measurement does not identify the precise eigenstate but specifies a subspace, which contains many microstates but is sufficiently specified for macroscopic needs.

3. OUR MODEL FOR THE MEASURING PROCESS

Consider the famous two-slit experiment. We want to observe whether the elctron passes through the upper slit. Thus we measure a position coordinate with only two values. My measuring apparatus consists of an atom together with the electromagnetic radiation field. The many degrees of freedom of the field make the apparatus behave as a macroscopic system and the emission of photons is an irreversible process. The metastability is obtained by assuming that the atom is in an excited state, which can decay only into the ground state through a forbidden transition. This atom is placed in the upper slit. When the electron passes through the slit it distorts the eigenfunctions of the atom in such a way that the matrix element between both states no longer vanishes and a decay of the atom, under emission of a photon, is triggered. The photon can be registered on a photographic plate. (One may consider the plate as part of the apparatus, but that is not necessary since the decayed atom contributes already a permanent record of the passage of the electron.)

The Hilbert space of the atom is spanned by the two states

$$\binom{1}{0} \text{ (energy } \Omega) \quad \text{and} \quad \binom{0}{1} \text{ (energy } 0) \ .$$

The field modes are labelled by their wave vector $\underset{\sim}{k}$, frequency $|\underset{\sim}{k}| = k$. The states of the radiation field are characterized by the occupation numbers of all these modes, but we shall only need the one-photon states $|\underset{\sim}{k}\rangle$ and the vacuum $|0\rangle$. They are all orthogonal to each other. The Hilbert space of our apparatus is the direct product of these two spaces, but we only need the space spanned by

$$\binom{1}{0} |0\rangle \quad \text{and} \quad \binom{0}{1} |\underset{\sim}{k}\rangle \quad (\text{all } \underset{\sim}{k}) \ . \tag{3}$$

The states Ψ of the apparatus combined with the object system are linear combinations of the states (3), with coefficients that are functions of the position $\underset{\sim}{r}$ of the electron:

$$\Psi = \varphi(\underset{\sim}{r}) \binom{1}{0} |0\rangle + \sum_{\underset{\sim}{k}} \psi_{\underset{\sim}{k}}(\underset{\sim}{r}) \binom{0}{1} |\underset{\sim}{k}\rangle \ . \tag{4}$$

The Hamilton operator of the total system is, in terms of the photon creation and annihilation operators and the Laplace operator with respect to $\underset{\sim}{r}$,

$$H = \Omega\begin{pmatrix} 1 & 0 \\ 0 & 0 \end{pmatrix} + \sum_{\underset{\sim}{k}} k a^{+}_{\underset{\sim}{k}} a_{\underset{\sim}{k}} - \tfrac{1}{2}\nabla^2$$
$$- u(r) \begin{pmatrix} 0 & 1 \\ 1 & 0 \end{pmatrix} \sum_{\underset{\sim}{k}} i v_k (a_{\underset{\sim}{k}} - a^{+}_{\underset{\sim}{k}}) \ . \tag{5}$$

The last term is the electric field multiplied with the operator of the dipole moment, whose strength depends on $\underset{\sim}{r}$ through a function $u(r)$. It is supposed that $u(r)$ is practically zero beyond a distance Δr from the atom. The coefficients v_k involve the normalization of the field modes, and in addition a form factor of the atom, which restricts the considered $\underset{\sim}{k}$ to those for which the dipole approximation holds.

The evolution of the wave function Ψ of the combined system is determined by the Schrödinger equation $i\dot{\Psi} = H\Psi$. Substituting (4) and (5) and separating the orthogonal components one obtains

$$i\dot{\varphi}(\underset{\sim}{r},t) = (\Omega-\tfrac{1}{2}\nabla^2)\varphi(\underset{\sim}{r},t) - u(r) \sum_{\underset{\sim}{k}} iv_k\ \phi_{\underset{\sim}{k}}(\underset{\sim}{r},t) \tag{6}$$

$$i\dot{\phi}_{\underset{\sim}{k}}(\underset{\sim}{r},t) = (k-\tfrac{1}{2}\nabla^2)\phi_{\underset{\sim}{k}}(\underset{\sim}{r},t) + u(r)\ iv_k\ \varphi(\underset{\sim}{r},t)\ . \tag{7}$$

Two-photon states have been cut out, which amounts to the rotating wave approximation. Spin, and exchange between the observed electron and the electrons of the atom, have been ignored, because they give only irrelevant complications. For the two-slit experiment the equations should be solved with boundary condition $\varphi = 0$, $\phi_{\underset{\sim}{k}} = 0$ on the screen, but we shall not do that explicitly [9].

Initially (that is for $t \to -\infty$) the atom is in its excited state, no photons are present, and the electron is described by some wave packet:

$$\Psi = \varphi(\underset{\sim}{r},t)\ \binom{1}{0}\ |0\rangle \qquad (t \to -\infty) \tag{8}$$

$$\varphi(\underset{\sim}{r},t) = \int C(\underset{\sim}{p})e^{i(\underset{\sim}{p}\cdot\underset{\sim}{r}-Et)}\, d\underset{\sim}{p}\ , \qquad E = \tfrac{1}{2}\underset{\sim}{p}^2 + \Omega\ . \tag{9}$$

On solving (6) and (7) one will find for $t \to +\infty$ a $\varphi(\underset{\sim}{r},t)$, which describes the probability density of the electron in case it has not been observed. Vulgarly speaking, the electron may either have passed through the lower slit, or through the upper slit without triggering the apparatus. Similarly $\phi_{\underset{\sim}{k}}(\underset{\sim}{r},t)$ for $t \to \infty$ refers to the case that the electron has betrayed its presence in the neighborhood u of the atom by causing the emission of a photon $\underset{\sim}{k}$.

An explicit solution, however, is not needed to recognize the relevant features of $\phi_{\underset{\sim}{k}}(\underset{\sim}{r},t)$. Set $\phi_{\underset{\sim}{k}} = e^{-ikt}\phi'_{\underset{\sim}{k}}$, so that

$$\dot{\phi}'_{\underset{\sim}{k}} - \tfrac{1}{2}i\nabla^2\phi'_{\underset{\sim}{k}} = e^{ikt}\ u(r)\ v_k\ \varphi(\underset{\sim}{r},t)\ . \tag{10}$$

Thus $\phi'_{\underset{\sim}{k}}(\underset{\sim}{r},t)$ obeys a free-particle Schrödinger equation with a source term. The source is confined to the neighbourhood of the atom and extends over a finite time, and moreover $\phi'_{\underset{\sim}{k}}(\underset{\sim}{r},-\infty) = 0$. Hence $\phi'_{\underset{\sim}{k}}(\underset{\sim}{r},t)$ is a wave function that originates from the position where the electron has been observed and at the time when the electron passes. The change from φ to $\phi'_{\underset{\sim}{k}}$ is the collapse of the wave packet resulting from the position measurement, as postulated by von Neumann. However, far from being basically different from the Schrödinger evolution, it is a consequence of the Schrödinger equation for object system and measuring aparatus together.

To summarize: When a successful measurement has occurred, the apparatus is left in a different state than before the measurement; the state vector of the combined system is therefore in a different subspace of the total Hilbert space; the projection $\phi_{\underset{\sim}{k}}$ thereof in the election space is a different function than the coefficient φ in the initial subspace. This is how the so-called collapse of the wave function comes about.

The solution of (7) can be written

$$\psi_{\underset{\sim}{k}}(\underset{\sim}{r},t) = v_k \int_{-\infty}^{\infty} e^{-ik(t-t')} dt' \int d\underset{\sim}{r}' \ G(\underset{\sim}{r},t;\underset{\sim}{r}',t') \ u(r') \ \varphi(\underset{\sim}{r}',t')$$

$$G(\underset{\sim}{r},t;\underset{\sim}{r}',t') = \theta(t-t')[2\pi(t-t')]^{-3/2} e^{-3\pi i/4} \exp[i(\underset{\sim}{r}-\underset{\sim}{r}')^2/2(t-t')] \ .$$

The integration over t' virtually extends over an interval Δt, the time during which the electron is in the range of u(r). Hence the k-values for which $\psi_{\underset{\sim}{k}}$ differs appreciably from zero have a range $\Delta k \sim 1/\Delta t$. Since Δk is the uncertainty in the energy of the photon, and hence also of the electron after observation, one has for the kinetic energy $E_m = \frac{1}{2}p^2$

$$\Delta E_m . \Delta t \sim 1 \ .$$

This is the familiar relation between the duration Δt of the measurement and the energy spread caused by it.

4. PROBABILITY DENSITY MATRIX, AND ENTROPY

Actually the situation is more complicated because the apparatus, being macroscopic, may be left in any one of a host of final states $\underset{\sim}{k}$. Accordingly the electron emerges in one of the many $\psi_{\underset{\sim}{k}}$ and the measurement does not tell which one. That is the reason why a density matrix is needed to describe the situation.

First note that the normalization of Ψ is, in view of the orthonormality of the apparatus states (3),

$$1 = \langle \Psi | \Psi \rangle = (\varphi | \varphi) + \sum_{\underset{\sim}{k}} (\psi_{\underset{\sim}{k}} | \psi_{\underset{\sim}{k}}) \ . \tag{11}$$

The angular brackets refer to the total Hilbert space, the round brackets to that of the object system alone, i.e., they indicate integration over the electron coordinate $\underset{\sim}{r}$. After the measurement (for $t \to \infty$) the separate terms in (11) are: the probability P_0 that the electron has not betrayed its presence, and the probabilities $P_{\underset{\sim}{k}}$ that it has, by emitting a photon $\underset{\sim}{k}$.

Let B be an observable <u>pertaining to the electron.</u> Its expectation value is

$$\langle \Psi | B | \Psi \rangle = (\varphi | B | \varphi) + \sum_{\underset{\sim}{k}} (\psi_{\underset{\sim}{k}} | B | \psi_{\underset{\sim}{k}}) \ . \tag{12}$$

There are no cross-terms between φ and the $\psi_{\underset{\sim}{k}}$ because they are coefficients of mutually orthogonal components of Ψ in the total Hilbert space [10].

As a special example take for B the operator that corresponds to the presence of the electron at $\underset{\sim}{r}_0$:

$$B(\underset{\sim}{r};\underset{\sim}{r}') = \delta(\underset{\sim}{r}-\underset{\sim}{r}_0)\delta(\underset{\sim}{r}'-r_0)$$

$$\langle \Psi | B | \Psi \rangle = |\varphi(\underset{\sim}{r}_0)|^2 + \sum_{\underset{\sim}{k}} |\psi_{\underset{\sim}{k}}(\underset{\sim}{r}_0)|^2 \ . \tag{13}$$

The first term shows that there is still some interference pattern on the screen, because φ is nonzero in both slits. The intensity of the pattern is reduced by the factor P_0, the probability that the electron is not observed. The second term does not give an interference pattern because each ϕ_k fans out from the upper slit alone. This implements the famous statement that the two-slit interference disappears as soon as the electron has been detected in one of the slits.

The function φ in (13) is the solution of the coupled equations (6), (7) taken at large t. It is therefore not identical with the incident wave packet, so that the reduced interference pattern is not just the original one multiplied by P_0. This shows that even if the result of the measurement is negative [11] the presence of the apparatus still has some effect, as pointed out by Heisenberg.

It is not possible to write (12) as an expectation value for some pure electron state, but only as an average Tr ρB over an electron density matrix

$$\rho(\underset{\sim}{r};\underset{\sim}{r}') = \varphi(\underset{\sim}{r})\varphi^*(\underset{\sim}{r}') + \sum_{\underset{\sim}{k}} \phi_{\underset{\sim}{k}}(\underset{\sim}{r})\phi^*_{\underset{\sim}{k}}(\underset{\sim}{r}') \ . \tag{14}$$

This density matrix describes an ensemble, which represents the case that I do not look at the result of the measurement. If I do look, I am dealing with a different ensemble (just as the ensemble for the cast die changes when I look at the result) [12]. If I see that there is no photon the ensemble that describes my knowledge has the density matrix

$$\rho_0(\underset{\sim}{r};\underset{\sim}{r}') = P_0^{-1}\ \varphi(\underset{\sim}{r})\varphi^*(\underset{\sim}{r}') \ . \tag{15}$$

If I look and find a photon I know that the electron is described by one of the $\phi_{\underset{\sim}{k}}$, but I don't know which one. In fact, an essential feature of a macroscopic system is that one cannot distinguish between its individual states, but merely between large subspaces. In our model that means that one can only observe whether or not a photon has been emitted. (Even if one would distinguish between various directions of the photon the subspaces would still be virtually infinite-dimensional.) Thus, having observed a photon I have to describe the electron by

$$\rho_1(\underset{\sim}{r};\underset{\sim}{r}') = (1-P_0)^{-1} \sum_{\underset{\sim}{k}} \phi_{\underset{\sim}{k}}(\underset{\sim}{r})\phi^*_{\underset{\sim}{k}}(\underset{\sim}{r}') \ . \tag{16}$$

It is left to the reader to construct the density matrices for the case that the emitted photon is detected with some efficiency less than 1.

In order to make the comparison with (1) it should be realized that our measured quantity A has only two eigenvalues: either the electron is not observed in the upper slit or it is. Thus our A as an operator in electron space is highly degenerate. For such an operator von Neumann obtains, even after reading the result, a density matrix similar to (16), but with the orthogonal eigenfunctions $\chi_n(\underset{\sim}{r})$ rather than our non-orthogonal $\phi_{\underset{\sim}{k}}(\underset{\sim}{r})$.

Entropy was defined by Gibbs as a functional of an ensemble or probability distribution, and depends therefore on our knowledge of the system. Only for the sharp distributions that are used to describe thermodynamic systems does it reduce to the Clausius entropy, which is (apart from an additive constant [13]) an objective function of the thermodynamic state variables. In quantum mechanics an ensemble is described by its density matrix ρ and the Gibbs entropy is

$$S = -\ \mathrm{Tr}\ \rho \log \rho \ . \tag{17}$$

It vanishes for a pure state and is otherwise positive.

In von Neumann's description of the measuring process (Sec. 2) the system is before the measurement in a pure state ϕ, and afterwards, if I don't look, in the ensemble described by (1) with entropy

$$S = -\sum_n |(\chi_n|\phi)|^2 \log |(\chi_n|\phi)|^2 . \tag{18}$$

In fact this increase of entropy was one reason for considering the measurement process as basically different from the Schrödinger equation. However, if I do read the result, the pure wave function ϕ reduces to a pure wave function χ_m with zero entropy change. The entropy increase (18) was merely due to the fact that I chose to ignore some of the information about the system.

Our model shows that actually a measurement does increase the entropy because of the inherent impossibility to know the precise state of the macroscopic apparatus. The combined system is always in the pure state Ψ, so that its entropy is zero and remains zero [14]. In the initial state (8) our apparatus is in a single eigenstate, that is, a one-dimensional subspace of its Hilbert space. Therefore the electron by itself could also be described by a pure wave function φ. Hence the entropy of the object system by itself is zero before measurement.

After the measurement, if I don't look whether the apparatus has been triggered, the electron is described by the density matrix (14) and has therefore positive entropy. If I look and find no photon the density matrix is (15) and the entropy equals zero. If I find that a photon has been emitted the density matrix is (16) and the entropy is positive. If I were able to determine the photon state $\underset{\sim}{k}$ as well, the entropy would again be zero, but this is precluded by the macroscopic nature of the measuring apparatus.

The entropy belonging to the density matrix (16) is

$$S_1 = -\sum_\nu \mu_\nu \log \mu_\nu , \tag{19}$$

where μ_ν are the eigenvalues of (16). It is easily seen that they coincide with the eigenvalues of the matrix

$$M_{\underset{\sim}{k}\underset{\sim}{k}'} = (1-P_o)^{-1} (\phi_{\underset{\sim}{k}}|\phi_{\underset{\sim}{k}'}) ,$$

apart from possible zero eigenvalues, which do not contribute to (19) anyway. The density matrix of the apparatus after a photon has been observed is found - by a similar argument as used in (16) - to be M^*. It has the same eigenvalues as M and therefore the entropy is again (19).

Summarizing: The entropy of the density matrix of the object system after observation is equal to the entropy increase of the apparatus due to its transition from the metastable initial state into the stable final state. The microscopic entropy expression (17) for the total system is zero at all times; but the coarse graining inherent in our inability to distinguish between the various $\underset{\sim}{k}$ creates an entropy increase, which affects the apparatus and the object system equally. The fact that the entropy can only increase is no surprise since we have chosen an initial state in which it had its minimum value.

REFERENCES

[1] For literature the reader is referred to M. Jammer, The Philosophy of

Quantum Mechanics (Wiley, New York 1974) ch. 11; B. d'Espagnat, Conceptual Foundations of Quantum Mechanics (2nd ed., Benjamin, Reading, Mass. 1976) part 4.

[2] J. von Neumann, Mathematische Grundlagen der Quantummechanik (Springer, Berlin 1931; Dover, New York 1943). Or for instance C.Cohen-Tannoudji, B. Diu, and F. Laloë, Mécanique Quantique I (Hermann, Paris 1973) p. 216 ff.

[3] F. London et E. Bauer, La Théorie de l'Observation en Mécanique Quantique (Hermann, Paris 1939); E.P. Wigner, Symmetries and Reflections (Indiana Univ. Press, Bloomington 1967) p. 171.

[4] N.G. van Kampen, Physica 20, 603 (1954); G. Ludwig, Z. Phys. 150, 346 and 152, 98 (1958); A. Daneri, A. Loinger, and G.M. Prosperi, Nuclear Physics 33, 297 (1962).

[5] As stated in L.D. Landau and E.M. Lifshitz, Quantum Mechanics (3d ed. Pergamon, Oxford 1977) p. 21.

[6] D. Bohm, Quantum Theory (Prentice-Hall, New York) ch. 4.

[7] The confusion about Schrodinger's cat is also due to the erroneous view that a cat has a two-dimensional Hilbert space with eigenvectors "life" and "death", rather than an enormous Hilbert space with two enormous subspaces. Even Wigner's friends have more states than just "yes" and "no".

[8] I owe this remark to J. Hamilton.

[9] That is, I rely on the Kirchhoff approximation of diffraction theory.

[10] The quantummechanical explanation of 1/f noise by P.H. Handel is based on such cross-terms and must therefore be rejected, see Th.M. Nieuwenhuizen, D. Frenkel, and N.G. van Kampen, Phys. Rev. (to be published); L.B. Kiss and P. Heszler, J. Phys. C (to be published).

[11] P.S. Epstein, Am. J. Phys. 13, 127 (1945); M. Renninger, Z. Phys. 158, 417 (1960). The presence of the apparatus modifies the wave function φ, but a collapse into a ψ_k occurs only if the measurement has a positive result.

[12] A wave function ψ describes a single system, just as the p,q in classical mechanics. A density matrix ρ represents an ensemble and is the analog of a classical probability distribution. Hence ρ depends on our knowledge. Those who regard ρ rather than ψ as the state of a single system inevitably get entangled in fruitless discussions about the subjective character of quantum mechanics.

[13] N.G. van Kampen, in : Essays in Theoretical Physics in Honour of Dirk ter Haar (W.E. Parry ed., Pergamon, Oxford 1984).

[14] Hence the act of measuring cannot be invoked as a separate cause of the increase of the entropy of the universe (if it is possible at all to define such an entropy).

DYNAMICAL SYSTEMS AND THE POSITIVE P-REPRESENTATION

Axel Schenzle

Fachbereich Physik
University of Essen
Essen, West Germany

ABSTRACT

The role of quantum fluctuations in dynamical systems can be described conveniently in terms of quasi-probabilities, since this concept bears a strong but formal analogy to classical statistical mechanics. At a closer look, however, only the method of the positive P-representation has the potential of associating the quantum mechanical problem with a classical stochastic process. Thereby it would become possible to simulate quantum statistics by classical stochastic processes. The price that has to be payed is a doubling of the dimension of phase space, by introducing additional degrees of freedom that have no classical counter part. By this extension of the phase space, the dimensions of the attractors of the associated deterministic system also double - a zero dimensional attractor of the classical system remains a fixed point, but a limit cycle turns into a two dimensional manifold attracting in the transverse directions and marginally stable inside. A strange attractor with its broken dimensionality also doubles its dimensions since the Lyapunov exponents of the deterministic evolution in the extended phase space come in pairs as well. The fluctuating forces of the full dynamical problem tend to distribute the probability density along the attractor, in case of the classical limit cycle over the entire two dimensional manifold which in general extends to infinity. A stationary solution therefore does not exist for the probability density in the doubled phase space. Transverse to the attracting manifold the transient distribution falls off rather rapidly, but nevertheless, the tails of the distribution eventually extend into regions of the phase space that are severely unstable and the simulation of the quantum process breaks down. These properties will be demonstrated using a simple tutorial model as well as a more realistic system taken from the field of quantum optics.

1. INTRODUCTION

Since the development of quantum theory, it has always been a point of interest or curiosity to search for processes that display in an obvious way the characteristic differences between classical and quantum mechanics. In order to understand the pecularities of quantum mechanics, one might want to demonstrate how quantum fluctuations emerge gradually when approaching the microscopic level. The conceptual differences between the classical and the quantum description of physical processes makes it difficult, however, to formulate this transition and to understand it on intuitive grounds. For this purpose, a formulation would be rather useful that uses the same language for classical as well as quantum processes. While classical mechanics in principle is deterministic, quantum mechanics is an inherently statistical theory. This suggests that an analogy - if it does exist - might be found by using the language of classical statistical mechanics. Wigner (1,2) has first given a quasiprobability formulation of quantum mechanics utilizing the concept of a classical probability density, and later Glauber and Sudarshan (3,4) formulated the diagonal P representation which has turned out to be an extremely valuable tool - conceptually and practically - e.g. in quantum optics. Recently it has been demonstrated that the elementary nonlinear processes in quantum optics display a rich variety of instabilities like transitions to limit cycles, bistability, perdiod doubling and chaos. Therefore it seems to be natural to investigate these nonlinear quantum systems in the neighbourhood of their points of instability. A closed system in its classical limit is described by a family of deterministic trajectories, while the associated quantum process - loosely speaking - is affected by intrinsic fluctuations. An open system displays already in its classical approximation thermal fluctuations about its deterministic paths, due to the coupling with an external reservoir, while the analogous quantum system is subject to both classical as well as quantum fluctuations. It should be noted here that it can be somewhat misleading to use the word noise or fluctuations for both cases of randomness, since classical and quantum noise are of quite different origin. This will become rather obvious in the subsequent chapters. The method of the quasiprobability, however, provides a consistent and joint formulation for both sources of random behaviour.

The concept of the quasiprobability distribution allows one to calculate expectation values of a quantum mechanical observable in the same way as ensemble averages are calculated in classical statistical mechanics, provided a definite ordering of the operators is imposed. The dynamic evolution as described by the quantum mechanical master equation, is turned over into an evolution equation for the quasiprobability. This equation is a partial differential equation, which in many cases of physical interest resembles the Fokker-Planck equation known from the classical theory of stochastic processes. In this way classical and quantum fluctuations are related intuitively. While the introduction of dissipation through an external thermal reservoir creates terms that are consistent with the interpretation as a genuine Fokker-Planck equation, nonlinearities of the Hamiltonian dynamics add contributions that cannot be interpreted as diffusion terms. A system dominated by thermal noise is therefore expected to behave more or less classically, and quantum features are suppressed. If, on the other hand, the internal nonlinearities are strong and dominate the dynamical evolution, then quantum fluctuations overcome the classical ones, and the corresponding evolution equation deviates essentially from the form of a Fokker-Planck equation, either by the appearance of higher order derivatives or by a negative definite diffusion matrix.

If the evolution equation can be solved in analytical terms then it is of no importance whether an analogy to the Fokker-Planck equation exists or not. However, in cases where the evolution equation resists an exact solution, one has to resort to other means in order to predict the properties of the process under investigation. While a genuine Fokker-Planck process can always be attacked by simulating the associated Langevin equation. This concept is statistically equivalent to the Fokker-Planck approach. This equivalence of partial differential and stochastic differential equation is lost, however, when the evolution equation deviates from the Fokker-Planck form. Since most relevant physical processes are nonlinear, and in case we are particularly interested in studying the quantum nature of the process, there is little hope that even an approximate analytical solution can be found by standard methods. We know from the classical theory of stochastic processes that a multivariate Fokker-Planck equation with nonlinear drift and multiplicative noise can only be solved in relatively rare cases (5,6) and we conclude that when the potential of simulating a nonlinear process through the associated stochastic differential equation is lost, then the quasiprobability formalism is of little help for describing quantum fluctuations.

Drummond and Gardiner have suggested a way out of this dilemma (7,8) by formulating the concept of the socalled positive P-representation. This concept is closely related to the definition of the nondiagonal coherent state representation of Glauber (9) but it has some fundamental advantage due to the analytic structure of the definition. the non-diagonal representation evolves on a phase space with twice the number of dimensions, containing thereby auxiliary information that has no physical counterpart. The corresponding probability density is not uniquely related to a given statistical operator, but the physical information one can draw from it, like the ensemble averages of the problem, are independent of the choice of the distribution. The necessity to deal with a problem with twice the number of degrees of freedom seems to be an unpleasant feature of the method, and definitely makes all analytical calculations rather difficult. The great practical advantage of the concept lies in the property that for low order nonlinearities it becomes possible to cast the dynamics of a dissipative quantum system into Fokker-Planck form, thereby relating the dynamics of a quantum system in n dimensions to a classical stochastic process in 2*n dimensions, thereby opening the possibility of numerical simulation.

It is the principle aim of this contribution to demonstrate on the one hand how this method works in practice by treating specific physical problems explicitly, and, on the other hand to shed light on the basic analytical properties of the deterministic diffusion process in the doubled phase space, which sets up the stage for the stochastic diffusion process. We will also show that the method bears some intrinsic and subtle difficulties when treating dynamical systems that undergo an instability, like a supercritical Hopf bifurcation. In this case large fluctuations occur, which only average out after an impossible large number of runs. We show that this has to do with the structure of the attractors in the doubled phase space, i.e. the attractors of the dynamical system associated with the purely deterministic part of the Fokker-Planck dynamics. A focus of the classical system remains a focus in the doubled phase space, while limit cycles of the classical dynamical system are turned into an entire ensemble of quasilimit cycles that trace out a two-dimensional manifold, embedded in the higher dimensional phase space. Due to the analytic structure of the theory, this manifold which can be shown to be an attractor for the dynamics in the doubled phase space, extends to infinity. The diffusion on this attracting manifold therefore resembles a random walk that eventually wanders off into arbitrarily large distance from the origin. A steady state distribution therefore does not exist. Furthermore, the deterministic evolution in the extended phase space is not necessarily bounded, when the corresponding classical evolution is well behaved. We demonstrate by using a simple model that the doubled phase space contains regions from which trajectories can escape to infinity even in a finite time. Certainly this region must be contained entirely in the unphysical section of phase space. However, it can be demonstrated explicitly that the boundaries of this dangerous region come asymptotically close to the attractor of the system. In this way, trajectories that have wandered off far enough on the attracting manifold finally come so close to the boundaries of the instabilities that even a minute transverse fluctuation can cause an enormous excursion or even a runaway in phase space. This is a rather unexpected and unfortunate observation since these fluctuations are practically impossible to deal with statistically as well as numerically.

If the classical dynamics eventually becomes chaotic by increasing e.g. the flux of energy through the system, then the asymptotic motion evolves on a strange attractor. The deterministic motion in the extended phase space is then chaotic as well, and the dimension of the attractor doubles when using e.g. the conjecture of Kaplan and York (10).

In spite of the serious problems that come along with this method, it seems that the concept of the positive P representation is the only quasi-probability approach that allows one to investigate problems of principle interest, concerning quantum noise in dissipative nonlinear systems that does not require approximations in the leading order of Planck's constant. Further work is therefore required to understand and, if possible, to find a way to eliminate these large fluctuations which do not indicate a physical instability, but a breakdown of the method.

2. THE POSITIVE P REPRESENTATION

We will now briefly introduce the concept of the positive P representation. The dynamical evolution of an open quantum system is characterized by the master equation for the statistical operator:

$$\dot{\rho} = -\frac{i}{\hbar}[H,\rho] + \left(\frac{\partial\rho}{\partial t}\right)_{irr} \tag{1}$$

where H is the Hamiltonian of the system, and $(\partial\rho/\partial t)_{irr}$ summarizes the influence of the reservoirs. In order to solve this equation, it becomes necessary to choose a suitable basis. A representation with a continuous parameter is obtained on the basis of the coherent states, and the most widely used quasiprobability description is the diagonal P representation (3,4):

$$\rho(t) = \int d^2\alpha \; |\alpha\rangle P(\alpha,\alpha^*,t)\langle\alpha| \tag{2}$$

where $P(\alpha,\alpha^*,t)$ is a nonanalytic c-number function of α and α^*. In general, P cannot be expected to be positive definite and therefore it does not qualify as the classical probability density of a stochastic process. Glauber (9) has introduced an offdiagonal representation years ago, but only little use has been made of this tool. As a consequence of the overcompleteness of the coherent states, the offdiagonal representation is not unique, and different c-number representations exist for the same physical state. Drummond and Gardiner (7,8) have made use of this ambiguity and have introduced the following representation:

$$\rho(t) = \int d^2\alpha\, d^2\beta \; |\alpha\rangle \frac{P(\alpha,\beta)}{\langle\beta^*\alpha\rangle} \langle\beta^*| \tag{3}$$

where

$$P(\alpha,\beta) = \frac{1}{4\pi^2} e^{-\frac{1}{4}|\alpha-\beta^*|^2} \langle \tfrac{1}{2}(\alpha+\beta^*)|\rho|\tfrac{1}{2}(\alpha+\beta^*)\rangle \tag{4}$$

The c-number function $P(\alpha,\beta)$ represents the statistical operator and is obviously real, positive and normalized to unity, due to the normalization of the statistical operator:

$$\mathrm{tr}\,\varrho = \int d^2\alpha\, d^2\beta\; P(\alpha,\beta) = 1 \tag{5}$$

From the knowledge of the master equation we can derive an evolution equation for the positive P representation $P(\alpha,\beta)$. This equation can be cast into the form of a Fokker-Planck equation if the nonlinearity of the dynamic system is given by a low order polynomial, or for more general potentials this would hold in first order in Planck's constant. In such cases it is always possible to choose the diffusion matrix positive due to the analytic properties of the P function. This is the essential advantage over the celebrated diagonal representation where there is no guarantee that the diffusion matrix turns out positive definite.

By this approach, the nonclassical evolution in the 2*n dimensional phase space of the diagonal representation is turned into a genuine classical diffusion process in 2*2*n dimensions. The evolution equation for the positive P function follows directly from the master equation of the process. However, in cases where evolution equation in the diagonal representation is already known, the equation for the positive P function is easily derived by the following transformation:

$$\alpha \to \alpha \quad ; \quad \alpha^* \to \beta$$

In order to illustrate this connection, we write down a typical problem as it is discussed e.g. in the field of nonlinear optics. The nonlinear dielectric properties of matter that become relevant for high field intensities, couple fields of different frequencies and typically lead to a set of equations of motion for the mode amplitudes

$$\dot{\alpha}(t) = (i\omega - \gamma)\alpha - \chi\,|\alpha|^2\cdot\alpha + \xi(t) \tag{6}$$

This equation assumes the form of a nonlinear Langevin equation in the presence of a finite temperature heatbath essentially - for zero temperature the noise vanishes. If we now quantize this problem in the language of the diagonal P representation, then the corresponding evolution equation for $P(\alpha,\alpha^*)$ deviates from the Fokker-Planck form. While the deterministic parts of the stochastic process determine the drift terms i.e. the terms connected with the first order derivatives, in the quasi Fokker-Planck equation, the classical fluctuations introduce dissipative terms with a positive diffusion matrix:

$$\sim \frac{\partial^2}{\partial\alpha\,\partial\alpha^*} \tag{7}$$

The nonlinearity of the Hamiltonian dynamics is a further source of noise, which is entirely of quantum mechanical nature, and the corresponding diffusion matrix is not positive definite:

$$\chi\left(\frac{\partial^2}{\partial\alpha^2}\alpha^2 + \frac{\partial^2}{\partial\alpha^{*2}}\alpha^{*2}\right) \tag{8}$$

It is just the presence of those terms that cause the difficulty with the statistical interpretation of the diagonal P representation. From here it is rather straight foreward to derive the Fokker-Planck equation for the positive P representation, by merely transforming $\alpha^* \to \beta$:

$$\partial P(\alpha,\beta)/\partial t = -\frac{\partial}{\partial\alpha}(i\omega-\gamma+\chi\alpha\beta)\alpha P - \frac{\partial}{\partial\beta}(-i\omega-\gamma+\chi\alpha\beta)\beta P + 2\gamma n_{th}\frac{\partial^2}{\partial\alpha\partial\beta}P + \chi\left(\frac{\partial^2}{\partial\alpha^2}\alpha^2 + \frac{\partial^2}{\partial\beta^2}\beta^2\right)P \tag{9}$$

Since equation (9) can be written in the form of a genuine Fokker-Planck equation, it defines a classical Markovian stochastic process, which can equivalently be formulated through a set of stochastic differential equations:

$$\begin{aligned} \dot{\alpha} &= (i\omega-\gamma+\chi\alpha\beta)\alpha + g_{\alpha,i}(\alpha,\beta)\,\xi_i(t) \\ \dot{\beta} &= (-i\omega-\gamma+\chi\beta\alpha)\beta + g_{\beta,i}(\beta,\alpha)\,\xi_i(t) \end{aligned} \tag{10}$$

where the fluctuating forces represent white noise e.g.

$$\langle \xi_i(t)\,\xi_j(0)\rangle = \delta_{ij}\cdot\delta(t) \tag{11}$$

The correlation matrix g_{ij} follows directly from the diffusion matrix of the Fokker-Planck equation and does not vanish in the limit of zero temperature.

In the Heisenberg picture, the nonlinear dissipative process would be described by a set of coupled operator equations driven by operator valued fluctuating forces. Nonlinear operator equations of that kind can hardly ever be solved and represent a mathematical nightmare. The same physical process is now replaced by a set of classical stochastic differential equations with multiplicative noise and these equations can at least be simulated numerically.

Before we turn our attention to specific physical examples and their numerical solution, it is very useful to discuss and to understand first the properties of the deterministic part of the associated Langevin equations, because they represent some classical dynamic process which bears a strong analogy to the corresponding classical process. The main difference between this deterministic part of the quantum mechanical Langevin equations and the classical dynamics lies in the different dimensions of the corresponding phase spaces. In the absence of noise, trajectories that initiate in the classical subspace $\alpha_i^* = \beta_i$ remain there, for all times, and the entire phase space is only traced out when fluctuations are present. Trajectories that start outside the classical subspace in general remain outside in course of time and have no classical counterpart.

3. THE ATTRACTORS OF THE DETERMINISTIC DYNAMICS IN THE DOUBLED PHASE SPACE

We now want to discuss the properties of the deterministic part of the Fokker-Planck dynamics and want to relate them to the properties of the corresponding classical dynamic system. The systems we want to discuss here are dissipative, and therefore their asymptotic dynamics evolves towards an attractor which can be a simple fixed point, a limit cycle or a chaotic attractor with a broken dimensionality. The corresponding classical system is a dissipative dynamic system as well which has its own attracting manifolds. The interesting question now is: In which way are these attractors of the classical and the quantum process related? We will investigate this question for the most typical types of attracting sets.

3.1 Fixed Points

The classical dynamics of a complex field may be given by:

$$\dot{\alpha} = K(\alpha, \alpha^*, \{c_i\}) \tag{12}$$

and we assume that it has a stable fixed point: $\alpha = \alpha_0$. Then the associated quantum mechanical process is characterized by a Langevin equation in the extended phase space, where the deterministic part is given by:

$$\begin{aligned} \dot{\alpha} &= K(\alpha, \beta, \{c_i\}) \\ \dot{\beta} &= K(\beta, \alpha, \{c_i^*\}) \end{aligned} \tag{13}$$

$\{c_i\}$ stands for the various parameters of the model.

Obviously, this set of c-number equations has a fixed point as well:

$$\alpha = \alpha_0 \, , \quad \beta = \alpha_0^*$$

for which we will show below that it is also a locally stable attractor.

3.2 A Line of Fixed Points

The most simple model that displays a symmetry breaking bifurcation in a two dimensional phase space is characterized by the following equation of motion:

$$\dot{\alpha} = \alpha (d - |\alpha|^2) \tag{14}$$

The stable fixed point of this model, which occurs for $d<0$ falls into the category above mentioned, and its local stability is not changed by doubling the phase space. For $d>0$ the attractor of the system is a one dimensional set of points:

$$\alpha \cdot \alpha^* = d \tag{15}$$

and each point on this circle is a stationary solution. A real equation for a complex variable defines a one dimensional manifold. When we now go over to the quantum mechanical case, we find the following deterministic equations of motion:

$$\begin{aligned} \dot{\alpha} &= \alpha (d - \alpha \cdot \beta) \\ \dot{\beta} &= \beta (d - \alpha \cdot \beta) \end{aligned} \tag{16}$$

the stationary solutions of which are located on a two dimensional manifold:

$$\alpha \cdot \beta = d \tag{17}$$

These are two equations reducing the four dimensional phase space to a two dimensional subspace, which in our case is a rotationally invariant hyperboloid embedded in four dimensions.

3.3 Limit Cycles

We will now consider a classical dynamical system that evolves asymptotically on a limit cycle in a two dimensional phase space, which we take as the complex plane of (α, α^*). The limit cycle is assumed to be defined by a single real equation of the form:

$$\mathcal{F}(\alpha,\alpha^*) = 0 \quad , \quad d\mathcal{F}/dt = 0 \tag{18}$$

An example of this kind is the previously discussed model with a complex bifurcation parameter $\alpha = \alpha' + i\alpha''$

$$\mathcal{F}(\alpha,\alpha^*) = 0 = \alpha' - |\alpha|^2 = 0 \tag{19}$$

In the extended phase space of the associated quantum case, the equation of the limit cycle now defines an invariant two dimensional manifold:

$$\mathcal{F}(\alpha,\beta) = 0 \quad , \quad d\mathcal{F}/dt = 0 \tag{20}$$

The classical limit cycle is part of this subspace, which was attractive with respect to transverse perturbations in the classical directions $\alpha^*=\beta$ and marginally stable when perturbed along the cycle. In the larger phase space of the quantum case, the limit cycle - so to speak - developes another direction of marginal stability transverse to the cycle itself. The two dimensional attracting manifold is made up by a continuous family of quasi-limit cycles, i.e. if the original limit cycle is embedded into the enlarged phase space, and is perturbed in transverse direction, then it relaxes to a new cycle in the neighbourhood of the old one. If one is repeating the perturbations, an entire family of quasi cycles is created that way, which traces out the two dimensional attractor. The dynamic flux is a local tangent vector to the attractor. To be able to define the local tangent plane we need one more linear independent direction. As it will be shown below, this direction can be found by multiplying the flux by an arbitrary complex number.

3.4 The Local Tangent Plane

A complex k dimensional vector space is isomorphic to a 2*k dimensional real space. A complex vector in the k dimensional complex vector space is equivalent to a two dimensional plane in the 2*k dimensional real space. The multiplication of the complex vector by a phase factor is then equivalent to a rotation in the plane of the real and imaginary parts.

In general we can define a limit cycle in the k dimensional complex vector space by a set of 2*k-1 real equations:

$$\mathcal{F}_\ell(\alpha_i,\alpha_i^*) = 0 \qquad i = 1,2 \ldots K \; ; \; \ell = 1,2 \ldots 2K-1 \tag{21}$$

where $\mathcal{F}_\ell(\alpha_i,\alpha_i^*)$ is invariant under the deterministic flow:

$$d\mathcal{F}_\ell(\alpha_i,\alpha_i^*)/dt = 0 \tag{22}$$

The two dimensional manifold that can be defined in the phase space of the associated quantum process by the identification $\alpha^* \rightarrow \beta$

$$\mathcal{F}_\ell(\alpha_i,\beta_i) = 0 \tag{23}$$

is invariant as well. In order to define this manifold locally, we need to find two linearly independent tangent vectors. The deterministic flux obviously is one of them, and we have to find the orthogonal direction on the manifold. Now any local difference vector that relates two points on the manifold is such a tangent vector. This connection is most easily seen, when we restrict ourselves to one complex field α i.e. a four dimensional phase space in the quantum case. The generalization to arbitrary dimensions is obvious. We assume now that eq. (23) can be solved for one of the fields, and we can characterize the manifold by the following complex equation:

$$\alpha = f(\beta) \tag{24}$$

Then any tangent vector can be written in the form of a local difference vector:

$$\underline{t} = \begin{pmatrix} \Delta\alpha \\ \Delta\beta \end{pmatrix} = \begin{pmatrix} f' \\ 1 \end{pmatrix} \cdot \Delta\beta \tag{25}$$

where $\Delta\beta$ is a complex parameter of small modulus. The two parameter family of vectors defines the tangent plane. Since the dynamic flux is part of this plain, it must be contained in this family. The dynamic flux of the model is locally directed along:

$$\underline{t}' = \begin{pmatrix} K(\alpha, \beta, \{c_i\}) \\ K(\beta, \alpha, \{c_i^*\}) \end{pmatrix} \tag{26}$$

By using the dynamic equations of motion, we easily convince ourselves that the following relation holds:

$$f'(\beta) = \left(K(\alpha,\beta)/K(\beta,\alpha)\right)\Big/_{\alpha = f(\beta)} \tag{27}$$

After inserting this relation into the definition of the local flux, we see that this vector is contained in the general tangent vector family for the special choice of $\Delta\beta$:

$$\Delta\beta = K(\beta,\alpha) \quad , \ \alpha = f(\beta) \tag{28}$$

As mentioned above, the multiplication of a vector in a complex vector space by a complex parameter only rotates this vector in the plain of its real and imaginary part, provided the modulus of that parameter is unity - otherwise it also stretches the modulus of the vector, which is of no interest here. Obviously, when we take the parameter:

$$\Delta\beta = i\,K(\beta,\alpha) \tag{29}$$

the tangent vector associated with this choice becomes orthogonal to the flux and both vectors together define uniquely the local orientation of the manifold:

$$\underline{t}' = \begin{pmatrix} K(\alpha,\beta) \\ K(\beta,\alpha) \end{pmatrix} \quad \text{and} \quad \underline{t}'' = i \begin{pmatrix} K(\alpha,\beta) \\ K(\beta,\alpha) \end{pmatrix} \tag{30}$$

To rotate the local flux into this new direction, we only have to multiply it by the imaginary unit. Now we remember that the evolution equations of the dynamic system are differential equations of first order in time, and do not contain time explicitly. Therefore this rotation into the linearly independent direction is equivalent to replacing the physical dynamics by an evolution along an imaginary time axis t→i*t. While the limit cycle can be viewed as a suitably deformed circle, the evolution into the imaginary time can be understood as a motion along a deformed hyperbola. Limit cycles and hyperbolas then define an interne, orthogonal coordinate net on the manifold. In nontrivial practical cases close to the bifurcation point, a limit cycle can only be found by perturbation theory, and the periodic orbit is obtained in the form of a Fourier series. In the two dimensional complex plain the motion on a limit cycle can be written in the form:

$$\alpha(t) = \sum_{n=0}^{\infty} C_n \cos(n\omega t + \varphi_n) \tag{31}$$

The evolution on the limit cycle of the classical analysis corresponds to the following dynamics in the doubled phase space of the associated quantum process:

$$\begin{aligned} \alpha(t) &= \sum C_n \cos(n\omega t + \varphi_n) \\ \beta(t) &= \sum C_n^* \cos(n\omega t + \varphi_n) \end{aligned} \tag{32}$$

Equation (31) can also be understood as the parameter representation of the limit cycle in the two dimensional complex plane:

$$\alpha = \alpha(t)\,,\ \alpha^* = \alpha^*(t) \quad \longrightarrow \quad \alpha = f(\alpha^*) \tag{33}$$

On the other hand, the relations in eq. 32 are not the parameter representation of the two dimensional object that corresponds to the limit cycle in the doubled phase space. In contrast to the single parameter t that can be understood as an internal coordinate of the limit cycle, a two dimensional manifold is traced out by a pair of internal coordinates. Time t can be taken as one of them, but where is the other parameter? At this point we remember that, when constructing the tangent plain, we realized that this plain can't be defined as the plain spanned by the dynamic flux, and a hypothetical flux along an imaginary time axis. This leads to the idea that a parameter representation of the two dimensional manifold is obtained by analytic continuation of time into the complex plain:

$$t \longrightarrow t' + i t''$$

Real and imaginary part of time are then the internal coordinates of the two dimensional manifold, and the eliminination of time now reduces the phase space of α and β down to the required two dimensions.

For practical purposes the parameter representation is much more convenient than the explicit form, where time has been eliminated. This can be seen as follows: close to the bifurcation point the limit cycle can well be approximated by a circle when the variables have been scaled property. This corresponds to the first term in the Fourier expansion. Further away from the point of instability the circle will be deformed continuously into a more complex object, and higher harmonics become of importance. Now when eliminating time, the circle is represented by a curve of second order, and higher harmonics inevitably lead to curves of increasing order, which in almost every case have t, remain in implicite form, since they cannot be solved for any of the variables explicitly. For the manifold in the doubled phase space, this observation has some striking consequences. A tiny deviation of the limit cycle from the pure circle, which may be totally invisible on a graphical plot, calls for a weak addition of higher harmonics, and therefore represents the limit cycle by a curve of higher order. The coefficients of the highest powers are very small, when we assume only a slight deviation from the circle. However, in the doubled phase space we can make these corrections as big as we wish, by going away from the classical domain and to large values of the variables. That means, far away from the classical limit cycle the two dimensional manifold always deviates drastically from the lowest order approximation, even very close to the bifurcation point, where the classical limit cycle is indistinguishable from a pure circle. From these considerations we conclude that the two dimensional manifold is a simple object only close to the classical trajectories, and becomes more and more entwined when moving out into the unphysical directions and to large values of the fields - even arbitrarily close to the bifurcation point.

3.5 Local Dynamic Properties

We may characterize the local dynamic properties by following the linearized dynamic evolution of a difference vector:

$$\dot{\alpha} = (C_{11} + iC_{12})\alpha + (C_{21} + iC_{22})\alpha^* \tag{34}$$

Expressed in real and imaginary part this corresponds to the following eigenvalue problem:

$$\underline{X} = \begin{pmatrix} \mathrm{Re}\,\alpha \\ \mathrm{Im}\,\alpha \end{pmatrix} : \qquad \dot{\underline{X}} = -\lambda \underline{X} = \underline{\underline{A}} \cdot \underline{X} \tag{35}$$

where the matrix $\underline{\underline{A}}$ is composed of the elements C_{ij} . The eigenvalues of this problem are easily found to be:

$$\lambda_{1,2} = C_{11} \pm \left(C_{21}^2 + C_{22}^2 - C_{12}^2 \right)^{1/2} \tag{36}$$

The dynamics of the corresponding quantum case is obtained from eq. (34) by replacing α^* by β :

$$\begin{aligned} \dot{\alpha} &= (C_{11} + iC_{12})\alpha + (C_{21} + iC_{22})\beta \\ \dot{\beta} &= (C_{11} - iC_{12})\beta + (C_{21} - iC_{22})\alpha \end{aligned} \tag{37}$$

After expanding again in real and imaginary parts, we obtain the following eigenvalue problem:

$$\dot{\underline{X}} = -\lambda \underline{X} = \underline{\underline{B}}\, \underline{X} \tag{38}$$

where $\underline{x}$ is the vector that represents the real and the imaginary parts of α and β . The corresponding matrix $\underline{\underline{B}}$ has no vanishing element and at first seems to couple all four components of $\underline{x}$ in a totally general way. A simple rotation among the real and imaginary parts, however, transforms $\underline{\underline{B}}$ into block diagonal form:

$$\underline{\underline{B}}' = \begin{pmatrix} \underline{\underline{A}} & 0 \\ 0 & \underline{\underline{A}} \end{pmatrix} \tag{39}$$

and $\underline{\underline{A}}$ is the matrix that characterized the classical dynamic flow in the two dimensional phase space. The eigenvalues of the associated quantum case are degenerated therefore and come in pairs:

$$\mathrm{Det}\left(\underline{\underline{B}} - \lambda \underline{\underline{1}}\right) = (\lambda - \lambda_1)^2 \cdot (\lambda - \lambda_2)^2 = 0 \tag{40}$$

These manipulations are easily generalized to an arbitrary number of dimensions. Since the transformation to block diagonal from is independent of the matrix A, all the steps simply repeat in higher dimensions.

The fact that the local dynamics in the doubled phase space does not contain any new elements i.e. no new time constants, verifies the statement above, that a stable fixed point retains its stability in the extended phase space. Since the properties of limit cycles and strange attractors are described quantitatively by the concept of the Liapunov exponent, we will use this approach also when we investigate the stability of those objects that emerge when extending the phase space to the quantum mechanical process.

3.6 The Liapunov Exponents

The Liapunov exponents describe how an infinitesimal neighbourhood of a given trajectory behaves asymptotically. For that purpose we follow a point x in phase space and a neighbouring one. The evolution equation, decomposed into real and imaginary parts, is written as:

$$\dot{x}_i = g_i(\{x_e\}) \tag{41}$$

The difference vector to an infinitesimal close by point then evolves in time according to:

$$\dot{\xi}_i = \frac{\partial}{\partial x_j} g_i(\{x_e\})\,\xi_j \equiv D\cdot g(\{x_e\}) \tag{42}$$

In general, the vector $\xi(t)$ will rotate in phase space and change its module while being transported along the original trajectory. The deformation of the entire n-dimensional neighbourhood is described by a vielbein, centered at x(t), which is transported along the dynamic trajectory. The rates by which the individual axis extend or shrink, defines the Liapunov exponents of the trajectory. If the chosen trajectory starts inside the basin of a stable attractor, then the Liapunov exponents define a unique set of numbers that are independent of the individual initial conditions and describe the stability of the attractor. The Liapunov exponents can be defined in the following way: We call Λ_i the real eigenvalues of the hermitian matrix:

$$L = L(T) = \hat{T}^+ e^{\int_0^T Dg^+ dt} \cdot \hat{T} e^{\int_0^T Dg\, dt} \tag{43}$$

then the Liapunov exponents are defined as the following limit:

$$\lambda_i = \lim_{T\to\infty} \frac{1}{2T} \ln \Lambda_i(T) \tag{44}$$

and $\hat{T}$ is the time ordering operator.

The dynamic flow in the phase space of the real and imaginary parts is isomorphic to a complex flow. Therefore a homogeneous rotation in the planes of the real and imaginary parts of the fields commutes with the dynamic flow and therefore with the matrix L. Therefore, if $\underline{e}$ is an eigenvector of the matrix L then any vector $\underline{e}'$ which is only rotated in the complex plane is an eigenvector as well. Therefore, all eigenvalues of the matrix L and all Liapunov exponents show up in pairs.

With the help of the Liapunov exponents we can also define the dimensionality of the attractor. The dimension of a simple attractor is given by the number of non-negative Liapunov exponents. For a strange or chaotic attractor, Kaplan and York (10) have given a definition which is slightly more involved; let $\{\lambda_i\}$ be a set of ordered Liapunov exponents, such that λ_1 is the largest one, and the index j is the largest index such that the following inequality holds:

$$\sum_{i=1}^{j} \lambda_i > 0$$

then one can associate a dimension with such a chaotic attractor through the following relation:

$$d = j - \left(\sum_{\ell=1}^{j} \lambda_\ell \right) / \lambda_{j+1} \tag{45}$$

We summarize our results on the properties of attractors in the doubled phase space and distinguish between the following cases:

- The attractor is a fixed point in the n dimensional phase space of the classical problem, and the dynamic equations can be linearized in the neighbourhood of the fixed point. Then we have n negative Liapunov exponents which turn into 2*n negative exponents after doubling of the phase space, and the attractor consequently remains stable.

- The attractor is a limit cycle in the n dimensional classical system. Then we find one vanishing Liapunov exponent, which reflects the marginal stability along the one dimensional manifold, n-1 negative exponents that indicate the attractive behaviour transverse to the limit cycle. After doubling of the phase space, the Liapunov exponents become degenerate, i.e. two of them vanish and 2*n-2 are negative. This proves the previous statement that the limit cycle turns into a two dimensional attracting manifold.

- The attractor is chaotic, then some exponents become positive, while the attractor itself remains a finite bounded object. The dimensionality of the attractor either defined by the number of positive exponents or the fractional dimension as defined by Kaplan and York double when doubling the phase space. For example, the two dimensional chaotic attractor, found by Savage and Walls (11) has four dimensions when extending it to the phase space of the quantum problem.

4. THE EXAMPLE OF SECOND HARMONIC GENERATION

The question of the parametric interaction of light fields in a non-linear medium seems to be an ideal problem for studying the quantum mechanics of nonlinear dynamic system - for two reasons:

- The classical dynamics of such a non-linear dissipative system exhibits a rather rich variety of instabilities and attractors - fixed points, limit cycles, chaotic attractors and even bistability of chaos and limit cycles that have undergone several period doubling transitions. Depending on the energy flux through the system all these transitions can be studied in terms of this rather simple system (12), (13).

- The nonlinearity that drives the second harmonic process is cubic and the evolution equation for the P representation that is derived from it contains only derivatives up to second order. The equation of motion for the positive P representation in that case can be cast into Fokker Planck form, which enables one to associate a classical stochastic process with the quantum process under consideration, without resorting to any approximations.

Stochastically equivalent to this Fokker-Planck process is a set of classical Langevin equations for the field amplitudes α_1 and α_2 and their counterparts in the extended phase space β_1, β_2:

$$\begin{aligned}
\dot{\alpha}_1 &= -(\gamma_1+i\delta_1)\alpha_1 + \chi\beta_1\alpha_2 + F_1 + g_{1\ell}\xi_\ell(t) \\
\dot{\alpha}_2 &= -(\gamma_2+i\delta_2)\alpha_2 - \tfrac{1}{2}\chi\alpha_1^2 \\
\dot{\beta}_1 &= -(\gamma_1-i\delta_1)\beta_1 + \chi\alpha_1\beta_2 + F_1 + g_{2\ell}\xi_\ell(t) \\
\dot{\beta}_2 &= -(\gamma_2-i\delta_2)\beta_2 - \tfrac{1}{2}\chi\beta_1^2
\end{aligned} \tag{46}$$

The inhomogeneities characterize real Gaussian white noise processes, and g is a suitable matrix that contains the strength and the field dependence of the noise forces:

$$\langle \xi_i(t)\,\xi_j(0)\rangle = \delta_{ij}\cdot\delta(t) \tag{47}$$

For weak intensities of the driving field $F_1 < 3$, in these scaled units, the system has a stable fixed point with finite amplitudes α_1 and α_2. At F=3 this point looses its stability in a super critical Hopf bifurcation, and a limit cycle becomes the attractor of the process. The directions of instability are the directions of the imaginary parts of α_1 and α_2, while the system remains attractive in the directions of the real parts. Infinitesimally close to the transition point, the limit cycle is an ellipsoid in the plain spanned by the imaginary parts of α_1 and α_2. The orbit is supposed to have a fundamental period of Ω^{-1}. When driving the system harder, the real parts also start to show excursions from the stationary values, and oszillate with a basic frequency of $2*\Omega$. The limit cycle can always be expanded into a Fourier series, the leading order of which represents the limit cycle by an ellipsoid. The higher order contributions show up gradually when moving away from the transition point, and deform the limit cycle into a more and more complex loop in the four dimensional phase space. Up to second order in the perturbative treatment, Mandel and Erneux (14) have derived the Fourier representation for this case analytically:

$$\begin{aligned}
\alpha_1(t) &= i\varepsilon C_{11}\cos(\Omega t+\varphi_{11}) + \varepsilon^2 C_{12}\cos(2\Omega t+\varphi_{12}) \\
\alpha_2(t) &= i\varepsilon C_{21}\cos(\Omega t+\varphi_{21}) + \varepsilon^2 C_{22}\cos(2\Omega t+\varphi_{22})
\end{aligned} \tag{48}$$

The parameter ε is a measure for the distance from the bifurcation point. All the parameters in eq. (48) are implicite functions of the system parameters and can be found in ref. (14). Here we are only interested in the structure of this solution and do not need the details.

As long as we remain in the immediate neighbourhood of the bifurcation point, the leading term dominates, and the cycle is represented by an ellisoid, which after suitable scaling of the variables is given by the following circle:

$$\begin{aligned}
&\mathrm{Re}(\alpha_1) = 0 \quad , \quad \mathrm{Re}(\alpha_2) = 0 \\
&(\mathrm{Im}(\alpha_1))^2 + (\mathrm{Im}(\alpha_2))^2 = 4
\end{aligned} \tag{49}$$

Now we want to demonstrate how this classical limit cycle extends into the two dimensional manifold, when doubling the phase space for the treatment of the quantum case:

$$\alpha_1 \, , \, \alpha_2 \quad \text{and} \quad \alpha_1^* \to \beta_1 \quad , \alpha_2^* \to \beta_2 \tag{50}$$

When using this identification, the attracting two dimensional manifold assumes the following form:

$$\begin{aligned} &\alpha_1 + \beta_1 = 0 \qquad \alpha_2 + \beta_2 = 0 \\ &(\alpha_1 - \beta_1)^2 + (\alpha_2 - \beta_2)^2 = -1 \end{aligned} \tag{51}$$

Decomposing the fields into real and imaginary parts, and rotating coordinates by 45 degrees:

$$\begin{aligned} x_i^{\pm} &= 2\,\mathrm{Re}\,(\alpha_i \pm \beta_i) \\ y_i^{\pm} &= 2\,\mathrm{Im}\,(\alpha_i \pm \beta_i) \end{aligned} \tag{52}$$

we obtain the following two dimensional object, embedded into the eight dimensional phase space of the quantum process.

$$\begin{aligned} &x_i^+ = 0 \qquad y_i^+ = 0 \\ &x_1^- y_1^- + x_2^- y_2^- = 0 \\ &(x_1^-)^2 + (x_2^-)^2 - (y_1^-)^2 - (y_2^-)^2 = -1 \end{aligned} \tag{53}$$

In order to visualize this manifold, we project it into the three dimensional subspace, spanned by the imaginary parts of α_1 and α_2 and the real part of α_2 . Since this object is a rotationally invariant hyperboloid, it is useful to introduce cylindrical coordinates:

$$\mathrm{Im}\,\alpha_1 = r\cos(\varphi) \; , \; \mathrm{Im}(\alpha_2) = r\sin(\varphi) \; , \; z = \mathrm{Re}(\alpha_2) \tag{54}$$

and the attractor assumes the following form:

$$z = \sin(\varphi) \cdot (r^2 - 1)^{+1/2} \tag{55}$$

In order to visualize this object, we intersect it with the plains φ = const. and $\varphi \neq n * \pi$ which produces a set of hyperbolas in z and r while the intersection with the plane z=0 generates a circle

$$z = 0 \quad , \; r = 1 \tag{56}$$

with two attached straight lines:

$$z = 0 \quad , \; \varphi = 0, \pi \quad , \; r > 1 \tag{57}$$

These straight lines are apparent intersections of the manifold with itself. We have to keep in mind, however, that these self-intersections are a consequence of the projection onto three dimensions, and it does not exist in the original eight dimensional phase space. A plot in Fig. (1 a-c) illustrates this behaviour.

As indicated above, the leading approximation to the limit cycle is expected to represent it well in the neighbourhood of the bifurcation point, nevertheless, the two dimensional attractor of the quantum case that follows from this approximation is only represented correctly close to the limit cycle itself, and gets poorer and poorer, when moving out in

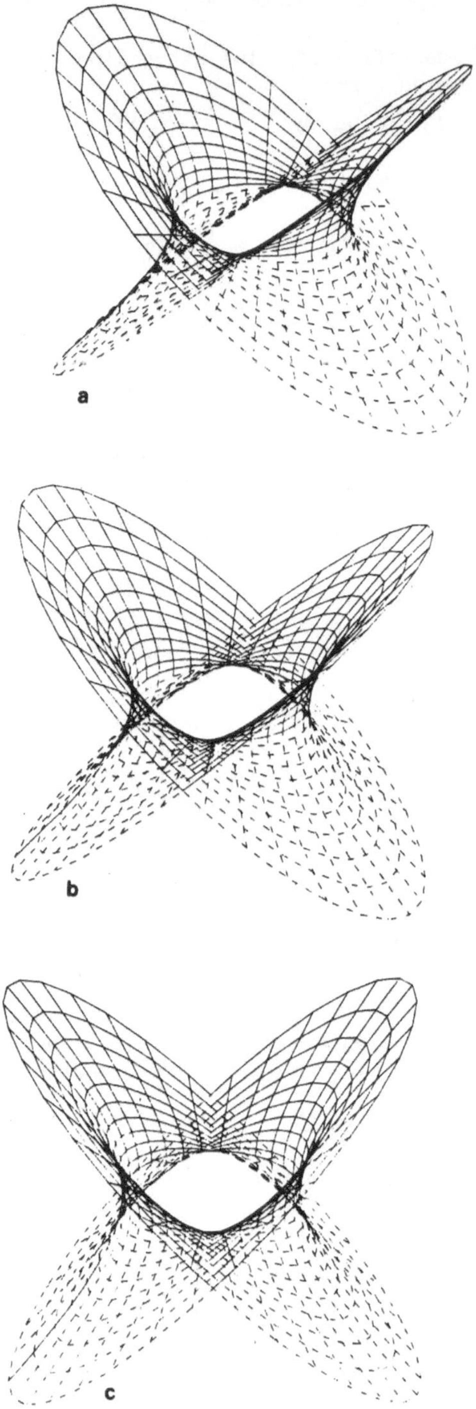

Fig. 1a-c Leading analytical approximation of the 2 dimensional manifold

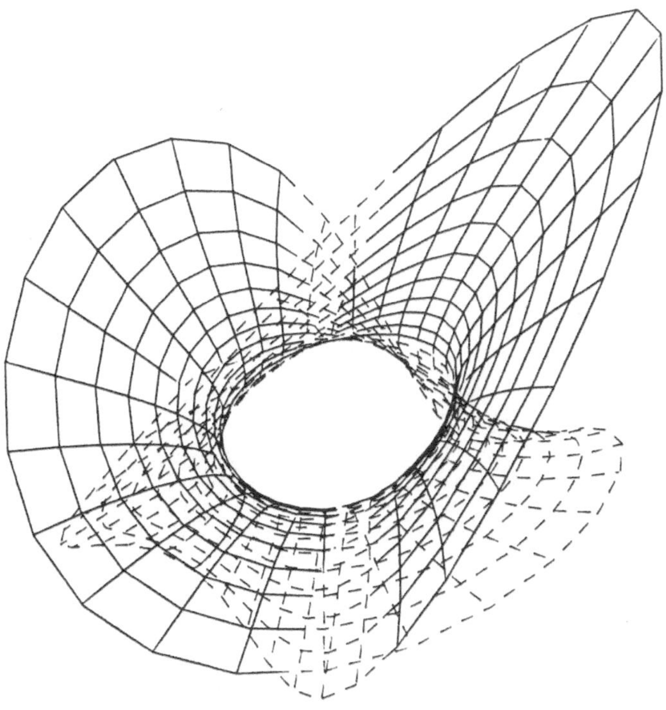

Fig. 2 Analytical approximation of the 2 dimensional manifold including the first harmonic correction

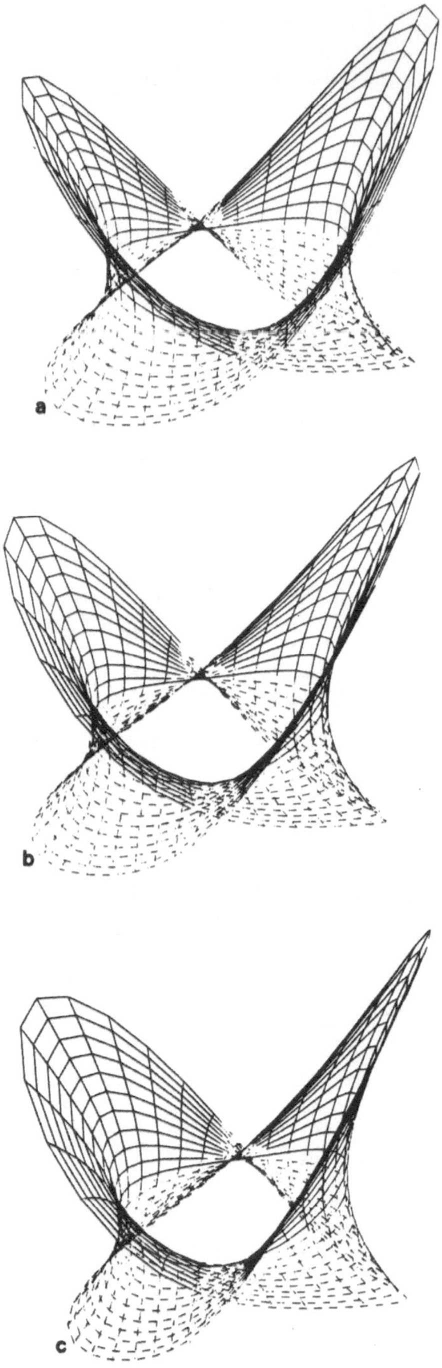

Fig. 3a-c Numerical integration of the attracting manifold

phase space. There even minute corrections to the leading order - i.e. the circle - become visible, and finally dominate the structure of the manifold. In other words, a finite approximation to the limit cycle can only represent a finite portion of the quantum attractor close to the classical attractor. If one wants to extend the region, even close to the transition point, one has to include higher and higher harmonics. The correction due to the second harmonic terms is shown in Fig. (2), where we have made use of the previous observation that the parameter representation of the two dimensional object is easily obtained by continuing the time into the complex plain.

These analytical results that are in principle based on the results of Mandel and Erneux, and which are expected to be the generic form of most Hopf bifurcations, can now be compared with the numerical simulations which we get from integrating the deterministic equations of motion of the quantum case. One set of internal coordinates is given by the deterministic flux, the other by the flux along an imaginary time. Since the latter is highly unstable and very sensitive to calculations with finite accuracy, we used the following procedure:

We start on the classical limit cycle and trace out the first coordinate line. Then we turn time into (i*t) and integrate for a short interval. This integration, due to the instability of that motion, takes us somewhat off the attractor. Then we integrate again along the physical flux, allowing it to relax to equilibrium, and trace out the next coordinate line. By continuing this procedure we trace out the attractor by a set of quasi limit cycles, that are densely distributed over the two dimensional manifold. For a special choice of the parameters ($\gamma_1 = \gamma_2 = 1$, $\delta_1 = \delta_2 = 0$, and $F = 3.5$) we have plotted the attractor in Fig. (3a-c) using the same set of axis as in the analytical approximation. A comparison of the two sets of pictures shows that the basic structure is identical. For a detailed comparison one has to keep in mind that the analytical result is drawn in scaled variables, and taken further away from the transition point, in order to demonstrate the influence of the higher harmonics. A very valuable byproduct of the numerical integration of the deterministic trajectories on the attractor is that it also allows one to determine the Liapunov exponents for that flow. Since the motion on the attractor is not ergodic, each quasi limit cycle in principle has its own Liapunov exponents. The numerical results, however, reveal that the exponents do not vary much from trajectory to trajectory. In agreement with our general discussion above we find that the Liapunov exponents precisely come in pairs, two vanish and six are negative, indicating the marginal stability inside the two dimensional manifold, and its stability - with respect to transverse perturbations. Typically we find:

$$\lambda_{1,2} = 0 \quad , \quad \lambda_{3,4} = -0.3 \quad , \quad \lambda_{5,6} = -1.7 \quad , \quad \lambda_{7,8} = -2.0$$

Since the flux of the dynamical model has constant divergence equal to -8, the sum of the Liapunov exponents must equal that number, which is a simple check for the reliability of the numerical integration.

5. QUANTUM DYNAMICS OF SECOND HARMONIC GENERATION

In the previous chapter we have set up the stage for the description of the quantum mechanical behaviour, by discussing the properties of the deterministic dynamics and its attractors. The full quantum mechanical formulation now requires that we also include the fluctuations that originate from the nonlinearity of the process. Since there is little, if any hope to be able to tackle the Fokker Planck equation in eight dimensions analytically, we describe the process of second harmonic generation by simulating the stochastic differential equations that are equivalent to the Fokker Planck description. In order to describe the transient buildup of the fields in the nonlinear interaction process, we numerically integrate the stochastic trajectories, form the moments required and average over a suitable number of runs. In this way we can evaluate for instance the following typical properties (15):

- Number of photons in the fundamental mode $I_1 = \langle \alpha_1 \beta_1 \rangle$
- Number of photons in the harmonic mode $I_2 = \langle \alpha_2 \beta_2 \rangle$
- Squeezing in the different quadratures

$$\begin{aligned} \langle \Delta^2 X_i^{\mp} \rangle = \; & 1 + 2\langle \alpha_i \beta_i \rangle \mp \langle \alpha_i^2 \rangle \mp \langle \beta_i^2 \rangle \\ & - 2\langle \alpha_i \rangle \langle \beta_i \rangle \pm \langle \alpha_i \rangle^2 \pm \langle \beta_i \rangle^2 \end{aligned} \tag{58}$$

In passing it may be worth noticing that the photon numbers found by only sampling a small number of trajectories are not - and cannot be expected to be - real or even positive, but the imaginary and the negative parts typically decay away rather rapidly with an increasing number of runs.

As an example of the simulation we have plotted in fig. 4 the transsient build up of the intensities and the amount of squeezing obtained in the different quadratures. The parameters have been chosen in a way that the system operates below the Hopf bifurcation, and the attractor is a simple fixed point. An interesting check of the reliability of the simulation is the calculation of the uncertainty product $\Delta^2 x_i^+ \cdot \Delta^2 x_i^-$; which has to satisfy the Heisenberg uncertainty relation. This product is also included in the figure, and comparison with the minimum uncertainty - the straight line - shows us that the quantum limits are never violated.

If we increase the intensity of the pump field above the critical value of F=3 the classical fixed point bifurcates into a limit cycle and the noisy trajectories are no longer trapped to the neighbourhood of the fixed point but can freely wander on the two dimensional manifold, which extends to infinity. Transverse to the attractor, the trajectories will be confined to a narrow region, depending on the strength of noise, due to the attractive property of the manifold.

The unboundedness of the attractor after bifurcating to the limit cycle, indicates that there might be an unexpected difficulty associated with this approach. The diffusive motion on the attracting manifold is analogous to a random walk in a plane. The probability diffuses out into the phase space, and a steady state density does not exist. Nevertheless, the ensemble averages must approach certain stationary values after a characteristic relaxation time of the process. And this is exactly what one observes in the numerical simulation close to the threshold of the instability.

Fig. 5 now shows a typical plot of a simulation further away from the bifurcation point, and there something unusual appears. Before becoming absolutely stationary, the transient moments of the fields develop rather unusual or unphysical bursts that indicate a break down of the method. When studying this phenomenon, we have made just as many runs, until the first drastic excursion appeared. Then we ran the simulation for twice as many runs and the spike went down in amplitude by a factor of two - which means that no other spike has appeared at the same point in time in the second half of the simulations. But at some other instant in time another spike appeared which again is reduced by averaging over more trajectories. So in principle one may think that these enormous excursions

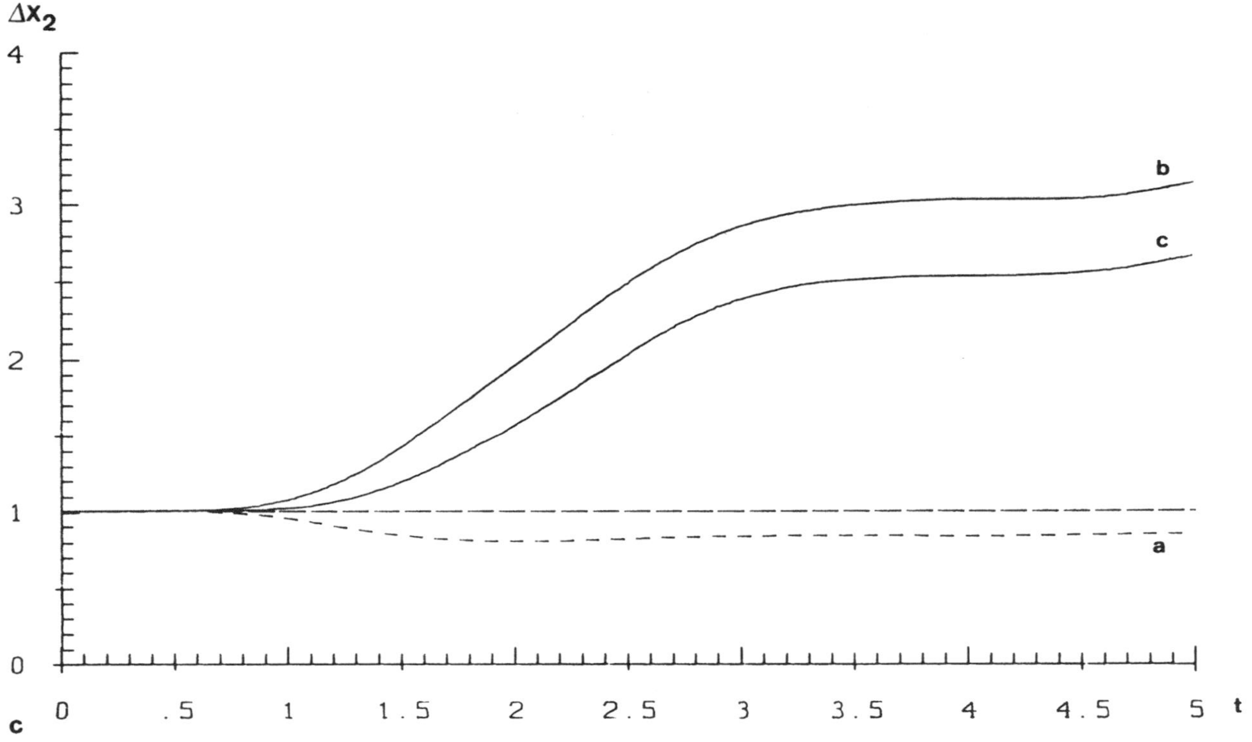

Fig. 4 Variances of the fundamental field
(F_1 = 2) a: $\Delta^2 x_1^+$ b: $\Delta^2 x_1^-$ c: $\Delta^2 x_1^+ \cdot \Delta^2 x_1^-$

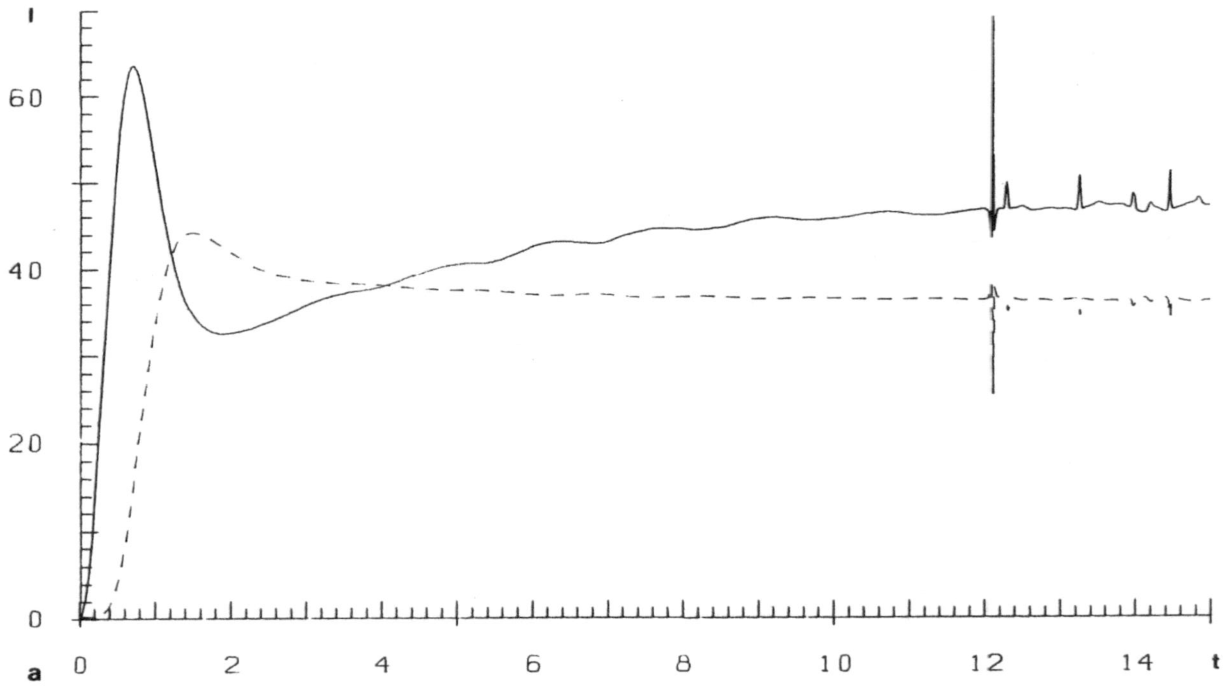

Fig. 5 Photon number of second harmonic generation above the threshold to the limit cycle $F_1 = 9$

from the mean values could be averaged out by averaging over an increasing number of trajectories. However, in practice this is no solution to the problem, since the longer sequence of runs is made, the greater the chance that a single trajectory appears deviating so much from the average that it totally dominates the average locally, and the number of run becomes astronomical to average this excursion out. Actually, the situation is even worse, since trajectories can eventually diverge to infinity and totally ruin the simulation. The positive P-representation has also been used to describe the quantum noise of absorptive optical bistability by Carmichael et.al. (16) and the laser by Sarkar et.al. (17) where similar problems have been observed.

The reason for the occurence of these large excursions can be illustrated quite easily by using a much simpler model that can be handled analytically. The most elementary model for a symmetry breaking instability, like the laser process, is characterized by the following model:

$$\dot{\alpha} = \alpha\,(d - |\alpha|^2) + \xi(t) \tag{59}$$

where α is a complex field amplitude and d is the bifurcation parameter. ξ(t) represents the spontanuous emission noise. For the present purpose we are only interested in the deterministic parts and solve eq. (59) in the absence of noise:

$$|\alpha(t)|^2 = d\,|\alpha(0)|^2 \cdot \left(|\alpha(0)|^2 + (d - |\alpha(0)|^2)\,e^{-dt}\right)^{-1} \tag{60}$$

This trajectory is stable for all initial conditions, and converges towards zero below threshold $d < 0$ and towards a finite value above $d > 0$. Now we translate this description of the deterministic classical evolution to the dynamics of the quantum system in the doubled phase space, by replacing $\alpha^* \rightarrow \beta$ and find the following coupled complex equations:

$$\dot{\alpha} = \alpha\,(d - \alpha\beta)\ , \qquad \dot{\beta} = \beta\,(d - \alpha\beta) \tag{61}$$

It is easily seen that the solution of this set of equation is made possible by solving for $\alpha\beta$:

$$\alpha(t)\beta(t) = I(0)\,d\,\left(I(0) + (d - I(0))\,e^{-dt}\right)^{-1} \tag{62}$$

where $I(0) = \alpha(0)\beta(0)$ is the initial condition. If the trajectory initiates in the physical subspace, i.e. $\beta = \alpha^*$ then it can easily be shown that the product $\alpha\beta$ remains finite for all times. Therefore, when starting in the physical domain, the deterministic trajectories are always well behaved, and average to:

$$\alpha(t)\beta(t) \longrightarrow d \tag{63}$$

for $d > 0$. More generally, the two dimensional manifold

$$\alpha \cdot \beta = d \tag{64}$$

is locally attracting, as proven in the previous chapters, even for trajectories that initiate in the nonphysical section of phase space. However, while the line $\alpha\alpha^* = d$ is globally stable in the two dimensional phase space of the classical dynamic system, the two dimensional manifold $\alpha\beta = \alpha$ is only locally but not a global stable in the extended phase space of the two complex variables α and β. Obviously, all trajectories that enter the region

$$\alpha \cdot \beta \leq 0 \tag{65}$$

which is the inner part of a cone embedded in four dimensions, diverge.

This cone becomes asymptotically identical with the attracting manifold itself, which demonstrates that the basin of attraction is squeezed down to the attractor when moving out in phase space. Therefore, all trajectories that may start initially in the physical domain are bound to exhibit large excursions sooner or later, when noise is included. This is seen most easily when we try to follow a trajectory that is affected by weak noise. The trajectory is supposed to start initially on the physical sector $\alpha^* = \beta$ and relaxes rapidly towards the attractor $\alpha\beta$ = d. Since the two dimensional manifold is locally attracting, in the transverse direction, we can assume that the trajectories are confined to its immediate neighbourhood. On the manifold itself the deterministic forces tend to drive the system on closed loops, while the noise causes an outward diffusion in a random walk fashion. Further out in phase space, the basin of attraction becomes more and more narrow and vanishes asymptotically. So if we go out far enough in phase space, then the transverse stability of the attractor is not strong enough to prevent the trajectories from crossing over into the region of deterministic instability, and large excursions are to be expected, even divergent behaviour, if the noise is so weak that the probability of kicking the trajectory back onto safe grounds becomes very small.

The large deviations of certain trajectories from the average values observed for the case of second harmonic generation can be traced back to three - probably generic - properties of this simple model. First, the extended phase space contains regions of deterministic instability, i.e. trajectories that emerge from there reach infinity in a finite time. Secondly, the deterministic attractor is unbounded, and the trajectories undergo a certain random walk that takes the trajectories far away from the classical limit cycle into the unphysical domain. The third and crucial property, however, is to be seen in the fact that the boundary of unstable behaviour and the attractor become identical for large values of the fields. These large values are accessible due to the random walk on the hyperbolic attractor.

Qualitatively this behaviour is identical to what has been observed in the numerical simulation of second harmonic generation. For a certain amount of noise, the trajectories remain well behaved for a certain time until the large deviations occur. When reducing the strength of noise, the trajectories stay well behaved for a longer time but eventually the problem sets in again. This is consistent with the picture we just created. When the noise is reduced, the random walk takes longer to reach the outer parts of the attractor, and the crossing over to the domain of instability is suppressed when the fluctuations are weaker. So for very weak noise, which actually is identical with the assumption of a small nonlinearity, the simulation can be expected to reach the stationary regime long before any irregularities occur. However, from a principle point of view, this singular behaviour which seems to be inevitable, must be understood as a serious problem of the concept, which needs further investigations.

The talk is based on results obtained in collaboration with M. Dörfle.

REFERENCES

1. Wigner, E.: Phys. Rev. 40,749 (1932)
2. Wigner, E.: Z. Phys. Chem. B19,203 (1932)
3. Glauber, R.: Phys. Rev. Lett. 10,84 (1963)
4. Surdarshan, E.C.G.: Phys. Rev. Lett. 10,277 (1963)
5. Schenzle, A., Brand, H.: Phys. Rev. A20,1628 (1979)
6. Schenzle, A., Graham, R.: Phys. Lett. A98,319 (1983)
7. Drummond, P.D., Gardiner, C.W.: J. Phys. A13,2353 (1980)
8. Gardiner, C.W. (ed.): Handbook of Stochastic Methods, in: Springer Series in Synergetics, vol. 13, Berlin, Heidelberg, New York: Springer 1983
9. Glauber, R.J.: Phys. Rev. 131,2766 (1963)
10. Kaplan, J., Yorke, J.: Functional differential equations and approximation of fixed points. In: Lecture Notes in Mathematics, Peitgen, H.-O., Walther, H.-O. (eds.), vol. 730, p. 204, Berlin, Heidelberg, New York: Springer 1979
11. Savage, C.M., Walls, D.F.: Opt. Acta 30,557 (1983)
12. Dörfle, M., Graham, R.: In: Optical instabilities, Boyd, R.W., Raymer, M.G., Narducci, L.M. (eds.), Cambridge: Cambridge University Press 1986
13. Drummond, P.D., McNeil, K.J., Walls, D.F.: Opt. Acta 28,211 (1981)
14. Mandel, P., Erneux, J.: Opt. Acta 28,7 (1982)
15. Dörfle, M., Schenzle, A.: Z. Phys. 65,113 (1986)
16. H.J. Carmichael, J.S. Satchell and S. Sarkar, Phys. Rev. A34,3166 (1986)
17. S. Sarkar, J.S. Satchell and H.J. Carmichael, J. Phys. A19,2765 (1986)

WORKSHOP REPORT ON QUANTUM CHAOS AND MEASUREMENT

William Firth, J.N. Elgin* and J.S. Satchell+

University of Strathclyde
* Imperial College, London
+ Royal Signals and Radar Establishment

Deterministic chaos has been such a stimulating and fruitful concept in classical mechanics that it is natural to ask to what extent, if any, chaotic behaviour survives the quantisation process. This was the main motivation for the workshop and hence for this report. Briefly, the verdict is that Quantum Chaos does not exist! We hasten to emphasise that this is true only for the strictest definition of chaos, and at infinite time scales: real systems can do display behaviour indistinguishable from their chaotic classical counterparts. Furthermore, there is a wealth of interesting and important science in the crossover region between fully classical ($h \to 0$ and/or short times) and fully quantum (h finite and $t \to \infty$) limits. We suggest "quantised chaos" as a succinct and graphic term to describe this "non-classical or semi-classical behaviour of quantum systems whose classical analogues show chaos" (M Berry).

The importance of time scales is vividly illustrated by bounded Hamiltonian systems. The Quantum energy spectrum is discrete and thus the dynamics necessarily quasiperiodic, even when the classical counterpart is fully chaotic, with exponential divergence of nearby trajectories and a broadband continuous spectrum. For finite times, however, the quantum dynamics necessarily has a continuous spectrum and, of course, quasiperiodic motion can be pseudorandom on short time scales. Some real physical systems are observable only on time scales so short that the quantum and classical evolutions are indistinguishable. Some of the work on the microwave ionisation of hydrogen reported by Bayfield at the workshop is in this category. Alternatively random external perturbations (classical noise) may continually "reset the clock" and maintain quasi-classical evolution indefinitely.

The realm of quantised chaos is entered when systems can be observed for long enough, with small enough external perturbation, that the intrinsic quantum limitations to chaotic evolution become appreciable. In quantum systems, a 2N-dimensional phase space can be discretised into Planck cells of volume h^N. This must wash out the self-similar structures and fractal dimensions characteristic of chaos below some inner scale determined by h. From a dynamical point of view, exponential divergence of nearby trajectories is possible only if "nearby" means "in separate Planck cells". The divergence between classical and quantum dynamics thus occurs when the system has evolved for long enough that the phase space points originating in the same Planck cell would be macroscopically distinguishable. This "break time" t_B is evidently controlled by h, so that we can envisage a diagram such as Figure 1.

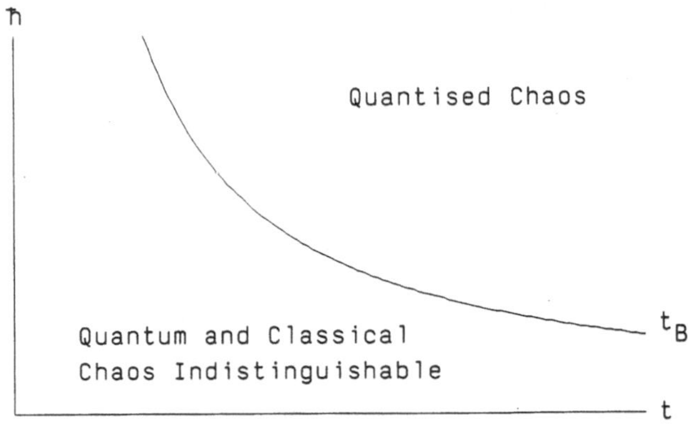

Figure 1. Quantum/Classical Break time

A plausible estimate for t_B is

$$t_B = \frac{1}{\lambda} \ln (S_0/h)$$

where λ is the largest positive Lyapunov exponent and S_0 is a characteristic action of the classical model of the system. Figure 2 shows the break time phenomenon in the widely studied kicked rotator model.

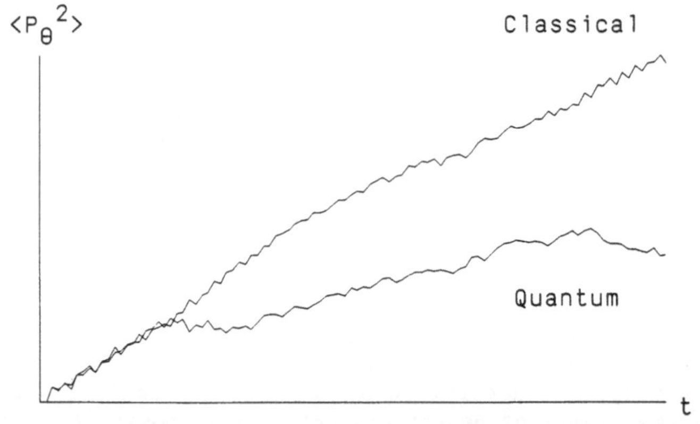

Figure 2. Energy growth for quantum and classical kicked rotators

In this particular case the chaotic diffusion in energy characteristic of the classical motion is arrested at t_B as the excitation becomes localised in energy space in a manner reminiscent of that in which Anderson localisation inhibits the spatial diffusion of electrons in amorphous solids. One interesting question in quantised chaos is the generality of this localisation phenomenon as a mechanism for quantum inhibition of classical chaos. It is known that the Anderson localisation mechanism is "weaker" in higher dimensional systems, and it may well be that in many degrees of freedom systems it is not important. The coupled spin system by Feingold and Peres has evidence of Anderson localisation being important. A similar observation can also be made about the kicked spin systems of Haake. These spin systems have a finite basis and so must, by construction, show only multi-periodic behaviour. Infinite basis systems with two or more degrees of freedom have not yet been studied, and this would be very interesting despite the formidable numerical difficulties.

The emphasis of normal quantum mechanics is very much on the energy spectrum, rather than the dynamics. Since fine spectral details require long times to manifest themselves, the corresponding branch of quantised chaos has large break times, and thus small h (ie $h \ll S_0$). This is the domain of the semi-classical mechanics of classically chaotic systems. There is evidence that the spectral properties of classically chaotic and classically integrable systems are qualitatively different in this semi-classical regime. A useful guide is the distribution function for the spacing of nearest neighbour energies. This is Poissonian for integrable systems, but chaotic systems exhibit level repulsion, and the distribution is well described by random matrix theory. If on the other hand one is interested in the dynamics of the system, this is more immediately related to the distribution of all energy level spacings rather than just the nearest neighbour ones. One important problem in such studies, however, is that the long time and small h limits do not commute:

$$\lim_{t\to\infty} \lim_{h\to 0} \neq \lim_{h\to 0} \lim_{t\to\infty}$$

Consequently, asymptotic classical results are never obtained as the semi-classical limit of a quantum system! It is worth noting that this difficulty is removed for dissipative systems.

The above considerations imply that contrary to the popular view, it is classical, rather than quantum, mechanics which is indeterministic and irreversible. Of course quantum irreversibility can also arise in the measurement process. A number of talks at the workshop addressed the quantum measurement problem, and there was a widespread feeling that its connection with quantised chaos may be important. More specifically several talks addressed the theory of the transfer of dynamical information from one quantum system to another as a prototype measurement stage. A number of examples were given but it was clear that the act of measurement inevitably introduces irreversibility and the language of dissipative quantum mechanics is the natural way to treat this. Introducing dissipation has important physical consequences. The long-time and semi-classical limits now commute and the system will always attain a statistical steady state. There was some speculation that dissipation would remove such phenomena as Anderson localisation, and restore a closer approach to the classical behaviour. However in the specific model of measurement for the kicked rotator suggested by Sarkar and Satchell this was not found. Instead they found the system approached a statistical steady state. This is significant for quantised chaos, because predictions about the behaviour of real quantum systems can only be tested by interrogating them via a measurement process, which ideally should be amenable to quantum analysis.

What then are the types of system that will most repay study? Transient experiments of the Bayfield type are clearly important, providing direct information on break time phenomena, albeit post hoc by destructive measurement. Driven, damped systems are particularly interesting, for the reasons mentioned above, especially in that they continually radiate diagnostic information. The micro-maser (a few microwave photons in a high Q single mode cavity, driven by an atomic beam of Rydberg atoms so weak that there is rarely an atom in the cavity) seems almost ideal amongst such systems. It is a fully quantised kicked system whose classical analogue is chaotic: furthermore the driving atoms also interrogate the cavity, so that downstream ionisation gives continuous, albeit statistical, information on the quantum dynamics over arbitrarily long times. Trapped-ion spectroscopy is, in a sense, dual to the micromaser, with one or a few ions in an electromagetic trap, driven by laser beams and interrogated by their own fluorescence. The classical electron orbits are chaotic for an ion like H_2^- and the detailed study of its spectrum may prove rewarding. Driven spin systems are a rather different, but very promising, class in which quantum effects will become significant if micro-magnetic particles with about one hundred spins can be fabricated. Superconducting devices such as the SQUID may also be an interesting class of systems with macroscopic quantum effects.

Finally one might ask what use if quantised chaos? Its importance for our fundamental understanding of the physical world is, we hope, clear from the above, but it may also have practical and technical importance. Progress in microfabrication techniques is leading to electronic devices which are more like molecules than crystals and it is probable that at least some will have dynamics in the quantised chaos domain. In fact it is conceivable that this domain may be one of the ultimate limits to miniaturisation of information processing devices!

CONTRIBUTORS

Dr S Sarkar
CENTRE FOR THEORETICAL STUDIES
RSRE
St Andrews Road
MALVERN
Worcs WR14 3PS UK

Professor W E Lamb, Jnr
848 N Norris Ave
Tucson
Arizona 85721
USA

Professor J Ford
School of Physics
Georgia Institute of Technology
ATLANTA
Ga 30332
USA

Professor M V Berry FRS
UNIVERSITY OF BRISTOL
Physics Dept
Tyndall Avenue
BRISTOL
BS8 1TL

Prof E Ott
Lab for Plasma & Fusion Energy
UNIVERSITY OF MARYLAND
College Park
MARYLAND 20742
USA

Dr Y Pomeau
S PhT ACEN Saclay
91191 Gif-sur-Yvette Cedex
FRANCE

Professor C M Caves
Theoretical Astrophysics 130-33
CALIFORNIA INSTITUTE OF TECHNOLOGY
PASADENA
California 91125
USA

Professor A Peres
Physics Dept
TECHNION
32000 HAIFA
Israel

Professor N G van Kampen
Institute for Theoretical Physics
UNIVERSITY OF UTRECHT
Princetonplein 5
3508 TA UTRECHT
The Netherlands

Professor F Haake
Fachbereich Physik
UNIVERSITÄT ESSEN
Essen
West Germany

Dr A Schenzle
Fachbereich Physik
UNIVERSITAT ESSEN - GHS
4300 Essen
West Germany

Professor G Casati
Dip Fisica
Università di Milano
Via Celoria 16
20133 Milano
ITALY

Professor R Graham
Fachbereich Physik
UNIVERSITAT ESSEN - GHS
4300 Essen
West Germany

Dr J Satchell
RSRE MOD (PE)
St Andrews Road
MALVERN
Worcs WR14 3PS, UK

Prof J Bayfield
UNIVERSITY OF PITTSBURGH
100 Allen Hall
3941 O'Hara Street
Pittsburgh PA 15260
USA

Professor E R Pike
Centre for Theoretical Studies
RSRE MOD (PE)
St Andrews Road
MALVERN, Worcs WR14 3PS, UK

Prof W Firth
UNIVERSITY OF STRATHCLYDE
Dept of Physics
John Anderson Bldg
GLASGOW G4 ONG

Dr J N Elgin
IMPERIAL COLLEGE
Dept of Maths
LONDON
SW7 2AZ

Dr S Graffi
Dip Matematica
Universita di Bologna
40127 BOLOGNA
Italy

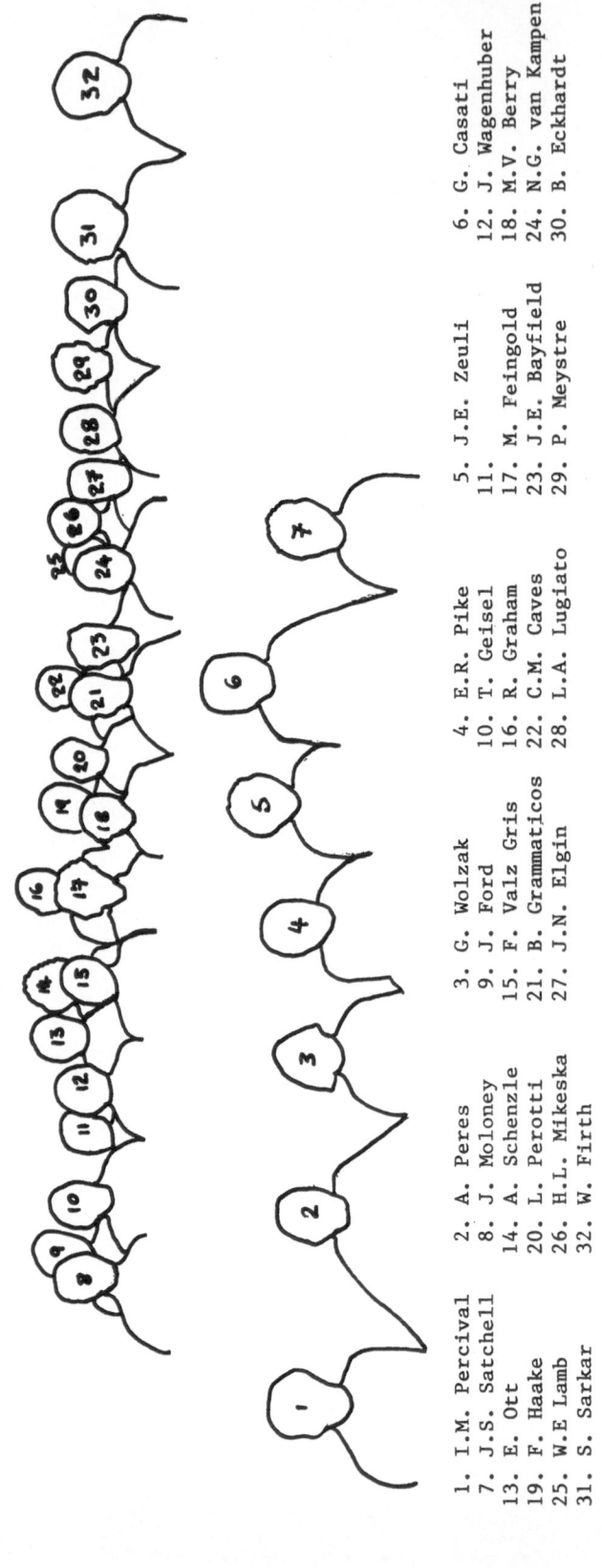

1. I.M. Percival	2. A. Peres	3. G. Wolzak	4. E.R. Pike	5. J.E. Zeuli	6. G. Casati
7. J.S. Satchell	8. J. Moloney	9. J. Ford	10. T. Geisel	11.	12. J. Wagenhuber
13. E. Ott	14. A. Schenzle	15. F. Valz Gris	16. R. Graham	17. M. Feingold	18. M.V. Berry
19. F. Haake	20. L. Perotti	21. B. Grammaticos	22. C.M. Caves	23. J.E. Bayfield	24. N.G. van Kampen
25. W.E Lamb	26. H.L. Mikeska	27. J.N. Elgin	28. L.A. Lugiato	29. P. Meystre	30. B. Eckhardt
31. S. Sarkar	32. W. Firth				

INDEX